Geoelectromagnetic Waves

Geoelectromagnetic Waves

A V Guglielmi and O A Pokhotelov

Institute of the Physics of the Earth
Russian Academy of Sciences, Moscow, Russia

Institute of Physics Publishing
Bristol and Philadelphia

British Library Cataloguing-in-Publication Data

A catalogue record for this book is available from the British Library.

ISBN 0 7503 0052 3

Library of Congress Cataloging-in-Publication Data

Guglielmi, A. V. (Anatol V.)
   Geoelectromagnetic waves / A. V. Guglielmi & O. A. Pokhotelov ;
translated from the Russian.
       p.   cm.
   Includes bibliographical references and index.
   ISBN 0-7503-0052-3 (alk. paper)
   1. Geomagnetic micropulsations. 2. Geophysics. 3. Earth
currents. I. Pokhotelov, O. A. (Oleg Aleksandrovich) II. Title.
QC809.M25G83 1996
550--dc20   95-44212
                                                                  CIP

Published by Institute of Physics Publishing, wholly owned by The Institute of Physics, London

Institute of Physics Publishing, Techno House, Redcliffe Way, Bristol BS1 6NX, UK

US Editorial Office: Institute of Physics Publishing, The Public Ledger Building, Suite 1035, 150 South Independence Mall West, Philadelphia, PA 19106, USA

Typeset in TEX using the IOP Bookmaker Macros
Printed in the UK by J W Arrowsmith Ltd, Bristol

Contents

Preface

This book is devoted to the theory of geoelectromagnetic oscillations and waves, i.e. waves of natural origin arising in the Earth's crust, in the ocean, in the atmosphere, in the ionosphere, in the magnetosphere and in the interplanetary medium in front of the magnetosphere.

We have sought to give a simple and clear presentation of the physical picture of the waves. When selecting the material special attention was paid to the theoretical inferences that can be easily compared with the observational data. Complicated and unwieldy calculations were omitted where possible. At the same time, where formulae have been presented without derivation, we have tried to explain their physical meaning.

The outstanding achievements of the many prominent scientists who contributed to the development of wave theory were used as one basis for this book. A certain portion of the material is based on the authors' work, most particularly the interpretation of geoelectromagnetic waves and the nonlinear theory. As far as we are aware this is the first use of the term 'geoelectromagnetic waves'. We consider that the term is more general and more appropriate than the common term 'geomagnetic pulsations' which is used in reference to the oscillations of the magnetosphere only. The abundance and the variety of the material resulting from generalization of the problem dictated that we should focus our attention on a number of specific areas which we considered to be most challenging and interesting.

In order to read this book, one should have a basic knowledge of electrodynamics (standard educational course) and a general understanding of the Earth and its environment. However, the book may be of interest to a less trained reader, although in this case some persistence and diligence will be required.

For convenience the book includes a commentary to the bibliography at the end of each chapter, a subject index and other references.

It was the authors' intention that the book should enable the reader not only to penetrate the physical problems of geoelectromagnetic waves and obtain the training essential to perform original work, but also to develop a clear identification of the unsolved problems and the trends in the field.

The book is written for a wide section of geophysicists—scientists, lecturers, post- and undergraduates as well as for physicists interested in geophysical applications of the electromagnetic wave theory. A certain portion of the material is given in the form of exercises supplied with solutions. This enables us to embrace a wider range of problems within the limits of the book and further to pursue didactic objectives.

The authors are grateful to Professors A Frazer-Smith, B Hultqvist, C F Kennel, R Lundin, J F McKenzie, C T Russell, K Schindler, L Stenflo, Drs P K Shukla, M Teague and K Yumoto for helpful discussions and to colleagues with whom some of the results described here were obtained.

Anatol Guglielmi
Oleg Pokhotelov
Moscow, May 1995

Notation

Electromagnetic field

The intensity and induction of the electric field are defined as E and D; the intensity and induction of the magnetic field are H and B respectively. Everywhere, unless indicated otherwise, $B = H$. A perturbation of the magnetic field in a wave is b. The current density is j.

The velocity of light $c = 3 \times 10^{10}$ cm s^{-1} and the elementary electric charge $e = 4.8 \times 10^{-10}$ cm$^{3/2}$ g$^{1/2}$ s^{-1} (i.e. 4.8×10^{-10} absolute units (abs. u.)). Here and in what follows numerical values are given within accuracy sufficient for geophysical applications.

The dielectric permeability is ε and the electroconductivity is σ or $\varepsilon_{\alpha\beta}$ and $\sigma_{\alpha\beta}$ respectively (in an anisotropic medium). The magnetic permeability is μ.

Condensed matter

The velocity is v; the pressure p; the density ρ; the displacement vector ξ; the strain tensor $u_{\alpha\beta}$; the stress tensor $\sigma_{\alpha\beta}$.

The modulus of dilatation is K; the shear modulus G; the Poisson coefficient α. In fact α changes within the limits from 0 to $\frac{1}{2}$.

The longitudinal sound velocity is c_l; the transverse sound velocity c_t. These velocities yield the inequality $c_l > c_t\sqrt{2}$. In a fluid $c_t = 0$.

The dynamic viscosity is η; the kinematic viscosity $\nu = \eta/\rho$.

Plasma

The electron mass is $m_e = 9.1 \times 10^{-28}$ g; the proton mass $m_p = 1.67 \times 10^{-24}$ g; the ion mass m_i. The charges of electrons and ions are $-e$ and ze respectively ($z = 1$ for the proton).

The concentration of electrons (ions) is $N_{e(i)}$ and the temperature $T_{e(i)}$; the plasma density ρ; the distribution function of the particles f; the collision frequency ν.

The plasma frequencies are

$$\omega_{0e} = (4\pi N_e e^2/m_e)^{1/2}$$

$$\omega_{0i} = (4\pi N_i z^2 e^2/m_i)^{1/2}.$$

The cyclotron frequencies are

$$\Omega_e = eB/m_e c \qquad \Omega_i = zeB/m_i c.$$

The thermal velocities of electrons and ions are

$$V_{T_e} = (T_e/m_e)^{1/2}$$

and

$$V_{T_i} = (T_i/m_i)^{1/2}.$$

The Alfvén velocity is

$$c_A = B/\sqrt{4\pi\rho}.$$

The ratio of plasma pressure to magnetic pressure is

$$\beta = 8\pi p/B^2.$$

Waves

The time dependence in a monochromatic wave is everywhere taken to be of the form $\exp(-i\omega t)$, and the spatial dependence in a plane wave is given in the form $\exp(i\mathbf{k} \cdot \mathbf{x})$.

The frequency ω; the wave vector $\mathbf{k}$, the period of a wave $T = 1/f$, $f = \omega/2\pi$; the wavelength $\lambda = 2\pi/k$; the polarization vector $\mathbf{a}$. The imaginary part of the frequency $\gamma = \mathrm{Im}\ \omega$, the growth rate (increment) $\gamma > 0$, the decrement $\gamma < 0$.

The refractive index $n = ck/\omega$.
The phase velocity $v_{ph} = \omega/k$.
The group velocity $v_g = \partial\omega/\partial k$.

Units

In theoretical deductions we use the Gaussian system of units. However, following tradition and being guided by efficacy considerations, we do not always follow this rule when physical values are expressed numerically. Thus the numerical values of $|\mathbf{B}|$ are sometimes given in gauss and sometimes in gamma. We recall that 1 G= 10^{-4} T, $1\gamma = 10^{-5}$ G i.e. $1\gamma = 1$ nT. The table to convert units of the Gaussian system into units of other systems may be found in Addendum A.1.

The time differentiation is sometimes replaced by a point over the symbol, e.g. $\dot{\rho}$ instead of $\partial\rho/\partial t$. The indices $\parallel$ and $\perp$ designate the longitudinal and the transverse vector components with respect to the external magnetic field. The signs $\pm$ in formulae like (4.4) are always arranged in such a way that the upper sign corresponds to the Alfvén wave and the lower one to the magnetosonic wave. Universal time is given as UT, and local time as LT.

In some cases one and the same symbol (γ for example) is used to designate different parameters. If the meaning of the symbol is not clear from the context then it is specially interpreted in the text.

Introduction

Electromagnetic waves of ultra-low frequency cascade over the Earth's surface from outer space. They are termed geomagnetic pulsations.

For more than a century now geomagnetic pulsations have attracted research interest because of the beauty of their shapes, and also because of their complex and puzzling behaviour. They show us an example of a self-consistent interaction of waves and particles, furnish information on the remote regions of the space environment of the Earth, and influence the course of geophysical processes to the point that they even determine individual elements of the large-scale structure of the magnetosphere. These pulsations exhibit diverse properties and constitute an important part of the world accessible to us.

The range of geomagnetic pulsations stretches from millihertz to several hertz. The waves arise in the magnetosphere of the Earth and in the interplanetery medium in front of the magnetosphere, penetrate through the ionosphere and reach the Earth's surface. But in the same frequency range 'terrestrial' sources of electromagnetic oscillations—lightning discharges, ocean waves and earthquakes—are active. The whole complex of waves of natural origin that perturb the electromagnetic field of the Earth will be termed geoelectromagnetic waves.

The longest waves are the standing Alfvén waves at the periphery of the magnetosphere. The frequency of such waves is of the order of $f \simeq c_A/2l$, where c_A is the Alfvén velocity and l is the length of the magnetic field lines. Typically, $c_A \simeq 3 \times 10^7$ cm s^{-1} and $l_{max} \simeq 1.5 \times 10^{10}$ cm and then we have $f_{min} \sim 10^{-3}$ Hz. We choose this frequency as the lower boundary of the geoelectromagnetic wave range.

The upper boundary is less clear. If we follow the traditional rule used for defining geomagnetic pulsations and consider 5 Hz to be the upper boundary, waves with frequencies of several hundreds of hertz, which are similar to pulsations but do not penetrate to the Earth as long as they are concentrated in the toroidal waveguide which surrounds the Earth, will disappear from view. Therefore, we choose 10^3 Hz as the upper boundary. In any case, the boundaries are somewhat arbitrary, i.e. set by convention. Our choice of the frequency range 10^{-3}–10^3 Hz seems reasonable and sufficient.

A few words on history

Attempts to find a general definition of a wave valid in every respect are fruitless, but we can indicate a characteristic feature. The wave can propagate, i.e. travel throughout space, and transfer energy, and possibly information, from one point to another. The propagation of a wave may be accompanied by a transfer of matter along the wave although it is not obligatory. For example, an electromagnetic wave is capable of propagating in a vacuum.

The formation of concepts on wave propagation can be traced back to great antiquity. Modern research has its beginnings in 1747 when d'Alembert deduced the equation that described the oscillations of a string. A year later, Euler obtained the solution known as the d'Alembert formula. In 1753 Daniel Bernoulli realized that the movements of a string may be described by a trigonometric series. Euler, however, objected to such an idea. It is curious that half a century later the Fourier theory was also severely criticized. Nevertheless, the theory of oscillations and waves developed successfully on the whole and at present it is a valuable source of ideas and methods for numerous applications.

Geoelectromagnetic waves as a subject of research emerged from the depths of classical geomagnetism. Evidently these waves were the first electromagnetic waves to be registered. They were observed by Celsius back in 1741 by comparing compass measurements with the pulsations of the aurora.

In the early stages, geomagnetic research was influenced directly and indirectly by the requirements of navigation, the safety of which depended on accurate magnetic direction finding. The ideological influence of Maxwell's electrodynamics is considered equally important. During the long period up to World War II experimental data were accumulated, and oscillation regimes were singled out and classified. The absence of a dominant concept or powerful economic stimuli impeded the development of this field of science, however.

After World War II, in the late 1940s and into the early 1950s, there emerged the conditions that predetermined the character and the tempo of geomagnetic wave research during the following years. By that time Alfvén had developed magnetohydrodynamics (MHD). The magnetohydrodynamic interpretation of the magnetospheric oscillations stimulated theoretical research. In addition, plasma physics had greatly advanced during this period as a result of fusion research. Space plasma physics had become a routine area in which to apply the ideas that were put forward in the course of this work. Furthermore, general scientific interest in space was stimulated by rocket and satellite research. Finally, geoelectromagnetic waves were used for sounding the Earth's crust as part of the search for oil and gas fields.

The problems with the theory

Despite the many aspects of this book, it is principally an attempt to answer just one question: how can geoelectromagnetic phenomena be understood within the theory of oscillations and waves? In order to grasp the simple and general laws underlying geoelectromagnetic phenomena, we shall study strongly idealized models. As is always the case, we risk obtaining a result that has nothing to do with reality. Therefore, the theoretical inferences should be thoroughly compared with the observational data.

The range of theoretical problems includes the following: the location of sources, the mechanism of excitation, the types of waves, the problems of propagation, the nonlinear effects, etc. The ways of solving these problems are not always clear. Most types of geoelectromagnetic waves are so complicated and intricate that there is at present no possibility of framing a unified theory which is applicable over a wide enough range of geophysical conditions. Instead, researchers try to estimate the dominant physical factors that control a specific electromagnetic process by means of observations and plausible reasoning. Having chosen these factors, they construct a model of the process which is as simple as possible. Then by combining the solutions of such local problems, introducing the necessary corrections prompted by a geophysical situation and adding general arguments and considerations, they try to synthesize the whole picture of the phenomenon.

Inexactness in our knowledge of the generation mechanism and the vagueness of the initial conditions requires the presentation of not just one but several scenarios for the origin and evolution of the waves.

Evidently, the effectiveness of practical applications depends on the progress of the physics of geoelectromagnetic waves. However, another aspect of the problem is worthy of attention. The Earth and its related environment, which is comparatively accessible for detailed study using ground and satellite observations, may be regarded as a model for the exploration of other planets in the solar system as well as remote space objects.

Time and again in the history of geomagnetic research it has transpired that the systematization and theoretical generalization of experimental data has led to important general cognitive and practical conclusions. The broad prevalence of low-frequency electromagnetic waves in the Universe and the difficulty of their direct observation everywhere except within the space environment of the Earth enables us to extrapolate (with certain reservations) some results of geomagnetic observations into the physics of remote space objects. Taking the above into account, it is most important to eliminate the slightest haziness in understanding the principal problems associated with the wave structure of the Earth's electromagnetic field.

Bibliography[1]

Information on low-frequency electromagnetic waves of natural origin may be found in the monographs by Helliwell (1965), Jacobs (1970), Guglielmi and Troitskaya (1973), Nishida (1978), Guglielmi (1979), Petviashvili and Pokhotelov (1992). The following reviews may also be useful: Campbell (1967), Troitskaya and Guglielmi (1967), Saito (1969), Dungey and Southwood (1970), Gendrin (1970), Hasegawa and Chen (1974), Southwood (1974b), Lanzerotti (1976), Guglielmi (1974, 1989), Rostoker (1979), Southwood and Hughes (1983), Guglielmi and Pokhotelov (1994).

The theory of geoelectromagnetic waves is based upon the electrodynamics of continuous media and plasma physics; magnetohydrodynamics turns out to be particularly useful in almost every case with some slight exceptions. To analyse the perturbations of a geoelectromagnetic field by seismic waves, additional information on elasticity theory and the physics of electrokinetic phenomena should be used. For this purpose, the handbooks on physics by Landau and Lifshitz (1970, 1984, 1988a), Lifshitz and Pitaevsky (1979) are recommended; however, first it is advisable to read the classical monograph by Alfvén and Fälthammar (1963) on magnetohydrodynamics.

The general theory of oscillations and waves may serve as a source of useful ideas particularly when studying nonlinear effects. The literature on this subject is voluminous. However, we single out the basic monograph by Whitham (1974). An introduction into the theory of oscillations and waves is given in the books by Gorelik (1959) and Pain (1979).

Information on the observation of magnetic pulsations in 1741 was borrowed from Buchert, Haerendel and Baumjohann (1990), who referred to Brekke and Egeland (1983). It is noteworthy that 250 years later a picture of waves observed by Celsius may be reconstructed, and using modern instrument data a model of this phenomenon may be created (see Buchert *et al* 1990). Information on the observations of the 19th century may be obtained from the monograph by Chapman and Bartels (1940).

Useful information on the development of the theory of oscillations and waves may be found in the booklet by Ostrovsky and Potapov (1988).

---

[1] In order not to interrupt the coherence of the text we have placed the bibliographic commentary at the end of each chapter.

Chapter 1

The Earth's crust

The elastic waves in the Earth's crust caused by earthquakes, explosions and so on excite oscillations of the geomagnetic field. In this chapter we shall consider the inductive and electrokinetic mechanisms of the generation of magnetic signals. The action of the first is based on Faraday's law of electromagnetic induction. The main point of the mechanism is the excitation of Foucault currents which oscillate in an external magnetic field in a conducting body. The action of the second is related to the movement of the fluid that fills pores and cracks of the rock. We shall deduce a simple equation that describes both these manifestations of seismomagnetism and show that the induction mechanism prevails at relatively low frequencies and the electrokinetic mechanism at relatively high frequencies.

Other forms of electromagnetic activity of the Earth's crust are also known. For example, indirect observations testify to the generation of electromagnetic impulses when the proliferation of cracks takes place during the preparatory period of an earthquake as well as at the very moment of the earthquake when the main discontinuity in the compactness of the rock is being formed. Many of these latter processes are still vague and enigmatic.

1.1 Seismic waves

The Earth's crust is separated from the mantle by the Mohorovičić discontinuity through which the velocities of elastic waves and the substance density undergo an abrupt increase. The effective crust thickness is typically 35 km; however, the thickness in the ocean differs considerably from that of mountain regions. The crust and the relatively solid upper part of the mantle is called the lithosphere.

The continental crust has a three-layer structure. The upper layer is formed of sedimentary rock. The thickness varies from hundreds of metres to 10–20 km (in large canyons). The intermediate layer consists of granites and gneisses and its thickness is 15–20 km over the plains and 30–35 km in mountainous regions. The lowest layer consists of basalt rock and may reach 30 km in thickness. The total thickness of the continental crust is 35–75 km. The relationship between the thickness of the crust and the relief is inverse in the sense that the crust under the plains is thinner than under the mountains.

The ocean crust is similarly formed in three layers. The uppermost layer is tens to hundreds of metres thick and consists of loose sediments. Unlike the continental crust, the ocean region does not possess a granite layer. Instead, the centre layer contains intermixed sheets of compressed sediments and basalt lavas. The thickness is of the order of 1–2 km. The lowest layer is formed of basalts and the compound thickness of the ocean crust is 5–10 km. The relationship between the crust and the relief is direct: that is the thinning of the crust corresponds to the rising of the bottom and vice versa.

The density and elastic properties of the rocks which form the crust determine the velocities of seismic waves. Three main types of seismic waves are distinguishable: longitudinal, transverse and surface waves. The velocities of longitudinal waves in various crust layers are as follows:

sedimentary rock (continent)	3–5 km s^{-1}
(ocean)	1.5–2 km s^{-1}
granite layer	5.5–6.5 km s^{-1}
basalt layer	6.5–7.5 km s^{-1}.

Transverse waves propagate more slowly than longitudinal waves. Consequently, longitudinal waves are sometimes termed primary and transverse waves, secondary waves. Therefore, these two types of waves differ from one another in polarization and propagation velocity. However, they have in common that they are both body waves and they may penetrate into the depths of the Earth's crust. In contrast, the surface waves propagate only along the Earth's surface. Their amplitude decreases exponentially with depth.

Surface waves may be of two types: Love waves and Rayleigh waves. They differ from each other in propagation velocity and polarization. The Love wave has transverse polarization. The oscillations of the ground take place in a horizontal plane in a direction perpendicular to the wave propagation. In contrast, the Rayleigh wave presents a superposition of longitudinal and transverse oscillations. The ground particles move along ellipses in a vertical plane drawn through the direction of propagation. Usually the Love wave is faster than the Rayleigh wave but both propagate more slowly than the body waves.

All types of seismic waves perturb the geomagnetic field. One basis for the task ahead lies in defining the magnetic field oscillations for the given field of elastic deformations. In this approach the ponderomotive forces which occur during the interaction between the main magnetic field and the electric currents induced by a seismic wave are not taken into account.

So, first of all, a mechanical problem concerning the propagation of elastic waves should be solved. We could immediately avail ourselves of known solutions but we shall reproduce the arguments briefly in order to provide a complete picture.

We shall consider the Earth's crust as an isotropic elastic body. Under adiabatic deformations the mechanical properties of the body in the linear approximation are completely determined by three parameters, namely the density ρ and two elastic constants such as the modulus of dilatation K and the shift modulus G. For example for granite the values of these parameters are: $\rho = 2.7$ g cm^{-3}, $K = 5 \times 10^{11}$ dyn cm^{-2} and $G = 2.4 \times 10^{11}$ dyn cm^{-2}.

In order to avoid misunderstanding, we note that these are just some typical values and are only presented here as a guide to the orders of magnitude of the values. The real values of parameters change noticeably with the mineral composition and the petrographic structure of the rock. It should also be noted that instead of K and G the Lame coefficients or the Young's modulus and the Poisson's coefficient are often used in the literature. Certain relationships exist between these three pairs of constants which allow the calculation of the unknown moduli on the basis of the two known ones, but we will not dwell further on this question.

The deformation of the body is described by the vector field $\boldsymbol{\xi}(\boldsymbol{x})$, where $\boldsymbol{\xi}$ is a displacement vector of a body point and $\boldsymbol{x}$ is the position of this point before the deformation. The change of the length element resulting from the deformation of the body is determined by the so-called strain tensor

$$u_{ik} = \tfrac{1}{2}\left(\frac{\partial \xi_i}{\partial x_k} + \frac{\partial \xi_k}{\partial x_i}\right). \tag{1.1}$$

Using u_{ik} and the moduli of elasticity K and G the Hookian stress tensor (Hooke's law) may be written in the form:

$$\sigma_{ik} = K u_{ll}\delta_{ik} + 2G(u_{ik} - \tfrac{1}{3}\delta_{ik}u_{ll}) \tag{1.2}$$

where δ_{ik} is the Kroneker delta symbol ($\delta_{ij}=0$ if $i \neq j$ and 1 if $i = j$) and u_{ll} is the sum of diagonal elements of u_{ij}.

Finally, using σ_{ik} the equation of motion of an elastic body is written in the form

$$\rho\ddot{\xi}_i = \frac{\partial \sigma_{ik}}{\partial x_k}. \tag{1.3}$$

Here, as well as in (1.1) and (1.2), the deformation is presumed small and accordingly only linear terms are retained.

Using (1.1) and (1.2) we rewrite the equation of motion (1.3) in the form

$$\ddot{\boldsymbol{\xi}} = c_t^2\nabla^2\boldsymbol{\xi} + (c_l^2 - c_t^2)\nabla(\nabla \cdot \boldsymbol{\xi}). \tag{1.4}$$

The following notations are introduced here

$$c_l = \left(\frac{3K + 4G}{3\rho}\right)^{1/2} \qquad c_t = \left(\frac{G}{\rho}\right)^{1/2}. \tag{1.5}$$

The parameters c_l and c_t may be used instead of K and G in order to describe the elastic properties of a medium. In particular, the elastic stress tensor (1.2) may be written as

$$\sigma_{ik} = 2\rho c_t^2 u_{ik} + \rho(c_l^2 - 2c_t^2)u_{ll}\delta_{ik}. \tag{1.6}$$

It follows from this formula that $c_l > (\frac{4}{3})^{1/2}c_t$. However, this inequality is deduced directly from (1.5) and the conditions $K > 0$, $G > 0$. We also note that a stronger inequality $c_l > c_t\sqrt{2}$ exists for all known bodies. This corresponds to the inequality $K > (\frac{2}{3})G$.

We now show that c_l and c_t are the velocities of longitudinal and transverse waves. For this purpose we shall imagine that an elastic body is homogeneous and infinite, i.e. it occupies the whole of space. We shall seek a solution to equation (1.4) in the form of plane monochromatic waves

$$\boldsymbol{\xi}(\boldsymbol{x}, t) = \boldsymbol{a}e^{i(\boldsymbol{k}\cdot\boldsymbol{x}-\omega t)} \tag{1.7}$$

where ω is the wave frequency, $\boldsymbol{k}$ is the wave vector and $\boldsymbol{a}$ is a constant complex vector (a complex wave amplitude).

Let us make a slight digression and explain the meaning of the notation $\boldsymbol{\xi}$ in the form of a complex value. Generally speaking, we are interested in real solutions of equation (1.4), that is, we suppose that a plane monochromatic wave has the form $\cos(\boldsymbol{k}\cdot\boldsymbol{x} - \omega t)$. The use of complex values like (1.7) is reasonable and permissible if the values undergo linear operations, but when the calculations are completed the real part of the corresponding value should be singled out.

The substitution of (1.7) into (1.4) leads to the replacement of the operators $\partial/\partial t$ and ∇ by $-i\omega$ and $i\boldsymbol{k}$ respectively. As a result, the differential equation (1.4) turns to an algebraic one

$$(\omega^2 - c_t^2 k^2)\boldsymbol{a} = (c_l^2 - c_t^2)\boldsymbol{k}(\boldsymbol{k} \cdot \boldsymbol{a}). \tag{1.8}$$

We can easily verify that the linear homogeneous equation (1.8) has nontrivial solutions only in the case that the wave frequency yields one of the relations

$$\omega = c_l k \tag{1.9}$$

or

$$\omega = c_t k. \tag{1.10}$$

The corresponding eigenvector solutions will be designated as $\boldsymbol{a}_l$ and $\boldsymbol{a}_t$. The vector $\boldsymbol{a}_l$ is parallel to the direction of propagation determined by the vector $\boldsymbol{k}$. The vector $\boldsymbol{a}_t$ lies in the plane perpendicular to this direction. Absolute values of the vectors $\boldsymbol{a}_l$ and $\boldsymbol{a}_t$ are arbitrary since equation (1.8) is homogeneous. The vector $\boldsymbol{a}$ may be presented in the form of the sum

$$\boldsymbol{a} = \boldsymbol{a}_l + \boldsymbol{a}_t \tag{1.11}$$

where the relations

$$k \times a_l = 0 \qquad k \cdot a_t = 0. \qquad (1.12)$$

hold.

A relation of either type (1.9) or (1.10) that determines the dependence of frequency on the wave vector is termed a dispersion law. It determines the phase $v_{ph} = \omega/k$ and the group $v_g = \partial \omega/\partial k$ wave velocities. In the case presented we have two waves independent from one another which propagate at phase velocities c_l and c_t. The group velocities are directed along the wave vector k. The moduli of the group velocities coincide with the phase velocities. The wave propagating at velocity c_l is termed longitudinal since the displacement ξ_l is directed along the wave propagation. The other wave propagating with velocity c_t is transverse since ξ_t is perpendicular to k.

The conclusion concerning the existence of two elastic waves propagating independently through an isotropic, homogeneous and infinite medium remains correct for the general case and not only for plane monochromatic waves. So let us return to the initial equation (1.4) and represent the displacement vector in the form of the sum

$$\xi = \xi_l + \xi_t. \qquad (1.13)$$

The longitudinal and transverse conditions analogous to (1.12) have the form

$$\nabla \times \xi_l = 0 \qquad \nabla \cdot \xi_t = 0 \qquad (1.14)$$

i.e. the vector ξ_l is potential and the vector ξ_t is solenoidal. Then (1.4) may be split into two independent wave equations

$$\ddot{\xi}_l - c_l^2 \nabla^2 \xi_l = 0 \qquad (1.15)$$

$$\ddot{\xi}_t - c_t^2 \nabla^2 \xi_t = 0 \qquad (1.16)$$

which describe the propagation of elastic waves with velocities c_l and c_t. Let us mention in passing that in granite typical velocities of longitudinal and transverse waves are $c_l = 5.5$ km s^{-1} and $c_t = 3$ km s^{-1}.

We now have to specify the model of the Earth's crust. Let an elastic solid body occupy a half-space, the free surface of which simulates the Earth's surface, considered to be plane. This idealization places some restrictions on the wavelengths of seismic waves in the sense that they must be small in comparison with, not only the Earth's radius, but also with the thickness of the Earth's crust.

For example, assume the body to occupy the half-space $z < 0$ (figure 1.1), and let a plane monochromatic longitudinal wave be incident upon the surface of the body at an angle θ_0. For simplicity we assume that the body borders on a vacuum[1]. Therefore, the boundary condition to be met on the surface of the body is

$$\sigma_{ik} n_k = 0 \qquad (1.17)$$

[1] In reality a small portion of a longitudinal seismic wave energy is refracted into the atmosphere in the form of an acoustic wave. Here we ignore this effect.

where n is the unit vector normal to the surface. Using equation (1.6) we can easily verify that condition (1.17) cannot be accomplished in the general case if the reflected wave is strictly longitudinal (the exception to the rule arises at $\theta_0 = 0$ only). When a longitudinal wave is incident at an arbitrary angle θ_0 the reflected wave field is composed of a superposition of longitudinal and transverse waves. The angles of reflection θ_l and θ_t are deduced on simple kinematic grounds: $\theta_l = \theta_0$, $\sin \theta_t = (c_t/c_l)\sin \theta_0$. The coefficients of reflection are complicated and are not given here but they may be determined by using the boundary conditions $\sigma_{xz} = \sigma_{zz} = 0$.

Similarly in the general case the reflected field incorporates both types of waves if a transverse wave is incident upon the surface of the body. The case where the polarization of the incident wave is perpendicular to the plane of incidence is exceptional with only one transverse wave being reflected in such a case.

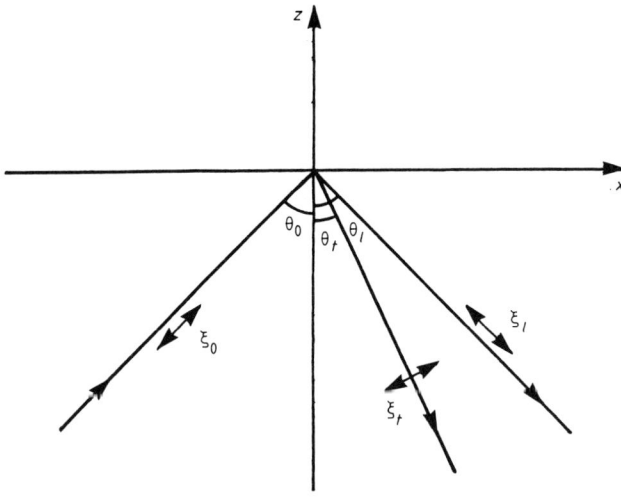

Figure 1.1. The incidence of the longitudinal elastic wave ξ_0 on the surface of the body, filling up the half-space $z < 0$. Longitudinal ξ_l and transverse ξ_t waves arise during the reflection. Double-headed arrows indicate the wave polarization.

Therefore, longitudinal and transverse elastic waves are not independent of each other in a bounded body (in contrast to a boundless body). On reflection they may be transformed into one another. Besides this distinction, the surface waves may propagate in a bounded body.

As above let a homogeneous body occupy a half-space $z < 0$ adjacent to a vacuum. Consider the monochromatic Rayleigh wave

$$\xi = \phi(z)\exp(ikx - i\omega t) \tag{1.18}$$

which propagates in the direction x and damps as $z \to -\infty$. The vector field $\xi(x, z, t)$ is expanded in longitudinal and transverse components governed by the wave equations (1.15) and (1.16) respectively. From the condition that oscillations vanish as $z \to -\infty$, and boundary condition (1.17), we find the dispersion relation for Rayleigh waves

$$\omega = c_R k \tag{1.19}$$

and nonzero components of the vector ϕ

$$\begin{aligned}
\phi_x &= \kappa_t u \exp(\kappa_t z) + k w \exp(\kappa_l z) \\
\phi_z &= -iku \exp(\kappa_t z) - i\kappa_l w \exp(\kappa_l z)
\end{aligned} \tag{1.20}$$

where

$$\kappa_t = k(1 - c_R^2/c_t^2)^{1/2}$$
$$\kappa_l = k(1 - c_R^2/c_l^2)^{1/2}$$
$$c_R = \nu c_t$$
$$u/w = -(1 - \nu^2/2)/(1 - \nu^2)^{1/2}.$$

The value ν increases monotonically from 0.874 to 0.955 with increase of Poisson's coefficient from 0 to $\frac{1}{2}$.

The velocity of a Rayleigh wave is somewhat lower than the velocity of a transverse wave. Usually we have $c_R < 0.92 c_t$. The waves have elliptical polarization as has already been mentioned.

The problem of the propagation of the Love wave is solved similarly. However, the stratified structure of the Earth's crust should be taken into account in this case. The velocity of the Love wave lies in a range, the lower border of which is equal to the velocity of transverse waves c_{t_1} in the uppermost layer and the upper border equals the velocity c_{t_2} in the lower-lying layers, so that

$$c_{t_1} < c_L < c_{t_2}.$$

There is no Love wave on the surface of a homogeneous body. It is also absent if $c_{t_1} > c_{t_2}$.

We have presented a reasonably complete picture of the mechanical motions of the Earth's crust during earthquakes and now, being fully conversant with the matter, we can proceed to the discussion of associated electromagnetic phenomena. Before that, however, we return to the problem of ponderomotive forces, i.e. electromagnetic forces exerted on a body.

Let us take into account the conductivity of the Earth's crust and add the ponderomotive force to the right-hand side of the equation of motion (1.3). Then the equation of motion and the equations of quasistationary electrodynamics will compose a self-consistent system analogous to the set of equations of magnetohydrodynamics (see exercise 1.1.1).

However, it is not necessary since a strong inequality

$$B \ll G^{1/2} \tag{1.21}$$

is established with ample margins in the Earth's crust (B is the value of the geomagnetic field)[2]. According to inequality (1.21), we may ignore the influence of the ponderomotive force on the motion of a solid body and solve the problem of the excitation of electromagnetic oscillations in the so-called kinematic approximation when the field $\xi(x, t)$ is considered given.

On the other hand, nothing prevents us from assuming as a basis a self-consistent set of equations of magnetoelastics and seeking a solution in the form of magnetoelastic waves similar to Alfvén waves. Then in the zero-order approximation to the small parameter $G^{-1/2}B$, solutions in the form of ordinary elastic waves would be obtained and, in the first approximation, perturbations of the electromagnetic field would be found. In other words, the magnetoelastic wave differs from the elastic wave in polarization. That is, the propagation of a magnetoelastic wave is accompanied not only by the deformation of the body but also by the oscillations of the electromagnetic field. From this point of view the problem that we take interest in is that of the polarization of magnetoelastic waves.

We shall take into account the inequality (1.21) from the very beginning and consider the problem in the given field approximation of an elastic wave.

Exercises

Exercise 1.1.1.

Find the dispersion law for the transverse magnetoelastic waves in a perfectly conducting solid body[3].

Solution 1.1.1.

The effective electric field $E + c^{-1}v \times B$ producing the conductivity current equals zero in an ideal conductor, i.e.

$$E = -\frac{1}{c}v \times B.$$

Here E is the intensity of the electric field, B is the magnetic induction and v is the velocity of the body motion, $v = \partial\xi/\partial t$. Substituting this into Maxwell's

---

[2] Inequality (1.21) is valid for all solid bodies known to us (see exercise 1.1.2).
[3] Some of the problems, including this one, have been compiled on the basis of other authors' works. In all such cases see the references to the original in the bibliographic commentary presented at the end of each chapter. The set-up of the problem and the method of the solution are not, as a rule, identical to the authors' originals.

equation

$$\nabla \times E = -\frac{1}{c}\frac{\partial B}{\partial t}$$

we find

$$\frac{\partial B}{\partial t} = \nabla \times (v \times B). \tag{1}$$

We will write the equation of solid body motion (1.3) in the form

$$\rho\frac{\partial^2 \xi}{\partial t^2} = \mu\nabla^2\xi + (\lambda + \mu)\nabla(\nabla \cdot \xi) + f. \tag{2}$$

The Lame coefficients $\lambda = K - (\frac{2}{3})G$ and $\mu = G$ are used here instead of the moduli of elasticity K and G. Furthermore, the volume density of electromagnetic force

$$f = c^{-1}j \times B \tag{3}$$

is added to the right-hand side of equation (2). In the quasistationary approximation the current density is expressed through the magnetic field intensity in the following way

$$j = \frac{c}{4\pi}\nabla \times H. \tag{4}$$

Suppose $H = B$, i.e. we shall consider that the magnetic permeability is indistinguishable from unity.

Equations (1)–(4) combined with the equation $\nabla \cdot B = 0$ form a closed self-consistent set of equations, analogous to the equations of magnetohydrodynamics. They are termed the equations of ideal magnetoelastics.

The body deformation is considered small. Consequently, it is reasonable to consider that the magnetic field perturbation is also small. Let us substitute $B(x, t)$ for $B + b(x, t)$ where B now is a constant magnetic field of external sources and b is a small perturbation. We shall obtain linearized equations of ideal magnetoelastics and seek their solutions in the form of plane monochromatic waves. For this we shall substitute the operators ∇ and $\partial/\partial t$ for ik and $-i\omega$ respectively. As a result we shall get a homogeneous set of linear algebraic equations. The matching condition of the system leads to the dispersion equation. Its three, essentially different, roots correspond to the three types of magnetoelastic waves that are analogous to magnetohydrodynamic waves. For transverse waves ($k \cdot \xi = 0$) the dispersion law has the form

$$\omega = (c_A^2 k_\parallel^2 + c_t^2 k^2)^{1/2}$$

where $c_A = B/(4\pi\rho)^{1/2}$ is the Alfvén velocity. In a fluid $c_t = 0$ and instead we have the known dispersion law for the Alfvén waves

$$\omega = c_A k_\parallel.$$

We see that shearing stress in a solid body leads to a situation where the wave does not disappear at $k_\parallel = 0$ as it is in a fluid where $G = 0$.

In solid bodies $c_A \ll c_t$ with ample margins. Therefore, with the greatest possible accuracy

$$\omega = c_t k$$

i.e. the magnetic field does not influence the wave propagation.

Exercise 1.1.2.

Evaluate the parameter $\alpha = G^{-1/2}B$ in the crust of pulsars.

Solution 1.1.2

The magnetoelastic waves analogous to the Alfvén waves possess interesting properties (see exercise 1.1.1). Naturally there arises a problem in searching for such waves. It is senseless to search for them over the Earth due to the fact that the parameter α is rather small (of the order of 10^{-6}). Maybe they exist over other space bodies? Possibly they do not exist at all.

Extremely strong magnetic fields ($B \simeq 10^{12}$–10^{13} G) are found in pulsars. A pulsar has a type of solid crust. At times 'pulsarquakes' happen in the course of which magnetoelastic waves could emerge. However, the shearing modulus in the pulsar's crust $G \simeq 10^{30}$ dyn cm^{-2}. Consequently, the parameter α does not exceed 10^{-2} even in a pulsar.

1.2   The inductive seismomagnetic effect

The conductivity of rock varies widely. The highest conductivity is intrinsic to moist stratified sandstone and shale. Lower conductivity occurs in dry gneisses and limestones and extremely low values are found in pure samples of rock-forming minerals. Quartz, for example, has $\sigma \simeq 10^{-3}$ s^{-1}.

Let us present some reference data on conductivity:

sedimentary rocks	10^7–10^{10} s^{-1}
metamorphic rocks	10^2–10^6 s^{-1}
igneous rocks	10^3–10^7 s^{-1}

We recall that 1 S m^{-1} = 9×10^9 s^{-1} (see also Addendum A.1).

The vertical profile σ is formed under the influence of many factors (humidity, temperature, pressure, mineral composition, petrographic structure etc). The upper water-saturated crust layer forms a highly conducting film. There is a thick layer of a low-conducting medium at a depth of 1.5–3 km which is 50–100 km thick in the continental regions and is termed a crystal base. The temperature increases with depth and partial melting of the rocks occurs at some level. New conductivity mechanisms are switched on here and

σ rapidly increases with depth. The surface that limits the crystal base from below is termed a highly conducting foundation.

The movement of the conducting layers of the Earth's crust in the magnetic field of the Earth's core during seismic wave propagation induces an alternating electromagnetic field. The effect is described by Maxwell's equations with a quasistationary approximation

$$\nabla \times E = -c^{-1}\partial B/\partial t \qquad \nabla \times H = 4\pi c^{-1}j$$
$$\nabla \cdot B = 0 \qquad j = \sigma(E_{\text{ef}} + c^{-1}v \times B). \tag{1.22}$$

Here E and H are the intensities of electric and magnetic fields respectively, $B = \mu H$ is the magnetic induction, μ is the magnetic permeability, σ is the conductivity, v is the medium motion velocity and c is the velocity of light. We shall specify the meaning of E_{ef} in section 1.3 and here we simply assume $E_{\text{ef}} = E$. This is enough to describe the inductive seismomagnetic effect.

The condition for the applicability of the quasistationary approximation has the form

$$\omega \ll \sigma/\varepsilon \tag{1.23}$$

where ω is the frequency of the field oscillations, and ε is the dielectric permeability of the rocks. This means that we may neglect displacement current when compared with the conductivity current. Inequality (1.23) is accomplished in the frequency range of seismic waves.

Equations (1.22) describe the electromagnetic field under the Earth's surface. Above the Earth's surface the second equation may be replaced by

$$\nabla \times H = 0. \tag{1.24}$$

Here the electroconductivity of air is disregarded. Moreover, equation (1.24) implies that the length of the seismic waves is much shorter than both the distance from the Earth's surface to the ionosphere (about 100 km) and the vacuum wavelength c/ω. The latter condition requires that the seismic wave velocity be much lower than the velocity of light, which is clearly the case inside the Earth's crust.

For simplicity and greater clarity we shall ignore the magnetic structure of the rock substance and assume $\mu = 1$, i.e. $B = H$. Yet it should be borne in mind that this assumption is conditional. It is justified merely by considerations of simplicity and convenience in the presentation of this material. The real rock possesses magnetic structure which causes piezomagnetic phenomena accompanying propagation of seismic waves.

Let us redefine $B(x, t) \rightarrow B + b(x, t)$, where B is the main magnetic field, and $b(x, t)$ is the inductive seismomagnetic signal. It is reasonable to consider the field B to be constant and homogeneous when analysing seismomagnetic signals. *A priori* $|b| \ll |B|$, then from (1.22) it follows that

$$\frac{\partial b}{\partial t} - \frac{c^2}{4\pi\sigma}\nabla^2 b = (B \cdot \nabla)v - B(\nabla \cdot v). \tag{1.25}$$

Here we have introduced a simplifying assumption that $\sigma = $ constant. This is a rough idealization since the electroconductivity of the Earth's crust changes with depth. The assumption may be partly weakened if we consider σ to be a piecewise function of depth. Then only the replacement $\sigma \to \sigma_n$, where n denotes the number of the Earth's crust layer with the given value of electroconductivity, should be made in equation (1.25).

The perturbation of the geomagnetic field over the surface of the Earth yields the Laplace equation

$$\nabla^2 b = 0. \tag{1.26}$$

Moreover, in and over the Earth the following condition holds

$$\nabla \cdot b = 0. \tag{1.27}$$

Equations (1.25)–(1.27) should be supplemented by the boundary conditions. The field b must be continuous at all separation boundaries, the surface of the Earth included. Furthermore, the field b must decrease when moving off the Earth's surface upwards since the source of the field is supposed to be located in the Earth's crust. If the source is a surface elastic wave, then the additional condition of decreasing b exists when descending into the depths of the Earth.

Now a velocity field $v(x, t)$ should be given in one form or another since the magnetic field damps inside the motionless Earth's crust under any initial conditions. The field $v(x, t)$ acts as a magnetic signal generator (see equation (1.25)). We shall determine $v = \partial \xi / \partial t$ according to the known displacement field $\xi(x, t)$ that emerges with the propagation of seismic waves of one type or another.

We have stated the problem of searching for the magnetic field of a seismic wave. On defining the magnetic field it will be possible to find the electric field and current according to the formulae

$$\begin{aligned} E &= j/\sigma - (1/c)v \times B \\ j &= (c/4\pi)\nabla \times b \end{aligned} \tag{1.28}$$

which follow from (1.22) with taking into account the assumptions we have made.

1.2.1 Body waves

Let us first consider a longitudinal elastic wave in a homogeneous infinite medium. Let the plane monochromatic wave

$$\xi \propto \exp[ik(x - c_l t)] \tag{1.29}$$

propagate in the x direction at velocity c_l. Here $k = \omega/c_l$ is the wavenumber and ω is the wave frequency. We know that the displacement vector ξ in a longitudinal wave yields the condition $\nabla \times \boldsymbol{\xi} = 0$ (see (1.14)). In a plane wave (1.29) the vector $\boldsymbol{\xi}$ is parallel to the direction of propagation x, i.e.

$$\boldsymbol{\xi} = (\xi, 0, 0). \tag{1.30}$$

From condition (1.27) it follows that the vector $\boldsymbol{b}$ lies in the plane which is perpendicular to the direction of propagation, that is $b_x = 0$. The orientation of $\boldsymbol{b}$ in this plane coincides with the orientation of the right-hand side of the vector equation (1.25). Without limiting the generality, let us direct the axes y and z in such a way that $\boldsymbol{B} = (B_{\parallel}, 0, B_{\perp})$. Then the vector $\boldsymbol{b}$ has only one component

$$\boldsymbol{b} = (0, 0, b). \tag{1.31}$$

From equation (1.25) and considering (1.29)–(1.31) we find the magnetic effect of a longitudinal elastic wave

$$b = k\xi B_{\perp}/(\mathrm{i} - kD). \tag{1.32}$$

Here we have introduced the designation

$$D = c^2/4\pi\sigma c_l. \tag{1.33}$$

The study of the induction mechanism of magnetic oscillation generation in the course of transverse elastic wave propagation is made similarly. The result is derived from (1.32) and (1.33) by a simple substitution $c_l \rightarrow c_t$, $B_{\perp} \rightarrow -B_{\parallel}$. The vector $\boldsymbol{b}$ is polarized just as $\boldsymbol{\xi}$ in this case. For example at

$$\boldsymbol{\xi} = (0, \xi, 0) \tag{1.34}$$

we have

$$\boldsymbol{b} = (0, b, 0). \tag{1.35}$$

The solution (1.32) contains two dimensionless parameters $k\xi$ and kD. Let us introduce a third parameter which equals the ratio of the first two and designate it $R_m = \xi/D$. This is the Reynold's magnetic number for the seismic waves.

Instead of the parameter kD we shall use the inverse value

$$L_n = 2T\sigma(c_l/c)^2. \tag{1.36}$$

Here $T = 2\pi/\omega$ stands for the oscillation period. The parameter L_n is analogous to the known Lundquist parameter. The magnetic field is 'frozen-in' into the substance at $L_n \gg 1$, and not at $R_m \gg 1$, as could have been assumed (which is really the case on some other occasions). For the seismic waves $R_m \ll 1$ always, whereas the parameter $L_n \sim 1$ and even may noticeably exceed unity.

For example, let $c_l = 5.5 \times 10^5$ cm s^{-1}, $\sigma = 5 \times 10^8$ s^{-1} and $T=3$ s. Then $L_n = 1$. The evaluation of the parameter L is useful and helpful when analysing all types of seismic waves. Only c_l should be substituted for the corresponding velocity of wave propagation in the formula (1.36).

Under the frozen-in condition ($L_n \gg 1$) the magnetic signal amplitude does not depend on the electroconductivity medium and equals

$$b \sim k\xi B. \qquad (1.37)$$

In the opposite limiting case ($L_n \ll 1$, diffusion regime) we have the following estimation

$$b \sim R_m B \qquad (1.38)$$

where

$$R_m = 4\pi\sigma c_l\xi/c^2 \qquad (1.39)$$

in the case of longitudinal waves. At $L_n \sim 1$ both formulae lead to a similar estimate.

The magnetic effect of body waves at fairly large depth under the Earth's surface is roughly the same as in the infinite medium. The effect becomes complicated due to the interference of the incident and reflected waves over the Earth's surface. Formulae analogous to (1.32) are not difficult to obtain but we will not dwell upon them. We just note that the magnetic signal on the Earth's surface is absent under the normal incidence of an elastic wave over the ground–air boundary. This is true for longitudinal as well as for transverse waves. The magnetic signal is also absent under inclined incidence of a transverse wave that is polarized perpendicularly to the plane of incidence. A wave of a type similar to the incident wave is reflected from the Earth's surface in all three cases.

1.2.2   The electromagnetic field of the Rayleigh wave

The body waves undergo rapid damping when moving away from the earthquake epicentre, and the Love wave, owing to its polarization pecularities, causes a negligibly small inductive magnetic effect over the Earth's surface. Therefore, the surface Rayleigh wave should be considered first far from the epicentre.

We shall consider the Earth's crust as a homogeneous elastic body that occupies the half-space. Let a monochromatic elastic wave of the form of (1.18)–(1.20) propagate along the plane surface of the body. Out of the body ($z > 0$) the magnetic field b yields the Laplace equation (1.26) and inside the body ($z < 0$) it yields the induction equation (1.25). At the body boundary ($z = 0$) the field b must be continuous. When moving off the boundary upwards ($z \to \infty$) and downwards ($z \to -\infty$) the field must vanish.

The initial conditions are not required since we are interested in stationary forced oscillations of the magnetic field. The dependence of b on t and x in this case is the same as for ξ. Then the solution of equation (1.26), assuming the boundary conditions at infinity, has the form

$$b \propto e^{-kz} \qquad (1.40)$$

at $z \geq 0$. Substituting this in (1.27) we obtain a polarization relationship

$$b_x = -ib_z. \tag{1.41}$$

Further on we shall reveal that the component b_y is negligibly small. Taking this into account, (1.41) signifies that magnetic oscillations have circular polarization in the vertical plane crossing the direction of propagation on the surface of the body and over it.

Now let us turn to equation (1.25). Considering (1.18)–(1.20) it takes the form

$$
\begin{aligned}
(\partial^2 - q^2)b_x &= p^2[u\kappa_t T(z) + w\kappa_l L(z)] \\
(\partial^2 - q^2)b_y &= -ip^2 w(k^2 - \kappa_l^2)B_y \exp(\kappa_l z) \\
(\partial^2 - q^2)b_z &= -ip^2[uT(z) + wL(z)]k
\end{aligned}
\tag{1.42}
$$

where

$$
\begin{aligned}
T &= (ikB_x + \kappa_t B_z)\exp(\kappa_t z) && L = (i\kappa_l B_x + kB_z)\exp(\kappa_l z) \\
p^2 &= 4\pi i\sigma\omega/c^2 && q^2 = k^2 - p^2 && \partial \equiv \partial/\partial z.
\end{aligned}
$$

The dependence on time and coordinate x in the form $\exp(ikx - i\omega t)$ is implied everywhere but it is not written explicitly. Equation (1.42) should be supplemented with the relation

$$b_x = (i/k)\partial b_z \tag{1.43}$$

which follows from (1.27).

Let us start by considering the equations for b_x and b_z. We solve them using the arbitrary constant variation method. As a result we obtain the amplitude of the magnetic field oscillations on the surface of the body ($z = 0$)

$$b_x = \left[\frac{uT(0)}{q + \kappa_t} + \frac{wL(0)}{q + \kappa_l}\right]\left(\frac{p^2 k}{q + k}\right). \tag{1.44}$$

Here $\mathrm{Re}\, q > 0$. Obviously we have $b_z = ib_x$. The dependence on t and x as well as on z at $z > 0$ is also clear. At $z < 0$ the dependence on z is rather complicated and it is not presented here.

Similarly, we seek b_y. Here the displacement current $c^{-1}\partial E/\partial t$ at $z > 0$ should be taken into account. On doing so we find out that b_y is approximately (σ/ω) times smaller than b_x and b_z.

The electromagnetic field components unite into two groups: b_x, b_z, E_y and E_x, E_z, b_y. It is natural to term the corresponding oscillations as oscillations of magnetic and electric types. We emphasize that both these types are not independent, and the six components form a united structure of the Rayleigh wave electromagnetic field. The electric component of magnetic-type oscillations at $z \geq 0$ equals

$$E_y = i(c_R/c)b_x. \tag{1.45}$$

Equation (1.45) is valid within the rest reference frame. In the reference system connected with the oscillating surface of the body, a corresponding kinematic term should be added to the right-hand side of (1.45).

We note that the polarization relationships (1.41) and (1.45) may be used to recognize the inductive seismomagnetic signal under interference conditions.

Within the ideal conductivity limits or infinitely small oscillation frequencies, equation (1.44) may be reduced to the form

$$b_x = -k[uT(0) + wL(0)].   (1.46)$$

The applicability criterion (1.46) has the form $L_n \gg 1$, where L_n is defined by the formula (1.36) provided c_l is substituted by c_R. Within the diffusional limits ($L_n \ll 1$) from (1.44) it follows that

$$b_x = \frac{p^2}{2}\left[\frac{uT(0)}{\kappa_t + k} + \frac{wL(0)}{\kappa_l + k}\right].   (1.47)$$

It will be useful to compare (1.46) and (1.47) with (1.37) and (1.38) respectively.

1.2.3   Comparison with observations

For more than a century now seismoelectromagnetic phenomena have been attracting the attention of geophysicists. Primarily these phenomena are of interest as long as they raise opportunities to find new ways to predict earthquakes. Secondly, the interest is justified by the fact that the observation of electromagnetic manifestations in seismic activity may present basic information on the physical processes in the depths of the Earth.

Seismomagnetic oscillations are difficult to observe. In the first place, an ordinary earthquake is too weak to cause an appreciable magnetic effect. In addition the observations are affected by the so-called seismographic effect caused by the motions of the magnetometer in a strong external magnetic field due to seismic wave propagation. This difficulty may be overcome if a quantum magnetometer is used, but a part of the information on the useful signal will be lost, since the quantum magnetometer registers only the modulus of total magnetic field vector.

Let us discuss the features of the magnetic signal recorded by the quantum magnetometer during a particular earthquake of extraordinary force.

The earthquake took place in Alaska on 28 March 1964 at 03 hours 36 min 10 sec UT. The epicentre was located in the northern part of Prince Williams Bay at 61°6′ N, 147°48′ W. The ripping of the rock started at a depth of 20–30 km. The magnitude of the earthquake was $M = 8.6$ on the Richter scale.

The seismomagnetic signal was recorded by a helium magnetometer installed in Berghen Park, Colorado. The geographical coordinates of the observation point were 39°42′ N, 105°22′ W. Magnetic oscillations started with the arrival of the surface wave. The period was $T \simeq 20$ s and the amplitude

$b \simeq 0.2\gamma$. The parameters of the seismic wave were the following: the velocity of vertical oscillations 0.7 cm s^{-1} the velocity of horizontal oscillations 0.4 cm s^{-1}. This corresponds to the displacement amplitudes $\xi_x = 1.3$ cm and $\xi_z = 2.2$ cm. The x axis here is directed along the wave propagation as before.

Let us make an attempt to explain the detected signal using the theses of the theory stated above. The horizontal and vertical components of the geomagnetic field equal $H = 0.22$ G and $Z = 0.52$ G at the observation point. The angle between the geomagnetic meridian and the direction of wave propagation equals $\theta = 48°$. Correspondingly, $B_x = H \cos\theta = 0.147$ G, $B_z = Z = 0.52$ G. A number of other parameters are required for the analysis but their precise values are unknown. Let us choose the typical values: $\nu = 0.92$, $c_t = 3 \times 10^5$ cm s^{-1}, $c_l = 5 \times 10^5$ cm s^{-1}. We shall estimate the magnetic signal amplitude using equation (1.44) in the frozen-in limits when it results in the maximum possible amplitude value. Furthermore, we should take into account that the device placed at Berghen Park recorded the oscillations of the total vector modulus of the geomagnetic field. According to the data on ω, ξ_x, ξ_z, B_x, B_z and elementary calculations we find $b = 0.1\gamma$.

We have estimated the contribution of the inductive mechanism in the formation of the seismomagnetic signal. The theory produces an amplitude at least a factor of two lower than the amplitude of the signal recorded at Berghen Park. Perhaps the influence of piezomagnetic and electrokinetic mechanisms should be taken into account? We shall continue this question in section 1.3.

Exercise

Exercise 1.2.1.

Find the magnetic field of the Rayleigh wave which propagates along the surface of a nonconducting body covered with a thin conducting film.

Solution 1.2.1.

It will be useful to apply the Levi-Cività boundary condition in the case of the thin conducting film. Let us introduce the integral conductivity

$$\Sigma = \int_{-h}^{0} \sigma(z)\, \mathrm{d}z$$

where h is the thickness of the conducting film and the z axis is perpendicular to the surface of the body that occupies a half-space $z < 0$. If conductivity is homogeneous then $\Sigma = \sigma h$. We make a limiting transition $h \to 0$, $\sigma \to \infty$, conserving the value of Σ as finite. Then, when crossing the surface $z = 0$ the tangential component of the magnetic field will undergo a finite jump, e.g.

$$b_x^+ - b_x^- = \frac{4\pi}{c} \Sigma E_y$$

where $b_x^+ (b_x^-)$ is the x component of the magnetic field above (below) the film.

In our case the film oscillates in the external magnetic field. Therefore, $\boldsymbol{E}$ should be substituted for $\boldsymbol{E} + c^{-1}\boldsymbol{v} \times \boldsymbol{B}$ in the Levi-Cività boundary condition. Let the Rayleigh wave propagate in the x direction. Due to the symmetry of the problem we have $b_x^- = -b_x^+$. The velocity $\boldsymbol{v} = -i\omega\boldsymbol{\xi}$ is found with the help of (1.20) at $z = 0$. Regarding (1.45) we obtain

$$b_x^+ = -\frac{2\pi i \Sigma \omega [uT(0) + wL(0)]}{c^2 - 2\pi i \Sigma c_R}.$$

The component b_z is determined by equation (1.41).

1.3   The inertial seismomagnetic effect

Hitherto we have considered the rock as a continuous one-component medium. This was an idealization sufficient to analyse seismomagnetic waves originating by induction. Now we shall take into account that the rocks that make up the Earth's crust contain pores filled with mineralized moisture. Such a medium is termed two-component as a result of the number of various substances contained in the medium.   The fluid that fills pores and cracks is sometimes termed underground water and the solid fraction of the rock is called a mineral skeleton.

The water filling the pores and cracks of the Earth's crust is a weak solution of electrolyte.   The surface which separates the solid phase from the liquid adsorbs electrolyte ions and this leads to the formation of a double electric layer. The outer part of the double layer is associated with the solid phase more or less weakly and may travel relative to it with the fluid.

Figure 1.2 shows a section fragment of a porous water-saturated body. A part of the mineral skeleton surface is shown on the left-hand side and the pore water is shown on the right-hand side.   Close to the solid surface, electrolyte anions and cations form a double electric layer.   The electric potential jump across the double layer of value ζ is shown in figure 1.2 as a bold line. The value ζ is termed the zeta-potential, electrokinetic potential or double-layer potential.

Let us expand the picture. The double electric layer emerges through the adsorption of electrolyte anions out of its solution. The process of desorption occurs simultaneously with adsorption.   An average time interval for the molecule (or ion) to be adsorbed is of the order of 10^{-6} s^{-1}. An electrokinetic potential is formed within approximately the same period of time.

The other specification concerns the spatial structure of the double electric layer. (This is shown in figure 1.2 as an approximate guide.) A monomolecular film of strongly combined water immediately adjoins the solid body surface.

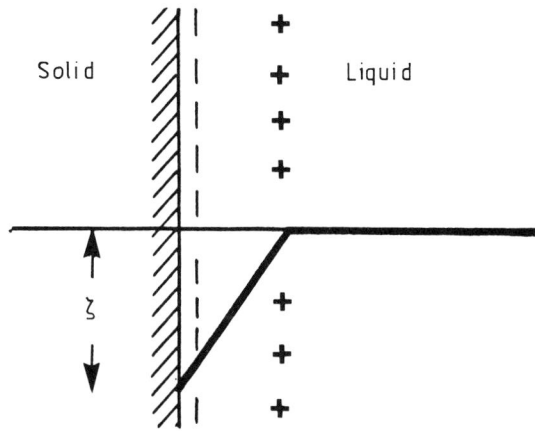

Figure 1.2. Sketch of the double layer.

The film moves along with the mineral skeleton, carrying away the adsorbed anions when the rock is deformed. The further the water molecule is from the pore wall, the weaker its connection is with the solid body. The conventional surface that separates a low-mobility water layer from the rest is termed a glide plane. The abundance of cations in mobile water forms the so-called diffusional double electric layer.

Under seismic wave propagation the fluid oscillates relative to the skeleton and these oscillations set in motion the electrokinetic mechanism which generates electromagnetic signals. We shall consider an isotropic porous medium. Suppose that there is no external magnetic field. If a plane longitudinal wave propagates in such a medium then the only distinguished direction is that of the propagation direction. In this case an excitation of electromagnetic oscillations is impossible. The symmetry of the system allows the excitation of potential electrostatic oscillations, but this does not include the excitation of magnetic field oscillations. This conclusion is true at any potential motion of an isotropic porous medium, and not just in the case of a plane longitudinal wave. Thus the generation of electromagnetic oscillations is likely only when the motion of a porous medium is characterized by vorticity.

Under real conditions the induction and electrokinetic mechanisms act simultaneously. We shall assume that this is the case, taking the set of equations (1.22) as a basis. However, the meaning of an effective electric field needs specifying.

On assuming earlier $E_{ef} = E$ we have lost the information on the most important effect. It turns out that in the Earth's crust, which is considered to be a porous water-saturated body, there arises something like the Tolman–Stewart effect under transverse seismic wave propagation. This results in excitation of magnetic field oscillations. In contrast to an ordinary Tolman–Stewart effect, a considerable 'weighting' of the charge carrier mass occurs in the Earth's crust. This enhances the effect and makes it quite amenable to experimental detection. We shall dwell upon this subject.

1.3.1 The Tolman–Stewart effect in the Earth's crust

Against the background of the remarkable scientific discoveries of 1916, the discovery of electron inertia in metals made by Tolman and Stewart slipped by almost unnoticed. The interpretation of the Tolman–Stewart effect was presented by Ch Darwin only in 1936. As was ascertained, the value of the effect is proportional to the ratio of the electron mass m_e to its charge $-e$. For example, the effective electric field equals

$$E_{\text{ef}} = E + \frac{m_e}{e}\dot{v} \qquad (1.48)$$

where $\dot{v}$ is the acceleration of the metal conductor.

In the Earth's crust, at least in its upper layers, not a metallic but an ion electroconductivity mechanism acts. In accordance with this fact, the electron mass m_e is to be replaced by ion mass m_i. This makes the gain 7×10^4 times, if the electroconductivity is accomplished by ions KCl, for example.

However, even enhanced, the Tolman–Stewart effect itself is negligibly small in the Earth's crust. To estimate the effect we employed extreme accelerations which emerge with the seismic wave propagation and ripping up of the break in the earthquake centre.

Using reasonable arguments we shall try to make the following statement convincing. Although the Tolman–Stewart effect in the Earth's crust is negligibly small in itself, nevertheless there exists a similar effect, and formally an effective electric field is defined by (1.48) if we substitute e for $-e$ and m_e for

$$m_{\text{ef}} = m_i + m_{\text{virt}}. \qquad (1.49)$$

Here the virtual mass of the charge carrier, m_{virt}, is the same order of magnitude as the mass of porous fluid contained in a disc the radius of which is of the order of a middle-sized pore and the thickness is of the order of the diameter of the ion contributing to the conductivity[4]. This new effect, similar to the Tolman–Stewart effect, is not small since $m_{\text{virt}} \gg m_i$. In the Earth's crust, which we are considering as a porous, moisture-saturated body, a weighting of the contribution of the ions occurs as a result of adding the virtual mass to the mass of the charge carrier.

---

[4] Relatively narrow pores are referenced here. In wide pores the radius of the disc indicated is of the order of a Debye radius.

1.3.2   Virtual mass

We would remind the reader of the origin of equation (1.48): if a metal moved in an accelerated manner, then a conductivity electron is influenced by the inertia $-m_e \dot{v}$ which is equivalent to the appearance of a supplementary electric field $(m_e/e)\dot{v}$. We shall try to apply this idea to an ion in the electrolyte solution which fills the pores of the Earth's crust.

Suppose the ion is a massive charged sphere. Let its mass, charge, and radius equal m_i, e and a respectively. The density, pressure and dynamic viscosity of the ambient fluid is denoted as ρ, p and η. If the solid skeleton is immobile and the porous fluid in the proximity of the ion moves at a speed w, then the ion is influenced by the force

$$e\boldsymbol{E} - 6\pi a \eta (\boldsymbol{u} - \boldsymbol{w}) \qquad (1.50)$$

where $\boldsymbol{u}$ is the ion velocity in the reference system associated with the solid skeleton. Here we made use of the Stokes formula for the resistance force acting on a spherical body in a flow of viscous fluid. But if the solid skeleton moves with acceleration $\dot{v}$, then the ion is influenced by the inertia $-m_i \dot{v}$ in addition to the force given by (1.50). It is equivalent to the electric field $-(m_i/e)\dot{v}$.

Until this point, the argument does not differ from that which leads to equation (1.48). However, we now take into account that $\boldsymbol{w}$ is modified in (1.50) under inertia and consequently an additional contribution to the effective electric field emerges.

Actually, the porous fluid was moving at a speed w relative to the immobile skeleton under pressure force $-\nabla p$. The accelerated skeleton motion results in the appearance of supplementary inertia $-\rho \dot{v}$ acting on the porous fluid. One may easily confirm that a supplementary force will change w by the value $-(\kappa/\nu)\dot{v}$, where $\kappa = k/m$, k is the permeability coefficient, m is the porosity of the solid skeleton and ν is the kinematic fluid viscosity[5]. This is equivalent to the superimposing of an electric field with intensity $-6\pi a \rho (\kappa/e)\dot{v}$.

We sum up both contributions to the effective electric field as

$$\boldsymbol{E}_{\text{ef}} = \boldsymbol{E} - \frac{m_{\text{ef}}}{e} \dot{v}. \qquad (1.51)$$

Here m_{ef} is determined by (1.49), and

$$m_{\text{virt}} = 6\pi a \kappa \rho. \qquad (1.52)$$

The drawbacks in deducing (1.52) are evident although the very existence of a strong effect in a porous body similar to the Tolman–Stewart effect is beyond doubt.

Let us transform (1.51) and (1.52) so that the microscopic parameters e, m_i and a do not appear in the subsequent equations. In the first place, we note

---

[5] The symbol m for porosity is introduced in respect to the existing tradition. This will not lead to confusion since we always supply the masses of real and virtual particles with subscripts.

that since the order of κ is equal to the square of a middle-sized pore, then $m_{virt} \gg m_i$ and m_{ef} in (1.51) may be substituted for m_{virt}. In the second place, we convert (1.52) into the form

$$\frac{m_{virt}}{e} = \frac{\varepsilon \zeta}{4\pi \sigma_w \nu} \tag{1.53}$$

and therefore we eliminate microscopic parameters from (1.52) completely. Here ε and σ_w are the dielectric permeability and the electoconductivity of the porous fluid and ζ is the electrokinetic potential.

When deducing (1.53) we used similarly structured arguments to the ones which led to (1.51) and (1.52). The general approach lies in an estimate of the kinetic coefficient and the density of volume charge in the porous space with regard to the results of the Debye–Hückel theory. A stricter base for (1.53) may be obtained, but only in the framework of a strongly idealized model, for example the model of a plane isolated capillary.

1.3.3 The equation of magnetic signal generation

Eliminating the field E from equation (1.22) and taking into account (1.51) and (1.53) we obtain

$$\frac{\partial}{\partial t}(b + \tilde{b}) = \alpha \nabla^2 b \tag{1.54}$$

where

$$\tilde{b} = \beta \Omega + \nabla \times (B \times \xi) \tag{1.55}$$

and

$$\alpha = c^2/4\pi m \sigma_w \qquad \beta = c\varepsilon\zeta/2\pi\sigma_w\nu \qquad \Omega = (1/2)\nabla \times v.$$

Here Ω is the vorticity, ξ is the deformation vector ($\dot{\xi} = v$) and B is the constant external magnetic field. When deducing (1.54) we considered the body to be homogeneous and stationary, i.e. m_{virt} and σ depend neither on time nor on coordinates ($\sigma = m\sigma_w$). The supplementary condition for the applicability of equation (1.54) is $v \ll l/\tau$, where l and τ are the characteristic scale and period of motion.

Equation (1.54) differs from the ordinary equation of magnetic induction in a moving conductor by the term $\beta\Omega$. One may easily see that this term is proportional to the difference between the vorticity filter flow of the porous fluid $m\Omega_w$ and the vorticity of the solid skeleton Ω.

We have assumed above that all the parameters of the medium are constant in time and space. In general, this leads to the requirement of motion incompressibility, i.e.

$$\nabla \cdot v = 0 \tag{1.56}$$

since otherwise the porosity m will be variable. With regard to (1.56), and (1.55) will take the form

$$\tilde{b} = \beta\Omega - (\boldsymbol{B} \cdot \nabla)\xi. \tag{1.57}$$

Equation (1.54) gives a uniform description to both manifestations of seismomagnetism. The first term of the right-hand side of (1.57) describes a new effect associated with the vorticity of a porous medium; the second term describes the ordinary effect of electromagnetic induction in a moving conductor. Both terms are mutually supplementary. The first (second) term prevails at high (low) frequencies. The boundary frequency equals

$$f_* = 2\sigma_w \nu B/c\varepsilon\zeta. \tag{1.58}$$

If $\varepsilon = 80$, $\sigma_w = 10^{10}$ s^{-1}, $\nu = 10^{-2}$ cm^2 s^{-1}, $\zeta = 50$ mV and $B = 0.5$ G, then $f_* = 0.25$ Hz.

At high frequencies ($f \gg f_*$) the generation of the magnetic field is described by the equation

$$\frac{\partial}{\partial t}(b + \beta\Omega) = \alpha\nabla^2 b. \tag{1.59}$$

1.3.4 The electromagnetic field of an elastic wave in a porous body

The vorticity $\Omega = 0$ in a longitudinal wave, and therefore only the induction mechanism of magnetic signal generation functions there. The displacement vector in a transverse wave yields the condition $\nabla \cdot \xi = 0$. Consequently, we may use (1.57). The vorticity Ω is not equal to zero. Both mechanisms of magnetic signal generation function simultaneously. However, equation (1.54) is linear and therefore these mechanisms may be considered independently.

The induction effect in a transverse wave has been described in the preceding section. To analyse the inertial (electrokinetic) mechanism we chose the wave polarization in the form (1.34). This does not limit the generality of consideration. So the vorticity is $\Omega = (0, 0, \Omega)$, where

$$\Omega = \omega^2\xi/2c_t. \tag{1.60}$$

Substituting (1.60) into (1.59) we get

$$b = i\omega \left(\frac{c_t}{c}\right) \left(\frac{\varepsilon\zeta m}{\nu}\right) \left[1 - \left(\frac{c_t}{c}\right)^2 \frac{4\pi i\sigma}{\omega}\right]^{-1} \xi. \tag{1.61}$$

Magnetic oscillations are polarized in the following way

$$b = (0, 0, b). \tag{1.62}$$

Thus, the inertial and induction signals are orthogonal to each other (see (1.35)). This follows directly from (1.57).

The electric field excited by a transverse elastic wave is purely solenoidal. The vector E is perpendicular to the vectors b and k. They form right-oriented three vectors. The oscillation amplitude of the electric field is equal to

$$E = (c_t/c)b. \tag{1.63}$$

The electric field of a longitudinal wave is strictly potential. (We will not consider the induction mechanism for the moment.) From the equation $\nabla \cdot j = 0$ and using equations (1.29), (1.30), (1.51) and (1.53) we find $E = (E, 0, 0)$, where

$$E = -\frac{\varepsilon \zeta \omega^2}{4\pi \sigma_w v} \xi. \tag{1.64}$$

However, this is an inaccurate formula since, when constructing the theory we did not take into account the compressibility of porous fluid and the porosity variation m under the volume deformation of the medium in a longitudinal wave field. We note that transverse wave propagation is not accompanied by volume deformation and, in this sense, (1.61) and (1.63) should be considered accurate.

In conclusion, let us refer to the magnetic effect of the earthquake of 28 March 1964 in Alaska. We recall that the theory gave an amplitude of $\simeq 0.1\ \gamma$, while according to the observation in Berghen Park the signal had an amplitude $\simeq 0.2\ \gamma$. The use of the electrokinetic mechanism cannot remove this discrepancy completely, since the oscillation frequency is lower than the boundary frequency (1.58). Nevertheless, we make an estimate using (1.61). Under typical values of the medium parameters we obtain $\simeq 50$ mγ. Even if this is correct there still remains the discrepancy 50 mγ. It could be explained by piezomagnetic oscillation but we really should admit that the theory only provides a qualitative explanation of the effect only. A more detailed calculation with regard to real experimental conditions is necessary for a quantitative comparison.

Bibliography

For section 1.1

An excellent introduction to seismology is given in the book by Bolt (1978). The theory of elastic wave propagation is expounded in the book by Landau and Lifshitz (1970). We constructed figure 1.1 using a similar picture from that book.

The concept of magnetoelastic waves in a solid body, which are analogous to magnetohydrodynamic waves in a fluid, was introduced by Knopoff (1955). Keilis-Borok and Monin (1959) developed the theory of magnetoelastic waves in connection with the problem of seismic-wave propagation near the Earth's core boundary. Exercise 1.1.1 was composed using these texts.

From the theoretical standpoint, magnetoelastic waves possess quite a number of interesting properties but they are not found in Nature since, not only in the Earth but also in the crust of pulsars $B \ll G^{1/2}$ with ample margins (Guglielmi 1992c). The estimate of G for the crust of pulsars is extracted from the text by Dyson (1971).

For section 1.2

The theory of electromagnetic induction in the Earth's crust in the given field approximation of a seismic wave is developed in the papers by Guglielmi (1986), Gorbachov and Surkov (1987), Guglielmi and Ruban (1990). They were partly engendered by the paper by Eleman (1965) dedicated to the response of magnetic instruments to the well-known 'Good Friday' earthquake with the epicentre in Alaska. The description of this earthquake, and the parameters of seismic and magnetic signals were borrowed from Folke Eleman.

A number of works are aimed at the search for magnetic signals outstripping the seismic wavefront. Gogatishvili (1984) as well as Levin *et al* (1988) report on the magnetic signals at fairly large distances from the epicentre (thousands of kilometres). Approximately an hour before the 'Good Friday' earthquake, at a distance of 440 km from the epicentre, Moore (1964) watched a series of short impulses with the amplitude 10–100 nT. Gokhberg *et al* (1989) detected a weak magnetic signal ($\simeq 10^{-3}$ nT) that had outstripped the elastic wavefront by 30 s, the distance from the epicentre being 220 km and the focus depth 30 km.

Let us dwell upon the paper by Belov *et al* (1974) where a magnetic signal with amplitude 0.4 nT which has outstripped the elastic wave by several seconds is described, the distance from the epicentre being 70 km and the focus depth being 70 km. The earthquake occurred in Kamchatka with magnitude $M = 6$. Taking into account the relatively short distance from the observation point to the epicentre, we may try to interpret the magnetic signal as a structure element of the elastic wavefront (Guglielmi 1991). However, the shunting influence of the separating ground–air boundary does not allow the observation of a magnetic signal which outstrips the plane front of an elastic wave over the Earth. If Belov *et al* (1974) observed the signal before the appearance of the elastic wavefront over the surface, then the curvature of the front should be taken into account. It is not impossible, however, that the signal was excited not before but after the appearance of the front over the Earth's surface. There are still too many vague points here and interesting discoveries are yet to come.

The idea of the boundary condition in the approximation of an infinitely thin film was proposed by Levi-Città in 1902 (e.g. Bateman 1955). We availed ourselves of this idea when composing the exercise for this section.

For section 1.3

The ideas of electrokinetics were conceived in the early nineteenth century. From 1809 to 1878 electro-osmosis, electrophoresis, flow potential and sedimentation potential were discovered by experiments with porous bodies and isolated capillaries. Within the context of this book the most important is the discovery made in 1859 when G Quincke found out that, when filtering fluid through a porous partition, differences of potential occurred. The understanding of this mechanism was achieved by the works of Helmholtz, Smolukhowski, Gouy, Stern and others. Regarding the modern state of the problem see, for example, the monographs by Delahay (1965), Bockris and Reddy (1970). The Debye–Hückel theory outlined in the monograph by Lifshitz and Pitaevsky (1979).

Seismoelectric signals were observed by Ivanov (1939, 1940), Martner and Sparks (1959), Long and Rivers (1975) and others. According to Ivanov (1939, 1940) the signal-generation mechanism is associated with the structural inhomogeneities of rock, i.e. with the presence of a system of pores and microcracks in the rock that are completely or partly filled with porous fluid. Frenkel (1944) estimated the idea of the electrokinetic origin of the signals and framed the corresponding theory. Later, Frenkel's theory was considerably developed, mainly in view of applications in seismology.

Frenkel (1944) confined himself to the analysis of curl-free (electrostatic) oscillations. There are no magnetic field oscillations in the classic Frenkel model. The idea of combining the electrokinetic mechanism with the Tolman–Stewart mechanism was first presented by Guglielmi (1992a,d) and Guglielmi and Levshenko (1993, 1994) and section 1.3 contains an extended reproduction of these papers. æ

Chapter 2

Ocean

Approximately 70 per cent of the Earth's surface is covered by ocean. Only a small portion of this vast expanse is smooth and almost everywhere water is agitated. In some places the ripples and small wave crests are seen. In other places, the crests start turning over, spume grows and, finally, white-caps appear and the wind begins tearing away the spume from the wave crests.

Salt water possesses fairly high conductivity. Therefore, ocean waves perturb the geomagnetic field noticeably and we face a magnetohydrodynamic problem. However, there is no need to use the whole set of magnetohydrodynamic equations since the real magnetic field of the Earth exerts practically no influence upon the motion of sea water. The situation here is similar to the one we dealt with when analysing the perturbation of the geomagnetic field by the Rayleigh seismic wave.

Usually, the theory is constructed in the following way. One solution of the hydrodynamic problem of wave propagation on the surface or in the depths of the water is chosen and then the solution of the equations of quasistationary electrodynamics with regard to water motion under respective boundary conditions is sought. In other words, the problem of the generation of magnetic oscillations is solved under the assumption that the motion of liquid does not depend on electromagnetic field.

Nevertheless, we start with magnetohydrodynamics. This is a beautiful and profound theory. It is useful to get to know it now as in the next chapters of the book it will play an important role. The magnetohydrodynamic approach is applicable for the analysis of linear and nonlinear long-period waves in the ionosphere and magnetosphere, waves in the solar wind in an upstream region before the bow shock, oscillations of the geomagnetic tail, etc.

Any redundancy in the whole set of magnetohydrodynamic equations will be examined after we proceed to the specific analysis of the geomagnetic field perturbations by the ocean waves.

2.1   Magnetohydrodynamics

2.1.1   Equations of magnetohydrodynamics

The subject of magnetohydrodynamics deals with the hydrodynamic movements in electroconducting fluids and gases with due regard for interaction of this motion with the magnetic field. The equations of magnetohydrodynamics contain the equations of the quasistationary electromagnetic field and the equations of ordinary hydrodynamics. The equations of the electromagnetic field have the form (1.22) if we suppose $E_{ef} = E$. Sea water and ionized gas in the upper atmosphere are diamagnetics, but their magnetic permeability differs from unity so insignificantly that we can suppose $H = B$. With this in mind, we shall write (1.22) in the form

$$\nabla \times E = -\frac{1}{c}\frac{\partial B}{\partial t} \tag{2.1}$$

$$\nabla \times B = \frac{4\pi\sigma}{c}\left(E + \frac{1}{c}v \times B\right) \tag{2.2}$$

$$\nabla \cdot B = 0. \tag{2.3}$$

We recall that the term placed in round brackets on the right-hand side of equation (2.2) is the electric field in the concurrent reference system, i.e. in the reference system where a given medium element rests at the present moment. For this, it is assumed that the hydrodynamic velocity is small in comparison with the velocity of light.

Equations (2.1)–(2.3) should be supplemented by a set of hydrodynamic equations, i.e. the equation of continuity, the equation of motion and the equation of heat balance. In a number of cases, the viscosity and thermal conductivity of the fluid may be neglected. Then, the laws of hydrodynamics are expressed by the equations

$$\frac{\partial \rho}{\partial t} = -\nabla \cdot (\rho v) \tag{2.4}$$

$$\rho\frac{dv}{dt} = -\nabla p + \frac{1}{c}j \times B \tag{2.5}$$

$$\frac{d}{dt}\left[\frac{p}{\rho^\gamma}\right] = 0. \tag{2.6}$$

Here ρ is the fluid density, p is the pressure, and γ is the ratio of specific heats, $d/dt = \partial/\partial t + (v \cdot \nabla)$. The second term on the right-hand side of equation (2.5) represents the Lorentz force that acts on unit volume of the liquid medium. If the fluid is also influenced by the gravitational field, then the right-hand side of (2.5) should be supplemented with the term ρg, where g is the acceleration due to gravity.

Using (2.2) we convert the equation of induction (2.1) and the equation of motion (2.5)

$$\frac{\partial B}{\partial t} = \nabla \times (v \times B) + \frac{c^2}{4\pi\sigma}\nabla^2 B \tag{2.7}$$

$$\frac{\partial v}{\partial t} + (v \cdot \nabla)v = -\frac{1}{\rho}\nabla p - \frac{1}{4\pi\rho}B \times (\nabla \times B). \tag{2.8}$$

When writing the second term on the right-hand side of equation (2.7) we suppose for simplicity that the electroconductivity σ does not change from one point to another.

The ponderomotive force (the second term on the right-hand side of (2.8)) may be converted by using the equality

$$B \times (\nabla \times B) = \tfrac{1}{2}\nabla B^2 - (B \cdot \nabla)B.$$

Then (2.8) takes the form

$$\frac{\partial v}{\partial t} + (v \cdot \nabla)v = -\frac{1}{\rho}\nabla\left(p + \frac{B^2}{8\pi}\right) + \frac{1}{4\pi\rho}(B \cdot \nabla)B. \tag{2.9}$$

When the equation of motion is written in this form, we can see that the force that influences the medium from the magnetic field is the sum of two terms. The first term represents the gradient of magnetic pressure (with the opposite sign)

$$-\nabla\left[\frac{B^2}{8\pi}\right].$$

The meaning of the second term

$$\frac{1}{4\pi}(B \cdot \nabla)B$$

is clearest if projected onto the normal n to the magnetic field line

$$\frac{1}{4\pi}n(B \cdot \nabla)B = \frac{B^2}{4\pi R}.$$

Here R is the radius of the field-line curvature. Hence, it follows that the second term is similar to the elastic force that appears when a stretched string is being distorted. The magnetic field line seeks to contract and become straight. At the same time the field lines 'push each other apart' under the influence of magnetic pressure.

The self-consistent set of equations (2.3), (2.4) and (2.6)–(2.8) describes the change of the values ρ, p, v, B in time and space. It is convenient to introduce the Reynolds magnetic number

$$R_m = \frac{vl}{\alpha} \qquad \alpha = \frac{c^2}{4\pi\sigma}.$$

It defines the relative part of the first term on the right-hand side of equation (2.7). Here v and l are the typical velocity and the scale of motion of the liquid medium and α is the coefficient of magnetic field diffusion. At $R_m \ll 1$ the perturbation of magnetic field due to fluid motion is small. At $R_m \gg 1$ the equation of induction may be approximately presented in the form

$$\frac{\partial B}{\partial t} = \nabla \times (v \times B). \tag{2.10}$$

In this case, we speak of the approximation to ideal magnetohydrodynamics. The physical implication of equation (2.10) lies in the fact that the field lines of the magnetic field are 'frozen in' the fluid, i.e. they move along with the fluid. The magnetic flux penetrating through any contour formed by liquid particles does not change in the process of motion. The electric field in the approximation of ideal magnetohydrodynamics is equal to

$$E = -\frac{1}{c}v \times B. \tag{2.11}$$

The applicability conditions of the approximation of ideal magnetohydrodynamics depend on the specific characteristics of motion and the criterion $R_m \gg 1$ may turn out to be too strict in reality. For example, in the case of propagation of Alfvén waves a weaker Lundquist criterion is necessary (see below, equation (2.20)).

Equally, there is no general answer to the question of whether the fluid can be considered incompressible. Sometimes a sufficient condition is that the dimensionless parameter (c_A/c_S), which is equal to the ratio of the Alfvén velocity

$$c_A = B/(4\pi\rho)^{1/2} \tag{2.12}$$

to the sound velocity

$$c_S = (\gamma p/\rho)^{1/2} \tag{2.13}$$

is small. However, even when $c_A \gg c_S$, the compressibility may also be disregarded. A trivial example is the propagation of transverse plane waves along an external magnetic field.

If the fluid may be considered incompressible in the process of motion, the equation of continuity (2.4) then becomes

$$\nabla \cdot v = 0 \tag{2.14}$$

and the equation of induction becomes

$$\frac{\partial B}{\partial t} + (v \cdot \nabla)B = (B \cdot \nabla)v + \frac{c^2}{4\pi\sigma}\nabla^2 B. \tag{2.15}$$

2.1.2 Magnetohydrodynamic waves

The equations of magnetohydrodynamics are nonlinear. Let us perform the linearization for the special case when, in an unperturbed state, the fluid is stationary and homogeneous. The unperturbed magnetic field is taken as homogeneous and constant.

Let us make redesignations $B \to B+b$, $\rho \to \rho+\delta\rho$, $p \to p+\delta p$, where b, $\delta\rho$ and δp are small perturbations of the same order. Using the adiabatic equation (2.6) we express δp through $\delta\rho$

$$\delta p = c_S^2 \delta\rho \tag{2.16}$$

and eliminate δp from the equation of motion (2.8). Then we obtain a linearized set of magnetohydrodynamic equations

$$\frac{\partial v}{\partial t} = -\frac{c_S^2}{\rho}\nabla\delta\rho - \frac{1}{4\pi\rho}B \times (\nabla \times b)$$

$$\frac{\partial \delta\rho}{\partial t} = -\rho\nabla \cdot v \tag{2.17}$$

$$\nabla \cdot b = 0$$

$$\frac{\partial b}{\partial t} = \nabla \times (v \times B) + \frac{c^2}{4\pi\sigma}\nabla^2 b.$$

Since the unperturbed medium is homogeneous and stationary, we may seek the solution of equations (2.17) in the form of plane monochromatic waves. From the equation of continuity we find the connection of $\delta\rho$ with v

$$\omega\delta\rho = \rho(k \cdot v) \tag{2.18}$$

and from the equation of induction, the connection of b with v

$$\left(\omega + i\frac{c^2 k^2}{4\pi\sigma}\right)b = (v \times B) \times k. \tag{2.19}$$

Suppose the electroconductivity is so high that the second term in round brackets in (2.19) can be neglected, i.e. we may proceed to the approximation of ideal magnetohydrodynamics. The condition of such transition

$$L = \frac{4\pi\sigma\omega}{c^2 k^2} \gg 1 \tag{2.20}$$

is termed the Lundquist criterion. Now instead of (2.19) we have

$$\omega b = (v \times B) \times k. \tag{2.21}$$

Using (2.18) and (2.21) we eliminate $\delta\rho$ and b from the equation of motion. As a result we obtain a set of three homogeneous algebraic equations relative to the components of the vector v. On equating the determinant of this set with zero we get the dispersion equation, the solutions of which determine the connection of ω with k.

The dispersion equation of the sixth power is relative to ω. However, in ideal magnetohydrodynamics there are no dissipative processes. This means that the wave processes are reversible in time. Therefore, the dispersion equation contains only even powers of frequency ω. In other words ω^2, and not ω represents the analytical function of k.

Taking this into account, it is clear that in the general case the dispersion equation has only three essentially distinct solutions $\omega(k)$, which correspond to the three types of magnetohydrodynamic waves. These are the Alfvén wave and two magnetosonic waves—fast and slow.

The dispersion law for the Alfvén wave does not depend on the elastic properties of the fluid

$$\omega = c_A k_{\parallel}. \tag{2.22}$$

Here $k_{\parallel}$ is the modulus of the vector projection k on the direction of the external magnetic field. In a wave of this type there are no fluctuations of density and pressure, that is $\delta\rho = 0$, $\delta p = 0$. The vector b is perpendicular to the planes of the vectors B and k; the vector E is perpendicular to B and lies in the plane B, k (figure 2.1). The velocity v is connected with the perturbation of the magnetic field b in the following way

$$v = \mp b/(4\pi\rho)^{1/2} \tag{2.23}$$

where the upper (lower) sign refers to the case $k \cdot B > 0$ ($k \cdot B < 0$).

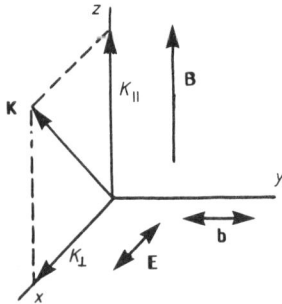

Figure 2.1. Polarization of the electric, E, and magnetic, b, field oscillations in the Alfvén wave.

The connection between ω and k for the magnetosonic waves depends essentially on the elastic properties of the fluid. In the general case, both v and b, in the same way as $\delta\rho$ and δp, are subject to oscillations. In a fast magnetosonic wave, the magnetic and the hydrodynamic pressures are added up and in the slow wave they are subtracted. We shall list the respective dispersion laws in two limited cases: a highly compressed medium ($c_S \ll c_A$) and an almost incompressible medium ($c_S \gg c_A$).

At $c_S \ll c_A$ we have

$$\omega = c_A k \tag{2.24}$$

for the fast wave and

$$\omega = c_S k_\parallel \tag{2.25}$$

for the slow wave. At $c_S \gg c_A$ for the fast wave

$$\omega = c_S k \tag{2.26}$$

for the slow wave

$$\omega = c_A k_\parallel. \tag{2.27}$$

We indicate that the dispersion law (2.27) does not differ from the dispersion law for the Alfvén waves (2.22). As a matter of fact at $c_S \gg c_A$ we have two waves: an ordinary acoustic wave (2.26) and the Alfvén wave with two independent directions of polarization.

2.2 Perturbation of the geomagnetic field by ocean waves

The density of water $\rho = 1$ g cm^{-3}. The typical value of the geomagnetic field $B = 0.5$ G. Hence it follows from equation (2.12) that the Alfvén velocity is $c_A = 0.14$ cm s^{-1}. This value is six orders lower than the velocity of sound $c_S = 1.5 \times 10^5$ cm s^{-1}. Therefore, the sea water may be considered almost incompressible. In such a medium, the ordinary acoustic waves propagate according to the dispersion law (2.26); possibly, this is also true for the Alfvén waves provided the criterion (2.20) is met.

Let us rewrite (2.20) using (2.22), drop non-essential numerical multipliers and for simplicity substitute k in place of $k_\parallel$

$$\frac{\sigma B \lambda}{\rho^{1/2} c^2} \gg 1. \tag{2.28}$$

Here $\lambda = 2\pi/k$ is the wavelength. When (2.28) is obeyed the damping of the Alfvén waves is weak. Otherwise, the Alfvén waves cannot propagate owing to the strong Joule dissipation.

The typical electroconductivity of sea water $\sigma = 3 \times 10^{10}$ s^{-1}. Substituting this value into (2.28) we see that the Lundquist criterion is not satisfied in seas and oceans.

So the Alfvén waves do not exist in sea water and sound waves propagate as if there were no geomagnetic field. Ponderomotive forces exert practically no influence upon sound waves since $c_S \gg c_A$.

Common sense indicates that the geomagnetic field exerts almost no influence on the movement of sea water. Therefore, we shall consider the problem of the excitation of magnetic field oscillations by sea waves within a given liquid motion approximation. For the corresponding criteria for water surface waves see exercise 2.2.2.

2.2.1   Waves on the surface of water

Gravity waves propagate along the surface of seas and oceans. If the amplitude of surface oscillations a is much shorter than the wavelength λ then the term $(v \cdot \nabla)v$ may be neglected as compared with $\partial v/\partial t$ in the equation of fluid motions and the linear approximation may be used. The motion may be considered to be potential, i.e. $v = \nabla\varphi$. Since we assume the fluid to be incompressible $(\nabla \cdot v = 0)$ then the potential φ yields the Laplace equation

$$\nabla^2\varphi = 0. \tag{2.29}$$

We now introduce a Cartesian system of coordinates in which the z axis is directed upwards and the fluid occupies the half-space $z \le 0$. We supplement the linearized equation of motion with the gravitational force and neglect, as we have agreed, the ponderomotive force. On integrating we find the pressure

$$p = -\rho g z - \rho\dot\varphi. \tag{2.30}$$

We shall denote vertical displacements of the fluid surface by $\zeta(x, y, t)$. It is evident that in the linear approximation $\dot\zeta = v_z$, i.e.

$$\frac{\partial\zeta}{\partial t} = \frac{\partial\varphi}{\partial z} \tag{2.31}$$

at $z = \zeta$. But on the surface $z = \zeta$, the pressure (2.30) must be equal to the constant atmospheric pressure. Therefore, on differentiating (2.30) with respect to time at $z = \zeta$ we use (2.31) and obtain the boundary condition

$$\left(\frac{\partial\varphi}{\partial z} + \frac{1}{g}\frac{\partial^2\varphi}{\partial t^2}\right) = 0 \tag{2.32}$$

at $z = 0$. (The application of condition (2.32) at $z = \zeta$ instead of $z = 0$ would lead us beyond the limits of the linear approximation.)

We shall consider a monochromatic wave whose wavelength is small compared with the depth of the fluid. Let the wave propagate along the x axis and damp when moving inwards away from the surface of the fluid

$$\varphi = \psi(z)\exp(ikx - i\omega t) \tag{2.33}$$

Here $\psi \to 0$ at $z \to -\infty$. Substituting (2.33) into (2.29) and noting (2.32) we derive the dispersion law

$$\omega = (kg)^{1/2} \tag{2.34}$$

and the distribution of velocities in a fluid

$$v_x = a\omega \exp[\mathrm{i}(kx - \omega t) + kz]$$
$$v_z = -\mathrm{i}a\omega \exp[\mathrm{i}(kx - \omega t) + kz].$$

(2.35)

Having integrated (2.35) by time we see that liquid particles move along circles, the radii of which decrease exponentially into the depth of the fluid.

Furthermore, we shall use the solutions to (2.34) and (2.35) to estimate the perturbation of the geomagnetic field by the gravity wave and, in the meantime, we proceed with the general consideration of the waves in a fluid.

If the wavelength $\lambda = 2\pi/k$ exceeds the depth of the reservoir H considerably, then the gravity wave is called long. For long waves the dispersion law is modified

$$\omega = (gH)^{1/2}k.$$

(2.36)

The character of the fluid motion also changes. The trajectories of the liquid particles have the form of ellipses vastly elongated in the direction of the wave propagation. Then the amplitude of the horizontal movement is almost constant with depth.

In the seas and oceans there exist horizontal surfaces that separate fluids of different density. If the density of the upper warmer and less dense layer is ρ_1 and that of the lower layer is ρ_2 then along the surface of separation there will propagate internal gravity waves similar to those described above but with the substitution of g for the effective parameter

$$g_{\mathrm{ef}} = g(\rho_2 - \rho_1)/(\rho_2 + \rho_1).$$

We draw your attention to the fact that the variations of density may correlate to changes in the concentrations of salts and consequently changes in the electroconductivity of the fluid. The jump of electroconductivity at the separation boundary should be taken into account when defining the magnetic field of the internal gravity waves.

Very short waves on the free surface of water are termed capillary since their existence depends on the surface tension. The lengths of capillary waves do not exceed 1–2 cm. These waves bring about ripples on the surface of the water. The amplitude of the waves is so small that they create a negligibly small magnetic effect.

The sea is a vast body, and the real picture of waves is rather intricate and unexpected. Usually waves of various scales exist simultaneously on the surface of water. The amplitudes and lengths of waves depend on the intensity and duration of the wind. Occasionally, the height of the waves exceeds 20 m and the length can be 500 m. When the wind subsides surge appears. This is a smooth choppiness devoid of sharp crests and white-caps. A surge may leave the winds that have generated it far behind and then (in windless weather) it is called a swell. The characteristic periods of surges are 10–20 s, the wavelengths are 150–600 m. In the open ocean where the depth is 2–3 km, a surge is satisfactorily described by the theory of gravity waves on the surface of deep water.

When waves of various directions meet or when a collision of waves with an air flow takes place at the shore of some cape, for instance, steep chaotic waves occur. They are called 'a crush' or 'whirlpool'. For example, when a high wind blows to meet the Laborador current a hard crush with a wave height exceeding 10 m arises to the south-east of the Newfoundland sand-banks.

A wind of gale force generates waves of great height. The waves are covered with capillary ripples and foamy white-caps. At the critical amplitude (approximately $a \sim k^{-1}$) wave crests on the surface of deep water sharpen (figure 2.2). When approaching a sloping shore the waves slow down, their forefront grows steeper and steeper and they rush over the shore creating surf.

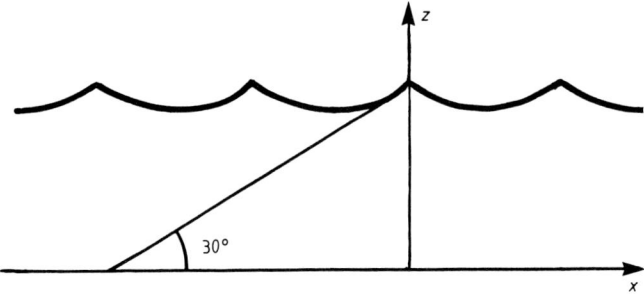

Figure 2.2. Stokes waves on the water surface. The angle at the sharp ridge of the wave is equal to 120°.

Extraordinarily big waves (tsunami) arise as an after-effect of submarine earthquakes and eruptions of submarine volcanos. In the open sea the waves are long and sloping. Here their lengths reach 100 km and their height is usually about 1 m. But when approaching the shore the waves shorten and grow higher. The height reaches dozens of metres. Tsunami can cause extensive destruction and loss of life.

All types of sea roughness are of interest for the geomagnetologist since they perturb the geomagnetic field.

2.2.2 Magnetic field of gravity waves

Let us use equations (2.34) and (2.35) to estimate the perturbation of the geomagnetic field by a surge on the surface of deep water. Ignoring the electroconductivity of air and taking into account the condition of quasistationarity $\omega \ll c/\lambda$, where λ is the wavelength, we shall describe the magnetic field over the water surface by the equations of the static field

$$\nabla \times \boldsymbol{b} = 0 \qquad \nabla \cdot \boldsymbol{b} = 0.$$

We shall treat the first of them with the operator $\nabla\times$ and use the second equation. As a result we get

$$\nabla^2 b = 0 \tag{2.37}$$

at $z > 0$. Here the z axis is directed upwards and the unperturbed water surface coincides with the plane x, y.

In the water ($z < 0$) the field is described by the equation of induction (2.15). We bring it to the form

$$\frac{\partial b}{\partial t} = (B \cdot \nabla)\, v + \frac{c^2}{4\pi\sigma}\nabla^2 b. \tag{2.38}$$

Here, besides the condition of water incompressibility ($\nabla \cdot v = 0$) and its homogeneity ($\sigma = $ constant), the condition of homogeneity of the external magnetic field ($B = $ constant) must also be added. Moreover, the terms $(b\cdot\nabla)v$ and $(v\cdot\nabla)b$ are neglected. Estimates reveal that these terms are of small order.

Let us specify the condition under which displacement currents in the sea may be neglected. The dielectric permeability ε of water is quite high. Therefore, the condition that the displacement current is small in comparison with the conductivity current must be written in the form $\sigma/\omega \gg \varepsilon$. The dielectric permeability of water approaches 80, but the electroconductivity of water is so high that the condition mentioned above is observed over the entire range of the frequencies that are of interest to us.

Now initial and boundary conditions should be imposed and the field of velocities of fluid motions should be given. We are interested in the oscillations of the magnetic field that are induced by the fluid motion under the propagation of the gravity wave (2.34) and (2.35) and thus we shall not need initial conditions. As far as the boundary conditions are concerned, they are given by the condition of continuity of the field b at the separation boundary $z = 0$ and by the condition of disappearance of the field b when moving in both directions away from this boundary.

We recall that the wave propagates along the x axis. Neither of the values depend on y. The dependence of the solutions $b(x, z, t)$ that are of interest to us on x and t is the same as the dependence of the velocities, i.e. it has the form $\exp(ikx - i\omega t)$. Taking all this into account we find the solution of the equation (2.37) at $z \geq 0$

$$b = b_0 \exp[i(kx - \omega t) - kz]. \tag{2.39}$$

Here, out of two independent solutions, we have chosen the one that does not increase upwards as the sources of the magnetic field are at the bottom. From the condition $\nabla \cdot b = 0$ regarding (2.39) we find the connection between the components of the magnetic signal

$$b_z = ib_x. \tag{2.40}$$

It is seen that the polarization of the magnetic field is circular in the vertical plane parallel to the wave propagation trace.

At $z < 0$ we have the equation

$$\partial^2 b - q^2 b = \beta m e^{kz} \tag{2.41}$$

that follows from (2.38), and a supplementary relation

$$b_x = \frac{i}{k} \partial b_z \tag{2.42}$$

ensuing from the condition of $\nabla \cdot b = 0$. Here we have introduced the designations

$$\partial = \partial/\partial z \qquad q^2 = k^2 - p^2$$
$$p^2 = 4\pi i\sigma\omega/c^2 \qquad \beta = p^2 ak(B_x - iB_z)$$
$$b = (b_x, 0, b_z) \qquad m = (-1, 0, i).$$

We find the solution to (2.41) using (2.42) that yields the conditions at $z = 0$ and $z = -\infty$

$$b_x = -ak(B_x - iB_z)\left[e^{kz} - \frac{2(1 - iL)^{1/2}}{1 + (1 - iL)^{1/2}}e^{qz}\right]$$
$$b_z = iak(B_x - iB_z)\left[e^{kz} - \frac{2}{1 + (1 - iL)^{1/2}}e^{qz}\right]. \tag{2.43}$$

Here $z \leq 0$, $\text{Re} q > 0$, $L = 4\pi\sigma g^2/c^2\omega^3$. The obvious dependence of b on x and t was omitted.

For $z > 0$ the solution of the problem has the form (2.39), with

$$b_{0x} = ak(B_x - iB_z)\left[\frac{1 - (1 - iL)^{1/2}}{1 + (1 - iL)^{1/2}}\right] \tag{2.44}$$

and $b_{0z} = ib_{0x}$.

The dimensionless parameter L is the Lundquist parameter (2.20) taking into account (2.36). It is small in comparison with unity in the range of surge periods[1]. The small magnitude of L signifies that the skin-length is much longer than the length of a sea wave. We shall avail ourselves of this fact in order to simplify equation (2.44) for the complex amplitude of the magnetic signal on the water surface

$$b_{0x} = -\tfrac{1}{4}R_m(iB_x + B_z) \tag{2.45}$$

where R_m is the Reynolds magnetic number for the gravity waves

$$R_m = \frac{4\pi\sigma g a}{c^2\omega}.$$

---

[1] We recall that for seismic waves $L \sim 1$, i.e. the Earth's crust 'carries' geomagnetic field lines better than the sea water.

The dependence of the field on x and t as has been given is identical over and under the water. The dependence on z over the water is given by (2.39). As regards the dependence of the field on z under the water, from (2.43) it follows that

$$b_x = b_{0x}(1 + 2kz)e^{kz}$$
$$b_z = ib_{0x}(1 - 2kz)e^{kz}$$

(2.46)

if we impose the restriction $|z| \ll c^2\omega/4\pi\sigma g$ to supplement the condition $L \ll 1$.

Let us make a numerical estimate. Let $\sigma = 3 \times 10^{10}$ s^{-1}, $T = 10$ s and $a = 50$ cm. Then $R_m \approx 3.2 \times 10^{-5}$. At high latitudes B_x may be neglected in comparison with $B_z \approx 0.5$ G. Then, according to (2.45), $|b_{0x}| \approx 0.4\gamma$. The amplitude damps rapidly when ascending over the water surface. With depth the amplitude reveals a somewhat more complicated behaviour (see (2.46)). The amplitude increases with the growth of the oscillation period of the water surface (see (2.45)).

We have defined the magnetic field. So far as the electric field is concerned it will be defined in the water by the magnetic field with the help of (1.28). E may also be defined over the water but this problem is of limited interest since the accomplishment of corresponding measurements is difficult.

The theory agrees with the observations well in all cases where the conditions of the theory applicability are met (remoteness from the sea shore, sufficient reservoir depth, quasisinusoidal form of the wave). The characteristic measured value of the spectral density of magnetic pulsations is of the order of $1\gamma^2$ Hz^{-1} at the intensity of the sea roughness 1 m^2 Hz^{-1} and at the frequency 0.1 Hz. (We recall that $1 \gamma = 1$ nT $= 7.96 \times 10^{-4}$ A m$^{-1} = 10^{-5}$ G.)

The advantage of the theory stated here lies in the fact that it gives a reasonable physical picture and a true estimate of the values of magnetic perturbations. But we repeat that because this is an idealized theory; it is inapplicable, for example, in the coastal zone where the finite water depth, the bottom relief and a considerable complication of the water motion should almost always be taken into account. These difficulties would seem surmountable if we use numerical methods of solution, and this is valid if we speak only of the solution of the electrodynamic problem. But before considering such a problem, a considerably more difficult hydrodynamic problem should first be solved.

The induction mechanism does not explain the whole variety of wave electromagnetic phenomena in the ocean. Close to the ocean bottom and at the boundaries between water layers with different salinity, the action of an electrochemical mechanism leads to the generation of an electromagnetic field. Over the rough water surface the atmospheric electric field is perturbed. If, over the quiet surface in clear weather, the field lines of the atmospheric electric field are directed vertically downwards, then when the sea is rough field lines are distorted so that the water surface should remain equipotential.

Exercises

Exercise 2.2.1.

Find the magnetic field oscillations due to the propagation of an acoustic wave in a homogeneous conducting fluid in the presence of an external magnetic field.

Solution 2.2.1.

Let a plane monochromatic wave propagate along the x axis. We orient the axes y, z of the coordinate system so that the external magnetic field should be parallel to the plane x, y. We assume $c_S \gg c_A$. Then the dispersion law (2.26) is valid and the wave is longitudinal, implying that $v = (v_x, 0, 0)$. Then it follows from (2.19) that $b = (0, b_y, 0)$ and

$$b_y = -R_m B_y$$

where the Reynolds magnetic number for the acoustic wave equals

$$R_m = \frac{4\pi \sigma c_S \xi}{c^2}.$$

Here ξ is the amplitude of the fluid displacement in the wave, $\xi = iv_x/\omega$. We have taken into account that $\omega \gg 4\pi\sigma(c_S/c)^2$. Under characteristic parameter values this condition is accomplished at frequencies considerably exceeding 1 Hz.

Using (2.16) and (2.18) the perturbation of the magnetic field may be expressed through the oscillations of the sound pressure

$$b_y = -\frac{4\pi i\sigma B_y}{\omega \rho c^2}\delta p.$$

Exercise 2.2.2.

Find the condition under which the influence of the ponderomotive force on the fluid motion under the propagation of surface waves may be ignored.

Solution 2.2.2

The ratio of the ponderomotive force to the inertia in the equation of motion (the first equation in the set (2.17)) is of the order of

$$\frac{kBb}{4\pi\rho\omega v}. \tag{1}$$

The perturbation of the magnetic field has the order

$$b \sim R_m B \tag{2}$$

where R_m is given by (2.9). Substituting (2) into (1) we find the required condition

$$\frac{\sigma B^2}{\omega \rho c^2} \ll 1. \tag{3}$$

For the gravity waves $\omega = (gk)^{1/2}$ and accordingly

$$\frac{\sigma B^2 \lambda^{1/2}}{\rho c^2 g^{1/2}} \ll 1 \tag{4}$$

where λ is the wavelength. For capillary waves $\omega = (\alpha/\rho)^{1/2} k^{3/2}$, where α is the surface strain (for water $\alpha \approx 70$ dyn cm^{-1}). Consequently condition (3) takes the form

$$\frac{\sigma B^2 \lambda^{3/2}}{(\alpha \rho)^{1/2} c^2} \ll 1. \tag{5}$$

Formula (4) should be used at $\lambda \gg \kappa$, and formula (5) at $\lambda \ll \kappa$, where $\kappa = (2\alpha/g\rho)^{1/2}$ is the capillary constant (for water $\kappa \approx 0.4$ cm).

Exercise 2.2.3.

The magnetic effect of the ocean waves is measured by an airborne quantum magnetometer. Find the connection between the solutions (2.39) and (2.45) and the indications of the device.

Solution 2.2.3.

Owing to the Doppler effect, a moving device will register oscillations at a frequency ω' which differs from the frequency ω in the initial reference system: $\omega' = \omega - k \cdot u$. Here u is the speed of the moving device. Considering the motion to be horizontal and using (2.34) we write $\omega' = \omega[1 - (\omega u/g) \cos \psi]$, where ψ is the angle between the direction of wave propagation and the direction of the aeroplane course.

The magnetic field does not change since $u \ll c$. For our purposes it should be assumed that the quantum magnetometer only registers the modulus of the total vector of the magnetic field $|B + b|$. Let us introduce the designation $b_\parallel = |B + b| - |B|$. Owing to the small order of magnitude of b we have $b_\parallel = (b \cdot B)/B$. Hence regarding (2.39), (2.40) and (2.45) we find

$$b_\parallel(t) = -\tfrac{1}{4} R_m B (\cos^2 I \cos^2 \theta + \sin^2 I) e^{-kz} \sin(\omega' t).$$

Here B is the modulus of the total vector of the geomagnetic field, I is the magnetic inclination and θ is the angle between the direction of the wave propagation and the magnetic meridian.

Exercise 2.2.4.

The atmospheric electric field over a calm sea surface in clear weather is directed vertically downwards. Find the oscillations of atmospheric electricity when the sea is rough. The water surface should be considered equipotential.

Solution 2.2.4.

We shall consider the electric field potential ($E = -\nabla\varphi$) and ignore the volume air charge over the sea surface. Then the potential yields the Laplace equation $\nabla^2\varphi = 0$. We write it in the form

$$\frac{\partial^2\varphi}{\partial x^2} + \frac{\partial^2\varphi}{\partial z^2} = 0$$

supposing that all the values depend only on x, z and the time t. The boundary conditions are $\varphi = 0$ on the sea surface and $\partial\varphi/\partial z = E_0$ at $z \to \infty$, where E_0 is the value of an unperturbed electric field (the z axis is directed upwards). Let us set the outline of the sea surface in the form $z = a\cos(kx - \omega t)$. The solution is a linear combination of $2D$ harmonic functions, the coefficients of which are found from the boundary conditions.

The problem is more interesting than it may seem on the face of it. For reasons of symmetry, it is obvious that over the crests and cavities the electric field is vertical. The solution of the problem gives the field value at these points $E_\pm = E_0(1 \pm ak)$. Here plus (minus) refers to the crest (cavity).

So the field is enhanced over the crest and is weakened over the cavity. With the growth of the wave amplitude the contrast deepens smoothly, but until when will this deepening develop? The answer is: it will develop until the instance when a fracture appears on the crest (see figure 2.2). This will happen at $a \sim 1/k$. At this moment the field over the crest is enhanced spasmodically.

As a matter of fact it is known from electrostatics that this is close to the rib wedge $E \propto r^{n-1}$, where $n = \pi/(2\pi - \theta)$, r is the distance to the rib and θ is the angle of the wedge flare. In the Stokes wave $\theta = 120°$ and $E \propto r^{-1/4}$.

In theory $E \to \infty$ as $r \to 0$, but of course this does not happen. The growth of E will be limited, for example, by the appearance of the coronal charge, an air hole, etc.

Bibliography

For section 2.1

Magnetohydrodynamics in the form of an independent branch of physics was created by Alfvén in the 1940s. He developed a set of magnetohydrodynamic equations, introduced the fundamental concept of the magnetic field 'frozen in' a highly conducting medium and discovered Alfvén waves. The principles of magnetohydrodynamics are presented in the books by Alfvén and Fälthammar (1963), Landau and Lifshitz (1984), Moffatt (1978) and Kadomtsev (1976). The criterion (2.28) was offered by Lundquist (1952).

For section 2.2

Concerning water surface waves, see the books by Lamb (1932), Landau and Lifshitz (1988a). Figure 2.2 was borrowed from the book by Lavrentiev and Shabat (1973).

Crews and Futterman (1962) calculated the oscillations of the geomagnetic field over the sea that are induced by the roughness of the wind. The calculations were carried out in two steps. First the Foucault currents induced by the given movement of the sea water were found, then according to the Biot–Savart law the magnetic field of these currents was defined through the estimation of the appropriate integral. Similarly Warburton and Caminiti (1964) defined the magnetic field under water. Weaver (1965) applied the Fourier method of seeking a solution of the problem of the magnetic field of gravity waves on the surface of seas and oceans. In this text we followed Weaver's paper (1965). In particular, exercise 2.2.3 is compiled on the basis of the results of this work. The magnetic field of internal ocean waves was analysed by Beal and Weaver (1970) in the framework of the two-layer model of a finite-depth ocean. Petersen and Poehls (1982) employed a more realistic model for this purpose. The work by Chave (1984) is also dedicated to the problem of electromagnetic field induced by internal waves.

It is not possible to present a full and complete description of the works which are dedicated to the *in situ* observations of electromagnetic effects caused by sea and ocean waves. Here we shall only point out the works by Maclure *et al* (1964) and Ochadlick (1989), which are directly connected with the Weaver theory (1965).

An important problem we have not touched upon at all consists in defining a magnetic field of sea waves in the coastal and shelf zones. Here the difficulty is the hydrodynamic problem, particularly if you take into account the unevenness of the bottom, the twisting character of the coastline, etc. In the same way, we refer to the problem of finding a magnetic field of long-period oscillations of the seiche type in closed reservoirs. In shallow water the nonlinearity of waves should always be taken into account, which leads to difficult problems. The corresponding estimations for cnoidal waves were made by Smagin and Savchenko (1986).

Chapter 3

Atmosphere and ionosphere

Thunderstorms, waterspouts, hurricanes as well as meteorite showers and eruption of volcanos and other cataclysms that perturb the atmosphere are accompanied by electromagnetic phenomena. In this chapter we shall concentrate on thunderstorm activity.

About 2000 thunderstorms rage over the Earth at a time. Every second about 100 lightning discharges occur in the atmosphere; the discharge current reaches tens and sometimes hundreds of kiloamperes. If we take into account that the length of the discharges can be several kilometres (and sometimes dozens of kilometres) then it is clear that thunderstorm activity is a powerful and constant source of electromagnetic waves.

A lightning discharge excites the electromagnetic field in a wide range of frequencies. At certain frequencies resonant enhancement of the field occurs. The properties of resonant oscillations, and even the mere possibility of their existence, depends on the global structure of the medium that includes an atmospheric layer limited by the Earth's surface at the bottom and by the ionosphere at the top.

A part of lightning discharge wave energy penetrates through the ionosphere and enters the magnetosphere. In addition, all kinds of magnetospheric waves that reach the Earth's surface inevitably pass through the ionosphere and atmosphere. Accordingly the need to consider the waves in both the atmosphere and ionosphere is obvious.[1]

3.1 Global resonances of the Earth–ionosphere cavity

While electromagnetic fields of sea waves and earthquakes weaken rapidly when moving away from their sources, the effect of a thunderstorm discharge is global. Electromagnetic oscillations emerge everywhere in the thin atmospheric layer that surrounds the Earth (figure 3.1). Such oscillations are termed global resonances or Schumann resonances after the author who proved their existence

---

[1] Some other aspects of the problem are mentioned in exercise 3.2.4.

in the early 1950s. Schumann was the first to calculate the oscillation spectrum, and to estimate the quality of the resonator and pointed out the thunderstorm discharges as the possible source of oscillations.

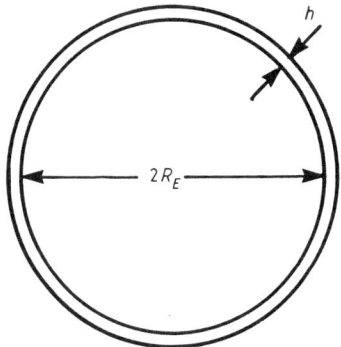

Figure 3.1. Bi-spheric Earth–ionosphere resonator on an arbitrary scale. In reality, the Earth's radius R_E is approximately 80 times greater than the spherical cavity thickness formed by the Earth's surface and the lower ionosphere.

A rough estimation of the frequency of resonances

$$f_n \simeq nc/2\pi R_E \tag{3.1}$$

is obtained from the condition that a whole number of wavelengths are kept within the circumference of the globe. Here R_E is the radius of the Earth, $n = 1, 2, \ldots$. The estimate gives $f_1 \simeq 7.5$, $f_2 \simeq 15$, $f_3 \simeq 22.5$ Hz, which are quite close to the experimental values $f_1 \simeq 8$, $f_2 \simeq 14$, $f_3 \simeq 20$ Hz.

3.1.1 Elementary theory

Let us make a more accurate estimation of the eigenfrequencies of the Schumann oscillations. First we shall consider an electromagnetic field in a vacuum between two concentric ideally conducting spheres. Then retaining the spherical symmetry we shall take into account the finite conductivity of the external resonator wall that imitates the ionosphere. At this stage we shall consider the Earth (the internal wall) to be ideally conducting, since its conductivity is much higher than the conductivity of the ionosphere. Thus, we can easily estimate the frequencies and quality of the resonances and see the field structure in the resonator. Later, we shall also take into account the finite conductivity of the Earth but only in order to discuss one version of electromagnetic sounding of the Earth's crust.

This programme is based on the idea of successive approximations. In this case it is justified by a series of strong inequalities

$$\delta_1 \ll \delta_2 \ll h \ll R_E \simeq \omega/c.$$

Here δ_1 and δ_2 are the depths of the field penetration into the Earth and ionosphere respectively and h is the distance from the Earth to the ionosphere. The relevant theory we call 'elementary' implying that a comparatively simple analytical scheme is used for its construction.

In a vacuum, arbitrary oscillations split up into sinusoidal $\exp(-i\omega t)$ oscillations for which Maxwell's equations have the form

$$\nabla \times E = ikb \qquad \nabla \times b = -ikE \qquad (3.2)$$

where $k = \omega/c$. Let us introduce the spherical coordinate system (r, θ, φ) and suppose for simplicity $\partial/\partial\varphi = 0$. Then for the electric-type oscillations $(b_r = 0)$ the components are E_z, E_θ and b_φ.

The next leading consideration will allow us to neglect E_θ and, in addition, to assume $\partial/\partial r = 0$. The inequality $h/R_E \ll 1$ gives a chance to affirm that a locally bi-spheric resonator is arranged almost like a gap between two ideally conducting planes. We recall that due to 'nonconnectedness' of the boundary in such a gap the TEM wave, or the so-called principal wave, may propagate at the velocity of light. It is a purely transverse wave and its field is homogeneous across the gap. The vector E is perpendicular and the vector b is parallel to the planes. It is natural to suppose that at $h/R_E \ll 1$ the field of electric-type oscillations in the Earth–ionosphere cavity has almost the same local structure. On the basis of this analogy, it follows from (3.2) that approximately we have

$$-\frac{\partial E_r}{\partial\theta} = iR_E kb_\varphi \qquad \frac{1}{\sin\theta}\frac{\partial}{\partial\theta}(\sin\theta b_\varphi) = -iR_E kE_r. \qquad (3.3)$$

Boundary conditions at the resonator walls in the given approximation are accomplished automatically. In the same approximation the obligatory condition $\nabla \cdot E = 0$ is met, which does not follow automatically from (3.3). On eliminating b_φ we get the equation

$$\frac{1}{\sin\theta}\frac{\partial}{\partial\theta}\left(\sin\theta\frac{\partial E_r}{\partial\theta}\right) + R_E^2 k^2 E_r = 0. \qquad (3.4)$$

This is the Legendre equation. The finite solutions of (3.4) for $0 \le \theta \le \pi$ are the polynomials $P_n(\cos\theta)$ under the condition that $R_E^2 k^2 = n(n+1)$. From this condition we obtain the spectrum of eigenfrequencies of Schumann oscillations

$$\omega_n = \frac{c}{R_E}\sqrt{n(n+1)}. \qquad (3.5)$$

The angular dependence of the field is described by the Legendre polynomials. Here we present $P_n(\cos\theta)$ for the first three harmonics

$$P_1 = \cos\theta$$
$$P_2 = \tfrac{1}{2}(3\cos^2\theta - 1)$$
$$P_3 = \tfrac{1}{2}(5\cos^2\theta - 3)\cos\theta.$$

In the case $\partial/\partial\varphi \neq 0$, the oscillations are described not by the Legendre polynomials but by spherical functions $Y_{nm}(\theta, \varphi)$, where $m = 0, \pm 1, \pm 2, \ldots,$ $|m| \leq n$. The spectrum has the form (3.5) as before. This means that the spectrum is $(2n + 1)$-fold degenerate.

The account of finite ionospheric conductivity leads to the shift of resonant frequencies and damping of oscillations. We shall refer to this problem below and here we shall simply consider that the ionosphere has an abrupt lower boundary placed at height h, and the conductivity σ is homogeneous and isotropic at $r \geq R_E + h$. We shall use the Leontovich's boundary condition

$$E_\theta = \zeta b_\varphi \tag{3.6}$$

at $r = R_E + h$. Here ζ is the surface ionospheric impedance defined by the relation[2]

$$\zeta = (1 - i)\sqrt{\omega/8\pi\sigma}. \tag{3.7}$$

The value ζ is considered to be small, $|\zeta| \ll 1$. Therefore, the account of the finite ionospheric conductivity almost does not change b_φ. We suppose that the b_φ component in (3.6) is obtained in a zero-order approximation (i.e. at $\sigma \to \infty$). As regards the E_θ component, it is equal to zero in the zero-order approximation and appears only in the first-order approximation with respect to ζ.

Energy leakage from the unit surface of the resonator upper wall in a unit of time is defined by the relation[3]

$$S = \frac{c}{8\pi}\mathrm{Re}(E_\theta b_\varphi^*).$$

Taking into account (3.6) this expression may be reduced to

$$S = \frac{c}{8\pi}|b_\varphi|^2\mathrm{Re}\zeta. \tag{3.8}$$

The flux from the total surface is obtained by integrating (3.8) over the sphere. Taking into account that the magnetic energy is equal to the electric energy the total energy may be evaluated as

$$\frac{R_E^2 h}{4}\int_0^\pi |b_\varphi|^2 \sin\theta \, d\theta.$$

---

[2] For more detail on impedance boundary conditions see Chapter 9.

[3] Here S is the vertical component of the Poynting vector averaged over the period $2\pi/\omega$. In section 1.1 it was pointed out that the use of a complex form for oscillating values is valid if the values are subject to linear operations only. At the end of the calculations, the real part of the corresponding value should be singled out. However, if the average from quadratic combinations of values is of interest one may use the rule according to which $\overline{(\mathrm{Re}A)(\mathrm{Re}B)} = (1/2)\mathrm{Re}(A^*B)$, where A and B depend on time as $\exp(-i\omega t)$ and the bar denotes the averaging over time.

Here we have taken into account that b_φ depends neither on r nor on φ and $h \ll R_E$. On dividing the total flow by the doubled total energy we get the decrement γ_n and from it we calculate the quality of the resonator $Q_n = \omega_n/2\gamma_n$. Using (3.7) it equals

$$Q_n = (h/c)\sqrt{8\pi\sigma\omega_n}. \tag{3.9}$$

Replacing $\mathrm{Re}\zeta$ by $\mathrm{Im}\zeta$ we find the frequency shift

$$\delta\omega_n = -\omega_n/2Q_n. \tag{3.10}$$

3.1.2   The gradient of the surface impedance of the Earth

We shall start the discussion and generalization of the elementary theory taking into account the conductivity inhomogeneity of the Earth.

The conductivity of the Earth considerably exceeds the ionospheric conductivity. So, the inclusion of the finite conductivity of the Earth is not required if we are interested in the global properties of the Schumann resonances (frequency, quality, mode field structure). However, the inhomogeneity of the conductivity distribution in the Earth's crust influences the local structure of the field. It may be used to elaborate on a version of magnetotelluric sounding of the Earth.

At a frequency of 8 Hz the depth of the field penetration into the Earth equals 530 m, if we assume as an example that $\sigma = 10^9$ s^{-1}. This means that Schumann resonances may be used for the sounding of the upper layers of the Earth.

At $r = R_E$ the impedance boundary condition is yielded

$$E_\theta = -\zeta b_\varphi. \tag{3.11}$$

Here ζ is the surface impedance of the Earth. An arbitrary but weak (in the scale of penetration depth) dependence on θ and φ is permissible. Owing to the small value of ζ, the magnetic field is taken in the zero approximation with respect to ζ, i.e. $b_\theta = 0$. Therefore, $E_\varphi = 0$. For the same reason $\partial b_\varphi/\partial\varphi = 0$, as has been assumed above when analysing an ideal resonator. Knowing this we substitute (3.11) into the r component of the induction equation and calculate b_r on the ground in the first-order approximation with respect to ζ

$$b_r = \frac{ic}{\omega R_E \sin\theta}\frac{\partial E_\theta}{\partial\varphi} = -\frac{icb_\varphi}{\omega R_E \sin\theta}\frac{\partial\zeta}{\partial\varphi}. \tag{3.12}$$

It is convenient to introduce a tangent b_τ and a normal b_ν components of the magnetic field on the surface of the Earth, where ν is the internal normal. Then using (3.12) and taking (3.5) into account gives

$$b_\nu^{(n)} = \frac{iR_E}{\sqrt{n(n+1)}}(b_\tau^{(n)} \cdot \nabla_\tau\zeta). \tag{3.13}$$

Here ∇_τ is the operator of the surface gradient.

In such form, the connection between $b_\nu^{(n)}$ and $b_\tau^{(n)}$ is valid not only for zonal harmonics ($m = 0$), but also for any other harmonic and for an arbitrary superposition of azimuthal harmonics. This can be verified by a straightforward calculation. But we may restrict ourselves by the assumption that connections of type (3.12) and (3.13) are a direct consequence of the induction equation, the impedance boundary condition and the transverse character of the oscillations. The latter condition means that

$$\nabla_\tau \cdot b_\tau = 0 \qquad (3.14)$$

and it is definitely accomplished for the Schumann oscillations in the zero approximation and evidently it also holds for TEM waves in the space between the conducting plates.

Equation (3.13) allows us to find the surface impedance gradient of the Earth at the frequencies of the Schumann resonances based on the observations of magnetic field oscillations at one point. Since the oscillations are global and the thunderstorm sources are constant, the measurement can be made at any point and practically at any time. The problems of measurement technique and the interpretation of the results require special consideration. However, we shall not dwell on them here.

As regards the applications of the method, we shall confine ourselves to pointing out the advisability of stationary observations of time variation of horizontal inhomogeneity of the Earth in the vicinity of the given point in the region under consideration. These observations may supply information on the development of unfavourable geological processes (earthquakes, landslides, etc). Great opportunities are offered by geoelectric profiling, as the definition of $\nabla_\tau \zeta$ from the observational data, even at one point, allows us to choose the route rationally when mapping a region.

3.1.3 Comparison with observations

All thunderstorm discharges excite background oscillations which are continuous. Against this background, from time to time a spike appears caused by a discharge of exceptional force. At distances of up to 10^3 km away from a discharge of sufficient force, one may observe a so-called 'flush' of electromagnetic oscillations.

Usually the vertical component of the electric field and (or) horizontal components of the magnetic field are registered. The resonant frequencies are found by the maxima of the spectral density of oscillations. The resonator quality is determined by the relative width of the spectral peak. As a rule not more than 3–5 resonances are registered confidently. The typical values of the frequencies observed and the qualities of the Schumann oscillations are as follows

$$f_1 = 7.9 \text{ Hz} \qquad\qquad Q_1 = 4.5$$

$$f_2 = 13.8 \text{ Hz} \qquad Q_2 = 5$$
$$f_3 = 20.1 \text{ Hz} \qquad Q_3 = 4.9.$$

The variations of Q_n significantly exceed the variations of f_n. So, for the first harmonic we have $f_1 = 7.5\text{--}8$ Hz, while $Q_1 = 3\text{--}5.5$. The intensity of oscillations is also highly variable. The characteristic spectral density values of magnetic field oscillations are of the order of $0.1\text{--}1$ $(\text{pT})^2$ Hz^{-1} and for electric field oscillations they are $0.01\text{--}0.1$ $(\text{mV})^2$ m^{-2} Hz.

Let us estimate resonant frequencies using (3.5): $f_1 = 10.6$ Hz, $f_2 = 18.3$ Hz and $f_3 = 26$ Hz. Evidently, this produces higher results than the observational data. Let us introduce a correction to (3.10) using experimental values for Q_n: $f_1 = 9.4$ Hz, $f_2 = 16.5$ Hz and $f_3 = 23.3$ Hz. The discrepancy becomes less but the result cannot be considered satisfying.

We shall now estimate the ionospheric conductivity from (3.9) using the experimental data on Q_n and f_n. We lack information on the height h of the ionospheric lower boundary but if we suppose, more or less arbitrarily, that $h = 70$ km then the error in the estimation of σ will not exceed an order of 1.5. For the first harmonic $f_1 = 7.9$ Hz, $Q_1 = 4.5$ and accordingly the ionospheric conductivity $\sigma = 3 \times 10^5$ s^{-1}. This is three to four orders less than the Earth's conductivity, but to what extent does it fit the reality?

To answer this question we have to make a slight digression. Let us discuss briefly the conductivity of partly ionized gas.

The conductivity is either measured during the experiment or evaluated in the framework of one or another model. Here we shall evaluate the conductivity in the framework of a cold isotropic plasma model. In other words we shall ignore the particle thermal motions, i.e. we shall use a so-called 'approximation of single particles' or as it is sometimes termed the cold-plasma approximation. Also, let us ignore the influence of the external magnetic field. Finally we shall consider the gas to be weakly ionized. All this corresponds approximately to the conditions in the atmosphere and the lowest layers of the ionosphere.

The motion of a free electron in a variable electric field is described by the equation

$$m\dot{v} = -e\mathbf{E} - \nu m v. \tag{3.15}$$

Here m is the electron mass, v is its velocity, e is an elementary electric charge and ν is the collision frequency of the electron with molecules. If the field $\mathbf{E}$ oscillates at frequency ω, then from (3.15) we find

$$v = \frac{e\mathbf{E}}{m(i\omega - \nu)}. \tag{3.16}$$

The current density induced by these oscillations $j = -eNv$, where N is the electron number density. (The ion current is negligibly small in the given approximation.) On the other hand $j(\omega) = \sigma(\omega)\mathbf{E}(\omega)$, where $\sigma(\omega)$ is the so-called high-frequency conductivity. (This is a generalization of Ohm's law for the variable electric field.) Comparing the two formulae for j and regarding (3.16) we find[4]

$$\sigma(\omega) = \frac{e^2 N}{m(v - i\omega)}. \tag{3.17}$$

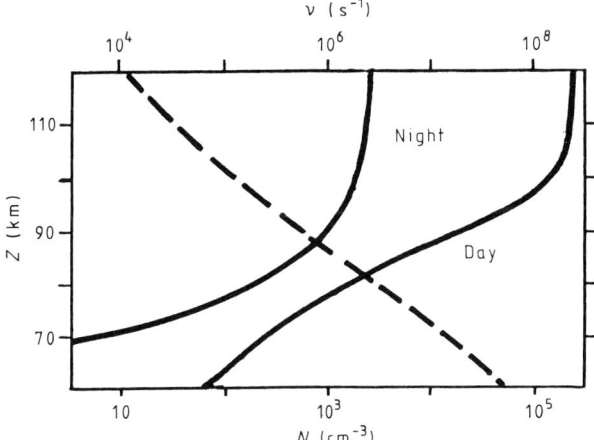

Figure 3.2. The approximate height distribution of electron density (solid line) and electron collision frequency (dashed line) at mid-latitudes during maximum solar activity.

In the lower ionosphere $v \gg \omega$ within the range of Schumann resonances (see figure 3.2). Therefore, $\sigma(\omega)$ may be substituted for the static conductivity, i.e. instead of (3.17) the following expression

$$\sigma = e^2 N / m v. \tag{3.18}$$

may be used.

If $\sigma = 3 \times 10^5$ s^{-1} as has been found for the harmonic $n = 1$, then $N/v = 1.2 \times 10^{-3}$ cm^{-3} s. This value corresponds to the values of N and v at heights of about 80–85 km. However, for the harmonics $n = 2, 3, \ldots$ we shall find other values of σ and we must select the values of N and v at other heights. This is certainly unsatisfactory. The theory should be improved and it is clear that the first step is to take account of the inhomogeneity of ionospheric conductivity distribution.

3.1.4 Ionospheric inhomogeneity

Instead of describing a medium by two real parameters—conductivity and dielectric permeability—as it has been done until now, it is convenient to

---

[4] Drude was the first to deduce this formula in 1900 in connection with his research on the high-frequency conductivity of metals.

introduce one complex parameter, namely the complex dielectric permeability $\varepsilon = \varepsilon' + i\varepsilon''$. This value will completely define the electrodynamic properties of the linear isotropic medium. In addition, we may introduce the notion of complex conductivity $\sigma = \sigma' + i\sigma''$, which also characterizes the medium. Between the complex values of ε and σ there exists a certain relationship. We shall present it for the case of monochromatic oscillations

$$\varepsilon(\omega) = 1 + \frac{4\pi i}{\omega}\sigma(\omega). \tag{3.19}$$

Considering (3.19) we shall write Maxwell's equation in the form

$$\nabla \times \mathbf{E} = ik\mathbf{b} \tag{3.20}$$

$$\nabla \times \mathbf{b} = -ik\mathbf{D} \tag{3.21}$$

where $k = \omega/c$ and

$$\mathbf{D}(\omega) = \varepsilon(\omega)\mathbf{E}(\omega). \tag{3.22}$$

Equations (3.20)–(3.22) describe the field in all the three media uniformly. The distinctions concern only the expressions for ε. In the Earth's crust $\varepsilon = 4\pi i\sigma/\omega$, where σ is the static conductivity of the rocks. In the air $\varepsilon = 1$ and the equations (3.20) and (3.21) turn into (3.2). In the ionosphere, from (3.17) and (3.19) we obtain at $\omega \ll \nu$

$$\varepsilon = 1 + i\omega_0^2/\omega\nu \tag{3.23}$$

where $\omega_0 = (4\pi e^2 N m^{-1})^{1/2}$ is the Langmuir frequency.

Let us introduce a spherical coordinate system (r, θ, φ) its centre being the centre of the Earth and we consider ε to be dependent only on r. In other words we shall retain the spherical symmetry of the model as it has been assumed in the elementary theory and we shall take into account the dependence $\varepsilon(r)$ more adequately. In addition, as above we shall discuss axially symmetrical oscillations for which all derivatives with respect to φ are equal to zero. Then the set of equations (3.20)–(3.22) will split into two independent subsystems, one of which describes the so-called electric-type oscillations with the components E_r, E_θ, b_φ, and the other magnetic-type oscillations with the components b_r, b_θ, E_φ. Electric-type oscillations will be of interest to us and we introduce the Hertz vector $\hat{\Pi} = (\Pi, 0, 0)$ expressed through the b_φ component

$$b_\varphi = ik\varepsilon^{1/2}\frac{1}{r}\frac{\partial\Pi}{\partial\theta}.$$

After introducing the Hertz vector, Maxwell's equations reduce to one scalar wave equation of the second order

$$\frac{\partial^2\Pi}{\partial r^2} + \frac{1}{r^2\sin\theta}\frac{\partial}{\partial\theta}\left(\sin\theta\frac{\partial\Pi}{\partial\theta}\right) + \left[k^2\varepsilon(r) - \sqrt{\varepsilon(r)}\frac{d^2}{dr^2}\left(\frac{1}{\sqrt{\varepsilon(r)}}\right)\right]\Pi = 0. \tag{3.24}$$

At $r = R_E$ the scalar function Π must match the boundary condition

$$\frac{\partial \Pi}{\partial r} = -ik\zeta \Pi \tag{3.25}$$

where $\zeta(\omega)$ is the surface impedance of the Earth. As $r \to \infty$ the condition of the Sommerfeld radiation should be accomplished. The latter means that we eliminate from consideration the waves that could penetrate the resonator from the outside.

The solutions of a homogeneous boundary problem (3.24) and (3.25) are sought using the separation of variables method

$$\Pi(r, \theta) = R(r)\Phi(\theta). \tag{3.26}$$

The angular function yields the equation

$$\frac{1}{\sin \theta} \frac{d}{d\theta} \left(\sin \theta \frac{d\Phi}{d\theta} \right) + \lambda \Phi = 0 \tag{3.27}$$

where λ is the division constant. The solutions to (3.27) that have no singularities are $\Phi(\theta) = P_n(\cos \theta)$ where the zonal quantum number $n \in \mathcal{Z}$ and $\lambda = n(n+1)$.

It is convenient to introduce the spherical impedance $Z_n(r)$ for each resonant mode with the number n

$$Z_n = \frac{1}{ik\varepsilon^{3/2} R_n} \frac{d}{dr} (\varepsilon^{1/2} R_n). \tag{3.28}$$

Then instead of seeking solutions of the second-order linear equation for the radial function $R_n(r)$ we may seek solutions of the Riccati equation for the spherical impedance

$$\frac{dZ_n}{dr} + ik\varepsilon Z_n^2 - ik - \frac{n(n+1)}{ikr^2\varepsilon} = 0. \tag{3.29}$$

We shall rewrite the boundary condition at the Earth's surface in the form

$$Z_n(R_E) = -\zeta. \tag{3.30}$$

The second boundary condition will be given at some spherical surface $r = r_1 > R_E$, using the Sommerfeld condition of radiation. For this we shall suppose that at $r \geq r_1$ the medium is homogeneous, i.e. $\varepsilon = $ constant. Then the radiation condition will be equivalent to the condition that at $r \geq r_1$ the radial functions are Hankel functions of the first kind. Next we note that in this region $n(n+1) \ll (kr)^2\varepsilon$ for low numbers n and use the asymptotic form of the Hankel function. Then we obtain

$$Z_n(r_1) = \varepsilon^{-1/2}(r_1). \tag{3.31}$$

In order to specify the problem, the values ζ, r_1 should be given and the profile $\varepsilon(r)$ within the range $R_E \leq r \leq r_1$ should be chosen. Usually it is assumed that $\zeta = 0$ without significant loss in accuracy. The choice of r_1 does not cause great difficulties either. The point is that the field penetrates the ionosphere not more than several dozens of kilometres within the range under consideration. If we take the initial value r_1 at a height that appreciably exceeds the depth of penetration (i.e. skin depth), then it will not be erroneous to use an arbitrary value $\varepsilon(r_1)$. Furthermore, we may choose the parameters so that at small displacement of r_1 the solution remains almost invariable.

The choice of radial dependence $\varepsilon(r)$ depends on the choice of the functions $N(r)$ and $v(r)$ (see (3.23)). The exponential approximation for v is often used, for example

$$v = v_0 \exp\left[(h_0 - h)/H\right]$$

where $h = r - R_E$ is the height over the Earth, $v_0 \simeq 10^7 \text{ s}^{-1}$ and $h_0 \simeq 70$ km. The alternative of choosing the model $N(r)$ is still left and it is used in order to make the result of calculations relevant to the experimental data in the best possible way.

The eigenvalues k_n in (3.29)–(3.31) are sought numerically using, for example, Newton's method of successive iterations. The positions of spectral maxima ω_n are found by $\text{Re}k_n$ and the relative qualities Q_n are found by $\text{Im}k_n$. Usually we succeed in reproducing the values ω_n, Q_n, known by observation of the first 3–5 resonances.

Further improvement of the theory is associated with the gyrotropy and horizontal inhomogeneity of the ionosphere. Gyrotropy depends on the geomagnetic field. It may explain splitting of spectral lines observed during the experiment. Daily inhomogeneity of the ionosphere partly accounts for the daily variation of the amplitude of Schumann resonances. Gyrotropy and horizontal inhomogeneity violates the spherical symmetry. This complicates the theory and requires us to draw on the methods of numerical analysis.

Finally the theory would not be complete unless field sources are taken into account. A vertical electric dipole that oscillates according to a harmonic law is chosen as the simplest source model (exercise 3.1.4). A vertical cloud-to-ground lightning discharge excites the Schumann resonator but this certainly is not a harmonic process but an impulsive one. An additional complication is associated with the fact that not one but a number of sources distributed over the planet are acting. A single source may be used to model the so-called Q-bursts, i.e. strong wave trains of Schumann oscillations with an amplitude one–two orders greater than the amplitude of background oscillations. Q-bursts are excited by a single lightning discharge of extraordinary force up to 10^6 A.

Exercises

Exercise 3.1.1.

Calculate the frequencies and qualities of the Schumann resonances for the planets of the solar system.

Solution 3.3.1.

Schumann oscillations exist if the planet possesses an ionosphere and a low-conducting atmosphere and if lightning discharges that excite the cavity between the planet's surface and the lower edge of the ionosphere arise. These conditions are satisfied over Venus, over Jupiter and probably over Mars. Thunderstorm discharges were observed on Venus and Jupiter. There were dust storms over Mars when charge separations, causing flashes of lightning, occur.

In the approximation of an ideal resonator the quality is infinite, and the frequencies are determined just by the radius of the planet. The frequencies calculated by (3.5) are presented in table 3.1.

Planet	Radius (km)	Frequency (Hz)		
		$n = 1$	$n = 2$	$n = 3$
Venus	6050	11.2	19.3	27.3
Earth	6370	10.6	18.3	26
Mars	3400	19.8	34.4	48.6
Jupiter	71000	0.95	1.67	2.33

Table 3.1. Frequencies of the Schumann resonances.

To calculate more realistic values we need the data on the lower ionosphere. At present sufficient information of this kind is available for the Earth and Venus. The appropriate calculation by the method of successive iterations gives for the Earth $f_1 = 7.4$ Hz, $Q_1 = 3$; for Venus $f_1 = 9$ Hz, $Q_1 = 5$. In the case of the Earth the observations show $f_1 = 7.9$ Hz, $Q_1 = 5$. For Venus we have no observational data at present.

Exercise 3.1.2.

Show how to use the *a priori* information on the structure of transverse resonances for the induction sounding of the Earth's crust.

Solution 3.1.2.

The Schumann resonances are sometimes called longitudinal. In addition to longitudinal resonances, there are also transverse resonances. The frequencies of transverse resonances of the Earth–ionosphere cavity are derived from the condition that a whole number of half-wavelengths is kept between the Earth and the ionosphere: $f_n \simeq nc/2h$, $n = 1, 2, \ldots$. At night the effective height of the ionosphere $h \simeq 90$ km and consequently $f_1 = 1.7$ kHz. At this frequency the depth of field penetration into the Earth is several dozens of metres, for example, 37 m if the conductivity is 10^9 s^{-1}. So the transverse resonances may be used for sounding the uppermost layers of the Earth.

Transverse resonances, as well as longitudinal, are excited effectively by lightning discharges. In contrast to the longitudinal, the transverse resonances are not global, but local or regional, since they are observed at distances of up to 2–3 thousand kilometres away from the source. This may be understood in the following way. TE and TM waves propagate horizontally away from the impulse source in the Earth–ionosphere waveguide. The group velocity becomes vanishing when the wave frequency approaches the transverse resonant frequency. Therefore, the absorption in the waveguide walls (mainly in the ionosphere) leads to damping of the field when moving away from the source. The nearer the wave frequency is to the resonant frequency the stronger is the damping.

Transverse resonances are observed at night in the form of so-called tweaks, that accompany spherics[5]. The dynamic spectrum of a tweak is non-stationary: in due course the carrier frequency of oscillations $f(t)$ asymptotically approaches the frequency of one of the transverse resonant harmonics f_n from above. It turns out that $\nabla \cdot b_\tau \to 0$ as $f \to f_n$. (For the TM wave this is evident and for the TE wave this follows from the theory of radio-wave propagation in the Earth–ionosphere waveguide.) Hence, for tweaks the formula

$$b_\nu^{(n)} = \frac{ih}{\pi n}(b_\tau^{(n)} \cdot \nabla \zeta)$$

is accomplished at least asymptotically and it is analogous to (3.13) for the Schumann resonances. This formula may be used to measure the gradient of the surface impedance using the observational data of the transverse resonances at one point only.

Exercise 3.1.3.

Estimate the the value of photon mass using the data on the observation of eigenfrequencies of Schumann resonances.

---

[5] Spheric is a variety of atmospherics, i.e. wide-band electromagnetic signals that accompany electric discharges in the atmosphere.

Solution 3.1.3.

The mass is introduced in electrodynamics by the following method. The first equation in (3.2) remains unchanged and the second one is replaced by

$$\nabla \times \boldsymbol{b} = -\mathrm{i}k\boldsymbol{E} - \mu^2\boldsymbol{A} \tag{1}$$

where $\boldsymbol{A}$ is the vector potential, $\boldsymbol{b} = \nabla \times \boldsymbol{A}$ and μ^{-1} is the Compton wavelength for a photon having mass

$$m_\gamma = \hbar\mu/c$$

where $\hbar$ is Planck's constant. The modification of Maxwell's equation (1) does not violate the Lorentz invariance of the theory and is permissible in this sense. However, the condition $m_\gamma \neq 0$ results in the violation of gauge invariance which leads to the violation of the charge conservation law and to electron instability. Thus, the problem of photon mass in electrodynamics has major importance.

Comparing the terms in equation (1) we may verify that the term $\mu^2\boldsymbol{A}$ may be significant in the case of large-scale fields. For the estimation of μ and m_γ let us use the fields induced by Schumann resonances which have typical wavelengths comparable with the Earth's dimensions.

Taking (1) into account, instead of (3.5), we obtain the following spectrum

$$\overline{\omega}_n = c\left[\frac{n(n+1)}{R_E^2} + \mu^2\right]^{1/2} \tag{2}$$

where the bar signifies that $\mu \neq 0$. If the theory, leading to the spectrum (2), is sufficiently rigorous and the measurements of the Schumann frequencies are accurate then using the values R_E, n, $\overline{\omega}_n$ one may calculate the value of μ with the help of equation (2). However, this is impossible due to the errors of spectral measurements and imperfection of the theory.

One may only make the estimation of μ from above

$$\mu < 2\sqrt{2}\pi c^{-1}(f_n\delta f_n)^{1/2}.$$

Here f_n is the observed resonant frequency and δf_n is the difference between f_n and the value of the frequency evaluated in the frame of ordinary ($m_\gamma = 0$) electrodynamics, or the error of measurement f_n if this error is more than the difference mentioned above. For the first harmonic, $f_1 \simeq 7.9$ Hz, we may use the value $\delta f_1 \simeq 0.5$ Hz. Then this leads to the estimation $\mu^{-1} > 1.7 \times 10^4$ km and $m_\gamma < 2 \times 10^{-47}$ g.

We compare the estimate with the result obtained by Schrödinger (1943) who used the fact of the existence of the Earth's magnetic field. Let us consider the case $\omega \to 0$ and define the potential of the magnetic dipole as follows

$$\boldsymbol{A} = \boldsymbol{M} \times \boldsymbol{r}\left[\frac{1+\mu r}{r^3}\right]\mathrm{e}^{-\mu r}$$

where M is the magnetic moment and r is the distance from the dipole. According to Schrödinger, the existence of the geomagnetic field leads to a certain limitation $\mu^{-1} > R_E$. This boundary approximately coincides with that obtained with the help of Schumann resonances. Stronger limitation follows from the observation of geomagnetic pulsations. We shall return to this problem in the next chapter.

Exercise 3.1.4.

Estimate the amplitude of the Schumann resonance excited by a monochromatic external source.

Solution 3.1.4.

Assume that an external current flows in a volume V inside the Earth–ionosphere cavity. Let $\mathbf{j}$ be the corresponding current density. According to the Poynting theorem the work of source done on the electromagnetic field per unit time equals $\int \mathbf{E} \cdot \mathbf{j}\, dV$. It is suggested that the volume V is small enough, and the frequency of current oscillations equals some resonance frequency (3.5). Then we are in a position to rewrite the integral as $E \int \mathbf{j}\, dV$. The integral $\int \mathbf{j}\, dV$ equals the time derivation of the electric dipole moment of the external current. Let us denote its vertical component by Il, where I and l are the effective total current and vertical length respectively. Let us denote also the vertical electric field of Schumann oscillations by E. Thereafter the rise of electromagnetic energy averaged over a period of Schumann oscillations equals $0.5EIl$. In the stationary state this value equals the energy leakage from the resonator upper wall. It follows that

$$E \simeq IlQ/cR_Eh$$

within a constant factor. Here we used equations (3.5) and (3.7)–(3.9). Closer inspection shows that

$$E_n = \frac{IlQ_n}{cR_Eh} \frac{2n+1}{\sqrt{n(n+1)}}.$$

3.2   The ionosphere as a reflector and transmitter

Here we shall consider the reflection of magnetospheric waves from the ionosphere and their transmission through the ionosphere and the atmosphere down to the surface of the Earth. The problem of reflection is interesting from the standpoint of the physics of magnetospheric waves, since the spectrum of magnetospheric oscillations depends on the coefficients of reflection. The problem of transmission arises in connection with the interpretation of the ground observations.

The ionosphere and the magnetosphere represent a plasma, i.e. a quasi-neutral totality of a large number of electrons and ions. In addition to charged particles there are neutral molecules, but their influence over the electrodynamic processes in the ionosphere and in the magnetosphere are quite different. The free path length in the magnetosphere is so great that we may neglect the collisions of particles with each other. In all the cases under consideration the plasma may undoubtedly be considered collisionless, starting with heights of, say, 1000 km. At this height we shall draw a conditional boundary that separates the ionosphere from the magnetosphere.

When approaching the Earth the gas density increases and collisions of particles begin to play an appreciable part. Collisions of electrons and ions with each other and with neutral molecules leads to the dissipation of wave energy and influences the phase and polarization characteristics of the wave field. Below 200 km the influence of collisions becomes dominant.

Another specific feature of electromagnetic wave propagation at ionospheric heights is connected with the strong vertical inhomogeneity of the medium. This inhomogeneity leads to the waveguide propagation of geomagnetic pulsations of Pc1 type along the ionospheric layers[6]. The centre of the ionospheric waveguide is located at a height of maximum F2 (approximately at 300 km)[7]. So-called ionospheric resonances appear at the critical frequencies of the ionospheric waveguide.

We shall leave the complicated problems of waveguide propagation in the ionospheric layers until the next section. Here we shall confine ourselves to the analysis of waves with frequencies much lower than the resonant frequencies of the ionosphere. At frequencies lower than approximately 0.1 Hz, which correspond to the geomagnetic pulsations of Pc2-5 and Pi2 types, the so-called thin-sheet model of the ionosphere may be used. This facilitates research and allows us to obtain a series of important results using analytical methods. In the general case, the numerical integration of the wave equation is the most appropriate method for solving problems of wave propagation in the ionosphere (see bibliographic commentary to this section).

3.2.1 Dielectric permeability of the magnetoactive plasma

Under the influence of the geomagnetic field the ionospheric and magnetospheric plasmas acquire anisotropic and gyrotropic properties. To describe these properties we shall introduce the tensor of complex dielectric permeability. At the same time we present some general equations of plasma physics. We shall deal with them below.

The electromagnetic field induces charges and currents in the plasma which in turn induce electromagnetic field. We shall suppose that the charge density $q(x, t)$ and the current density $j(x, t)$ are averaged over small volumes that contain a large number of charged particles. This also refers to the electric $E(x, t)$ and magnetic $B(x, t)$ fields.

---

[6] For the classification of geomagnetic pulsations see Addendum A.2.
[7] For brief information on ionospheric layers see Addendum A.1.

Using Maxwell's equations

$$\nabla \times \boldsymbol{B} = 4\pi c^{-1}\boldsymbol{j} + c^{-1}\partial \boldsymbol{E}/\partial t \qquad \nabla \cdot \boldsymbol{B} = 0$$
$$\nabla \times \boldsymbol{E} = -c^{-1}\partial \boldsymbol{B}/\partial t \qquad \nabla \cdot \boldsymbol{E} = 4\pi q \tag{3.32}$$

we obtain the charge conservation equation

$$\frac{\partial q}{\partial t} = -\nabla \cdot \boldsymbol{j}. \tag{3.33}$$

Hence, the influence of the medium on the electromagnetic field is characterized by one value, i.e. by the induced smoothed current $\boldsymbol{j}$. If the current is known then the induced charge may be found from (3.33).

It is convenient to introduce the vector of electric induction

$$\boldsymbol{D}(\boldsymbol{x}, t) = \boldsymbol{E}(\boldsymbol{x}, t) + 4\pi \int_{-\infty}^{t} \boldsymbol{j}(\boldsymbol{x}, t')\mathrm{d}t'$$

that also characterizes the influence of the medium completely. Then equations (3.32) will take the form

$$\nabla \times \boldsymbol{b} = \frac{1}{c}\frac{\partial \boldsymbol{D}}{\partial t} \qquad \nabla \cdot \boldsymbol{b} = 0 \tag{3.34}$$

$$\nabla \times \boldsymbol{E} = -\frac{1}{c}\frac{\partial \boldsymbol{b}}{\partial t} \qquad \nabla \cdot \boldsymbol{D} = 0.$$

Here we have made a replacement $\boldsymbol{B} \rightarrow \boldsymbol{B} + \boldsymbol{b}$, where $\boldsymbol{b}(\boldsymbol{x}, t)$ is the variable magnetic field of the wave. The constant external magnetic field $\boldsymbol{B}(\boldsymbol{x})$ is known.

In order to close the system it is necessary to find the dependence $\boldsymbol{D}(\boldsymbol{E})$ or $\boldsymbol{j}(\boldsymbol{E})$. Assuming the fields to be small, we can use the linear approximation. Then the connection between $\boldsymbol{j}$ and $\boldsymbol{E}$ has the form

$$j_\alpha(\boldsymbol{x}, t) = \int_{-\infty}^{t} \mathrm{d}t' \int \mathrm{d}\boldsymbol{x}' \, \sigma_{\alpha\beta}(\boldsymbol{x}, \boldsymbol{x}'; t, t')E_\beta(\boldsymbol{x}', t'). \tag{3.35}$$

This relation is the most general for the linear functional $\boldsymbol{j}(\boldsymbol{E})$.

Now we suppose the medium to be homogeneous and stationary. Then the core of the integral operator in (3.35) has the form of a finite-difference function of coordinates and time: $\sigma_{\alpha\beta}(\boldsymbol{x} - \boldsymbol{x}'; t - t')$. Using the condition of homogeneity and stationarity we shall expand all the values into the Fourier integrals, for example

$$\boldsymbol{E}(\boldsymbol{x}, t) = \int \boldsymbol{E}_{\omega k} e^{\mathrm{i}\boldsymbol{k}\cdot\boldsymbol{x} - \mathrm{i}\omega t} \, \mathrm{d}\omega \, \mathrm{d}\boldsymbol{k} \tag{3.36}$$

$$E_{\omega k} = (2\pi)^{-4} \int E(x, t) e^{-ik \cdot x + i\omega t} \, dx \, dt. \tag{3.37}$$

It follows from (3.35) that

$$j_{\alpha \omega k} = \sigma_{\alpha \beta}(\omega, k) E_{\beta \omega k} \tag{3.38}$$

where

$$\sigma_{\alpha \beta}(\omega, k) = \int\limits_0^\infty dt \int dx e^{-ik \cdot x + i\omega t} \sigma_{\alpha \beta}(x; t).$$

Similarly we have

$$D_{\alpha \omega k} = \varepsilon_{\alpha \beta}(\omega, k) E_{\beta \omega k} \tag{3.39}$$

where

$$\varepsilon_{\alpha \beta}(\omega, k) = \delta_{\alpha \beta} + \frac{4\pi i}{\omega} \sigma_{\alpha \beta}(\omega, k). \tag{3.40}$$

In the general case, $\varepsilon_{\alpha \beta}$ depends on the frequency ω (frequency dispersion) and on the wave vector k (spatial dispersion).

Consequently, the medium in the linear approximation is characterized completely by the tensor of dielectric permeability $\varepsilon_{\alpha \beta}$ (or the conductivity tensor $\sigma_{\alpha \beta}$). The advantage of such a phenomenological description lies in the fact that it permits the determination of some general ratios for the field without specifying the type of the tensor $\varepsilon_{\alpha \beta}$. First of all it corresponds to the dispersion relations (exercise 3.2.1). Furthermore, using $\varepsilon_{\alpha \beta}$ we may write compactly the equation of wave energy transfer (Chapter 5) and the expression for the ponderomotive force (Chapter 7). Finally, in terms of $\varepsilon_{\alpha \beta}$ one may express general stability criteria for the medium (Chapter 6).

So, $\varepsilon_{\alpha \beta}$ is the most vital characteristic of the medium. But can an explicit expression for $\varepsilon_{\alpha \beta}$ be found?

The values of $\varepsilon_{\alpha \beta}$ are either measured experimentally or calculated within the framework of one or other specific model of the medium. The first method is used for the condensed matter (sea water, Earth's crust); the second one is for the plasma (ionosphere, magnetosphere). Let us outline the derivation of $\varepsilon_{\alpha \beta}$ within the framework of the cold plasma model.

The motion of a single particle with a charge e and a mass m is described by the equation

$$\frac{d^2 x}{dt^2} = \frac{e}{m} \left[E(x, t) + \frac{1}{c} \frac{dx}{dt} \times B(x) \right] \tag{3.41}$$

where $x(t)$ is the trajectory of a particle. We have ignored the term $(e/mc)\dot{x} \times b$ on the right-hand side of (3.41), assuming $|b| \ll |B|$. Moreover, we have temporarily ignored the collisions of the given particle with other charged and neutral particles.

Introducing a notation $\Omega = eB/mc$ we may rewrite (3.41) in the form

$$\dot{v} + \Omega \times v = (e/m)E. \tag{3.42}$$

Here $v = \dot{x}$. The value $|\Omega|$ is termed the cyclotron frequency or gyrofrequency of a particle. Let us introduce an additional designation $\Omega = e|B|/mc$. We do not specify the type of particles by an index, but imply that $\Omega = \Omega_a = e_a|B|/m_a c$. It is obvious that $\Omega < 0$ for the electrons and $\Omega > 0$ for the positive ions. At the same time we agree that if we mean the electrons ($a = e$) or ions ($a = i$) then we shall write $\Omega_e = e|B|/m_e c$ or $\Omega_i = ze|B|/m_i c$, where e is an elementary charge, ze is the ion charge, i.e. $\Omega_e > 0$, $\Omega_i > 0$ (for positive ions). These simple rules facilitate the writing of rather complicated formulae for the tensor of dielectric permeability.

We consider E and b to be small values of the same order. The same order of magnitude is intrinsic to the particle velocity v, since at $E = 0$ and $b = 0$ the particle is supposed to be immobile. But this means that the inhomogeneity of E and B in the equation of motion may be neglected. Then (3.42) reduces to an algebraic equation

$$-i\omega v + \Omega \times v = (e/m)E \tag{3.43}$$

if E changes in due course according to $\exp(-i\omega t)$. When solving equation (3.43) we find the dependence $v(E)$.

The current density, induced by the wave, is obviously

$$j = \sum eN v(E). \tag{3.44}$$

Here the summation is made over the particle species. The value N is the number density of charged particles of given type and the condition of quasi-neutrality takes place

$$\sum eN = 0. \tag{3.45}$$

Now we shall compare (3.44) with (3.38), taking into account (3.40), and find the dielectric permeability tensor of a cold collisionless plasma

$$\varepsilon_{\alpha\beta} = \begin{pmatrix} \varepsilon_\perp & ig & 0 \\ -ig & \varepsilon_\perp & 0 \\ 0 & 0 & \varepsilon_\parallel \end{pmatrix} \tag{3.46}$$

where

$$\varepsilon_\perp = 1 - \sum \frac{\omega_0^2}{\omega^2 - \Omega^2} \qquad g = -\sum \frac{\omega_0^2 \Omega}{\omega(\omega^2 - \Omega^2)} \qquad \varepsilon_\parallel = 1 - \sum \frac{\omega_0^2}{\omega^2}.$$

Tensor (3.46) is given in a Cartesian system of coordinates (x, y, z) with the z axis being directed along the external magnetic field B. The value

$$\omega_0 = (4\pi e^2 N/m)^{1/2}$$

is termed the plasma or Langmuir frequency for particles of the given type.

The tensor (3.46) depends on the frequency ω, but it does not depend on the wave vector k, i.e. there is no spatial dispersion. This is a direct consequence of the fact that we have ignored the thermal motions of the particles. Hence it follows that we may use (3.46) to study the waves in an inhomogeneous medium. (The requirement for the stationarity of an unperturbed medium is still conserved.) Furthermore, the tensor (3.46) is Hermitian ($\varepsilon_{\alpha\beta}^* = \varepsilon_{\beta\alpha}$), i.e. the dissipation of the wave energy is absent in the plasma. This is natural since the plasma is considered collisionless.

Collisions of charged particles with neutrals in a partly ionized gas (ionosphere) may be taken into account approximately if we add friction force $-\nu v$ to the right-hand side of (3.43). This will lead to the requirement to make a formal substitution in all the components of the dielectric permeability tensor (3.46)

$$m \rightarrow m(1 + \mathrm{i}\nu/\omega). \tag{3.47}$$

Here ν is the frequency of collisions between charged particles of the given type and the neutrals.

The approximation of a cold plasma, which has the advantage of simplicity, is widely employed in the theory of geoelectromagnetic waves. However, the applicability of this approximation is restricted by a number of conditions: (1) the wave phase velocity along the external magnetic field must be much higher than the average thermal particle velocity; (2) in the direction, perpendicular to the external magnetic field, the wave field must change slightly within a distance of the order of the Larmor radius $v_T/|\Omega|$ of particles that move at the average thermal velocity; (3) the wave frequency must not be close to the gyrofrequencies of particles or to the harmonics of these gyrofrequencies.

3.2.2 The wave equation and boundary conditions

Eliminating b from the third equation of the system (3.34) and with the help of the first equation, we obtain the wave equation for E

$$\nabla^2 E - \nabla(\nabla \cdot E) + \frac{\omega^2}{c^2} D = 0. \tag{3.48}$$

Here the dependence of the field on time is taken in the form $\exp(-\mathrm{i}\omega t)$. The field, which depends on time arbitrarily, may be represented as a superposition of monochromatic waves. The connection of D with E may be represented in the following way

$$D = \varepsilon_\perp E + (\varepsilon_\parallel - \varepsilon_\perp)\tau(E \cdot \tau) + \mathrm{i}E \times g \tag{3.49}$$

where $\tau = B/B$ is a unit vector along the geomagnetic field and $g = g\tau$ is the so-called gyration vector.

We want to carry out an analytical study and so we must inevitably introduce a number of strong idealizations while the problem is being set. We shall consider the medium to be plane-stratified (figure 3.3). This imposes a restriction on the wavelengths in the horizontal direction: they have to be much shorter not only than the radius of the Earth but also than the typical dimensions of both the horizontal plasma and magnetic field inhomogeneities. Let us suppose the field lines of the geomagnetic field to be strictly vertical. This condition occurs approximately only at high latitudes. We shall direct B downwards as is the case in the northern hemisphere.

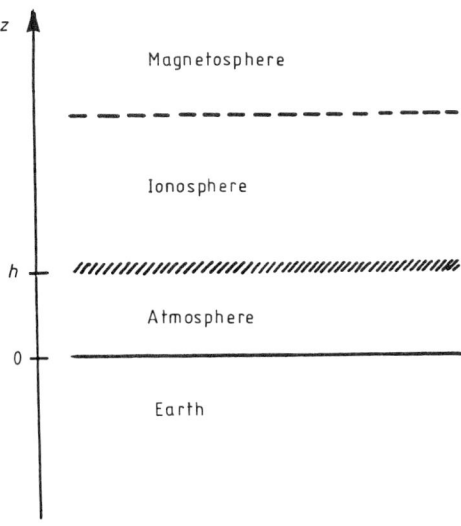

Figure 3.3. Plane model of the medium. The so-called Hall (gyrotropic) layer is indicated in the lower ionosphere at an altitude of $h \approx 100$ km above the Earth's surface.

Let us direct the z axis of the Cartesian system of coordinates vertically upwards and make a spatial Fourier expansion of the wave field using the coordinates x and y. The equations for the Fourier components of the field E are obtained from (3.47) by substitution

$$\partial/\partial x \rightarrow ik_x \qquad \partial/\partial y \rightarrow ik_y \qquad \nabla^2 \rightarrow d^2/dz^2 - k_\perp^2$$

where $k_\perp = (k_x^2 + k_y^2)^{1/2}$. The field, which arbitrarily depends on x and y, may be represented in the form of superposition of spatial harmonics.

Without restricting the generality we may assume $k_y = 0$. Then the wave equations (3.48) and (3.49) will take the form

$$\left[\frac{d^2}{dz^2} + k_0^2 \varepsilon_\perp \right] E_x = igk_0^2 E_y + ik_\perp \frac{dE_z}{dz} \tag{3.50}$$

$$\left[\frac{d^2}{dz^2} - k_\perp^2 + k_0^2 \varepsilon_\perp \right] E_y = -igk_0^2 E_x \tag{3.51}$$

$$E_z = \frac{ik_\perp}{k_0^2 \varepsilon_\parallel - k_\perp^2} \frac{dE_x}{dz} \tag{3.52}$$

where $k_0 = \omega/c$. The magnetic field is expressed through the electric field by means of the formulae

$$b_x = \frac{i}{k_0} \frac{dE_y}{dz} \tag{3.53}$$

$$b_y = -\frac{k_0}{k_\perp} \varepsilon_\parallel E_z \tag{3.54}$$

$$b_z = \frac{k_\perp}{k_0} E_y. \tag{3.55}$$

If the medium is not gyrotropic ($g = 0$), then the set of equations would fall into two independent subsystems. Equations (3.50), (3.52) and (3.54) would describe a so-called E-wave with the components E_x, E_z, b_y. It is also termed the transverse magnetic wave (TM), since the component b_y is perpendicular to the direction of propagation. Equations (3.51), (3.53) and (3.55) would describe the b-wave with the components E_y, b_x, b_z (transverse electric or TE wave).

Gyrotropy is negligible in the Earth's crust and in the air. In the magnetosphere, where there are practically no collisions between particles, $g \simeq (\omega/\Omega_i)\varepsilon_\perp$. Therefore, the gyrotropic term may be neglected at fairly low frequencies (see exercise 3.2.2). As regards the ionosphere, owing to collision, gyrotropy does not disappear at whatever low frequencies. But the ionosphere is strongly inhomogeneous along the height. It turns out that at $\omega \ll \Omega_i$ gyrotropic properties of the ionosphere show only in a comparatively thin layer located in the lower ionosphere (exercise 3.2.3). This gives us reason to substitute the gyrotropic layer for the gyrotropic film when analytical modelling of low-frequency wave propagation is being carried out.

Generally speaking (but not in this particular case), gyrotropy is not necessarily accompanied by anisotropy or dissipation, these are the three independent properties of substance. Here an anisotropic highly absorbing layer of the ionosphere is situated almost at the gyrotropic layer. This layer is also usually modelled by an absorbing film. In order to simplify the analysis we shall place both films at one and the same height (about 100 km).

Let us specify the model of the medium. It has been assumed above that the medium is horizontally homogeneous, and the external magnetic field is vertical. We shall also consider the Earth to be ideally conducting. Hence, the boundary conditions on the Earth's surface ($z = 0$) will take the form

$$E_x = E_y = b_z = 0. \tag{3.56}$$

In the range $0 < z < h$ we shall place a vacuum gap that imitates the atmosphere ($\varepsilon_{\alpha\beta} = \delta_{\alpha\beta}$). At height $z = h$ we shall place a gyrotropic absorbent film. Its properties will be described by the tensor of integral conductivity

$$\Sigma_{\alpha\beta} = \begin{pmatrix} \Sigma_\perp & -\Sigma_H \\ \Sigma_H & \Sigma_\perp \end{pmatrix} \tag{3.57}$$

$$\Sigma_\perp = \int \sigma_\perp(z)\,dz \qquad \Sigma_H = \int \sigma_H(z)\,dz$$

where $\sigma_\perp = (\omega/4\pi i)(\varepsilon_\perp - 1)$, $\sigma_H = (\omega/4\pi)g$. The indices α, β assume the values x, y. The integration is made over a gyrotropic absorbent layer of the ionosphere.

We have already dealt with boundary conditions of the film type (see exercise 1.2.1 in section 1.2). The idea is that the horizontal components of the electric field are continuous, and the horizontal components of the magnetic field are subject to discontinuity when passing through a thin conducting film. Let us find the connection between E_t and b_t over the film.

Let us first introduce the impedances

$$Z_E(z) = -E_x(z)/b_y(z) \tag{3.58}$$

$$Z_b(z) = E_y(z)/b_x(z) \tag{3.59}$$

for E-waves and b-waves in the air. From the solution of the wave equation (3.48) at $\varepsilon_{\alpha\beta} = \delta_{\alpha\beta}$ and using (3.56) we find

$$Z_E(z) = i(k_\perp/k_0)\tanh(k_\perp z) \tag{3.60}$$

$$Z_b(z) = -i(k_0/k_\perp)\tanh(k_\perp z). \tag{3.61}$$

Here it has been taken into account that $k_0 \ll k_\perp$. This inequality is obeyed with ample margins in the range $f < 0.1$ Hz that is of interest to us, since *a priori* $k_\perp \gg R_E^{-1}$.

At the height $z = h$ under the film we have

$$\begin{aligned} b_x^- &= Y_b(h)E_y \\ b_y^- &= -Y_E(h)E_x \end{aligned} \tag{3.62}$$

where $Y(z) \equiv Z^{-1}(z)$. At the same height, but above the film, after straightforward calculations we get

$$\begin{aligned} b_x^+ &= 4\pi c^{-1}\Sigma_H E_x + Y_b(h)E_y \\ b_y^+ &= 4\pi c^{-1}\left(\Sigma_H E_y - \Sigma_\perp E_x\right). \end{aligned} \tag{3.63}$$

When deriving (3.63) the inequalities

$$k_0/k_\perp^2 h \ll (4\pi/c)\Sigma_\perp \ll 1/k_0 h$$

are used. We have also made a simplifying supposition that the thickness of the film is small compared with h and $k_\perp^{-1}$.

Let the magnetosphere be located immediately above the film. We shall consider it to be non-gyrotropic, homogeneous, and infinite, and describe it by a diagonal tensor

$$\varepsilon_{\alpha\beta} = \varepsilon_\perp \delta_{\alpha\beta} + (\varepsilon_\parallel - \varepsilon_\perp)\tau_\alpha \tau_\beta \tag{3.64}$$

where the indices α, β assume the values x, y, z; $\tau = B/B$.

It is necessary to introduce the sources and take into account the condition of radiation at infinity ($z \to \infty$). Let us suppose, that the sources are located fairly high above the ionosphere. Then they may be given in the implicit form of a wave which is incident upon the ionosphere. The total field, in this case, will certainly not satisfy the condition of radiation. It will be yielded only by the diffracted (scattered) field.

According to Mandelshtam any idealization will sooner or later revenge itself. A number of applicability conditions of our ideal model are shown in the explicit form here, a number of others are implied. The most important condition is the absence of field oscillations inside the Hall layer, as well as inside the ionosphere above this layer, which we simply disregarded. On the face of it, this condition is provided by the limitation $f < 0.1$ Hz, but this is deceptive. We know that the frequency of the Alfvén wave is almost independent of the transverse structure of the wave field to a certain extent, i.e. f is almost independant of $k_\perp$. But this means that the condition $f < 0.1$ Hz does not limit the value $k_\perp$ strictly enough. Now one can easily imagine what will happen if the direction of the vector B deviates even slightly from the vertical. Then, for sufficiently large values of $k_\perp$ the oscillations, which we seek to avoid, will certainly arise, despite the condition $f < 0.1$ Hz.

However, the model is so interesting that we will pursue its analysis.

3.2.3　Alfvén waves

Let an Alfvén wave of unit amplitude be incident upon the ionosphere. From (3.50) and with the help of (3.64) at $k_\perp^2 \ll k_0^2 \varepsilon_\parallel$ we obtain

$$E_x = e^{-ik_\parallel(z-h)} \qquad b_y = -n_A E_x \qquad (3.65)$$

where $n_A = c/c_A$, $k_\parallel = \omega/c_A$ and $c_A = c\Omega_i/\omega_{oi}$. The reflected wave, which yields the condition of radiation, splits into an Alfvén wave at $z > h$

$$E_x = R e^{ik_\parallel(z-h)} \qquad b_y = n_A E_x \qquad (3.66)$$

and a magnetosonic one

$$E_y = T e^{ik_z(z-h)} \qquad b_x = -n_z E_y. \qquad (3.67)$$

Here R is the coefficient of reflection, T is the coefficient of transformation, $n_z = (n_A^2 - n_\perp^2)^{1/2}$, $n_\perp = ck_\perp/\omega$, $k_z = k_0 n_z$.

Substituting (3.65)–(3.67) at $z = h$ into boundary conditions (3.63) we obtain

$$R = \frac{\Sigma_w - \Sigma_\perp + iS}{\Sigma_w + \Sigma_\perp - iS}. \qquad (3.68)$$

Here $\Sigma_w = cn_A/4\pi$ and

$$S = \frac{k_\parallel}{2k_\perp} \frac{\Sigma_H^2}{\Sigma_w}(1 - e^{-2k_\perp h}).$$

Two new restrictions are assumed: $k_\parallel \ll k_\perp$, $k_\parallel h \ll 1$. Instead of (3.68) it is more appropriate to use the expression

$$R = \frac{\Sigma_w - \Sigma_\perp}{\Sigma_w + \Sigma_\perp} \tag{3.69}$$

which follows from (3.68) with the proviso that $S \ll \Sigma_w$.

Similarly we obtain the coefficient of transformation

$$T = -\frac{4\pi \Sigma_H (1 + R)}{c[Y_b(h) + n_z]} \tag{3.70}$$

or at $k_\perp h \ll 1$

$$T = \frac{2ik_\parallel h \Sigma_H}{\Sigma_w + \Sigma_\perp}. \tag{3.71}$$

Let us discuss these solutions. The coefficient of reflection does not depend on gyrotropy (see (3.69)). With the disappearance of the absorbent layer ($\Sigma_\perp \to 0$) it approaches $+1$. Thus the electric field E_x has an antinode at the height $z = h$, while the magnetic field b_y has a node at the same height.

On the other hand, the coefficient of transmission is proportional to the gyrotropy Σ_H. Generally speaking the coefficient is rather small. For example, at $-\Sigma_H \simeq \Sigma_\perp \simeq \Sigma_w$, $h \simeq 10^7$ cm, $c_A \simeq 10^8$ cm s^{-1} and $\omega \simeq 0.1$ s^{-1} we have $T \simeq 10^{-2}$. However, the transmission should be taken into account, since it provides the penetration of the field into the surface of the Earth.

At height $z = h$ under the film, according to the condition of continuity of the horizontal components of the electric field $E_x = 1 + R$ and $E_y = T$. For the magnetic field we have (see (3.62))

$$b_x^-(h) = Y_b(h)T \qquad b_y^-(h) = -Y_E(h)(1 + R).$$

Extending the fields downwards to the surface of the Earth we have $E_x(0) = E_y(0) = 0$ and

$$b_x(0) = \frac{2\Sigma_H}{\Sigma_w + \Sigma_\perp} e^{-k_\perp h} b_y^{inc}(h) \tag{3.72}$$

where b_y^{inc} is the amplitude of the incident Alfvén wave. The component $b_y(0)$ is negligibly small in the given approximation.

Apparently, the very possibility of observing Alfvén waves on the Earth's surface is associated with the gyrotropy of the ionosphere. The field at the Earth's surface vanishes at $\Sigma_H \to 0$. It also vanishes as $k_\perp \to \infty$ (at $k_\perp h > 1$ the field at the Earth's surface is exponentially small). Finally, it is evident from (3.72), that after transmission through the ionosphere, the vector of the magnetic perturbation turns in the horizontal plane by $\pi/2$.

Let us estimate the reflection coefficient and the transmission coefficient. In the daytime $\Sigma_w \simeq 10^{12}$ cm s^{-1}, $\Sigma_\perp \simeq 10^{13}$ cm s^{-1}, $\Sigma_H \simeq -2 \times 10^{13}$ cm s^{-1}. Accordingly we have $R \simeq -0.8$. At night $\Sigma_w \simeq 3 \times 10^{11}$ cm s^{-1},

$\Sigma_\perp \simeq -\Sigma_H \simeq 10^{11}$ cm s^{-1} and $R \simeq +0.5$. The ratio $b_x(0)/b_y^{inc}(h)$ will be estimated under the condition $k_\perp h \ll 1$. In the daytime this ratio is of the order of 3.5, and at night is of the order of 0.5.

3.2.4 Magnetosonic waves

The incident magnetosonic wave has the components

$$E_y = e^{-ik_z(z-h)} \qquad b_x = n_z E_y. \qquad (3.73)$$

Then the reflected magnetosonic wave has the components

$$E_y = R e^{ik_z(z-h)} \qquad b_x = -n_z E_y \qquad (3.74)$$

and the Alfvén wave

$$E_x = T e^{ik_\parallel(z-h)} \qquad b_y = n_A E_x \qquad (3.75)$$

where R and T are the coefficients of reflection and transformation.

Substituting (3.73)–(3.75) into (3.63) we obtain R and T. Here the condition $kh \ll l$ is natural. Then we have $R \simeq -1$. This means, first, that the gyrotropy influences the coefficient of reflection weakly. Secondly, the field E_y has a node at height $z = h$ (and on the Earth's surface $z = 0$). Accordingly, the field b_x has an antinode.

The coefficient of transformation approximately equals

$$T \simeq -2ik_z h \frac{\Sigma_H}{(\Sigma_w + \Sigma_\perp)} \qquad (3.76)$$

and is rather small. The gyrotropic absorbent layer of the ionosphere weakly influences the reflection and the transmission of magnetosonic waves.

Exercises

Exercise 3.2.1.

Use $\varepsilon_{\alpha\beta}(\omega, k)$ to derive the dispersion equation and to expand the arbitrary field in a homogeneous medium by normal waves.

Solution 3.2.1.

The equations of electrodynamics (3.34) in the linear approximation written for the Fourier components with regard (3.39) have the form

$$M_{\alpha\beta}(\omega, k) E_{\beta\omega k} = 0 \qquad (1)$$

where

$$M_{\alpha\beta} \equiv n^2 \delta_{\alpha\beta} - n_\alpha n_\beta - \varepsilon_{\alpha\beta}.$$

Here $n = ck/\omega$. Owing to the superposition principle, the arbitrary wave field $E(x, t)$ has the form of the Fourier integral (3.36), in which $E_{\omega k}$ represents the solution of equation (1). The condition of the existence of nontrivial solutions (1) leads to the dispersion equation, which connects the frequency ω with the wave vector k

$$\text{Det}[M_{\alpha\beta}(\omega, k)] = 0.$$

The roots of the dispersion equation ω_k^σ correspond to the different normal waves in the linear homogeneous medium (σ signifies the type of the normal wave).

The Fourier components may be presented in the form

$$E_{\omega k} = \sum_\sigma a_k^\sigma E_k^\sigma \delta(\omega - \omega_k^\sigma).$$

Hence we obtain the expansion of the arbitrary field by the normal waves

$$E(x, t) = \sum_\sigma \int a_k^\sigma E_k^\sigma \exp(ik \cdot x - i\omega_k^\sigma t)\, dk.$$

Here E_k^σ is the amplitude of the σ-type wave. The polarization vectors a_k^σ are defined by the equation

$$M_{\alpha\beta} a_\beta^\sigma = 0$$

in which it is supposed that $\omega = \omega_k^\sigma$. If the tensor $\varepsilon_{\alpha\beta}$ is Hermitian, then the normalized polarization vectors yield the condition of orthogonality

$$a_\alpha^{\nu*}(\delta_{\alpha\beta} - n_\alpha n_\beta / n^2) a_\beta^\sigma = \delta_{\sigma\nu}$$

and the ratio

$$n^2 \delta_{\nu\sigma} = a_\alpha^{\nu*} \varepsilon_{\alpha\beta} a_\beta^\sigma.$$

Here it is also assumed that $\omega = \omega_k^\sigma$. The asterisk stands for the complex conjugate. Other ways of normalization of polarization vectors are also possible.

The finding of eigenfrequencies ω_k^σ and polarization vectors a_k^σ is one of the principal problems of the linear theory of waves in a homogeneous infinite medium.

Exercise 3.2.2.

Find the solution of the dispersion equation, derived in the previous exercise, within the framework of a cold plasma model at $\omega \to \infty$ and at $\omega \to 0$.

Solution 3.2.2.

Instead of first determining the general solution and then making respective limiting transmissions, we shall make these transmissions in (3.46). This is not quite correct, but reduces the length of the solution considerably. As $\omega \to \infty$ we have $\varepsilon_{\alpha\beta} = \varepsilon \delta_{\alpha\beta}$. In a collisionless plasma we have

$$\varepsilon = 1 - \omega_{0e}^2/\omega^2$$

where ω_{0e} is the electron Langmuir frequency. (The limiting transition here is not brought to its end. If $\omega \to \infty$, then $\varepsilon = 1$.) In the high-frequency limit the plasma is isotropic and non-gyrotropic, and the ions do not influence the propagation of waves. The dispersion equation splits into three independent equations. One of them, $\varepsilon(\omega) = 0$, corresponds to the longitudinal Langmuir oscillations ($\omega = \omega_{0e}$, $\boldsymbol{E}\|\boldsymbol{k}$). The other two are similar: $n_{1,2}^2 = \varepsilon$. They correspond to transverse waves ($\boldsymbol{E} \perp \boldsymbol{k}$) with two independent polarizations. The dispersion relation for the high-frequency transverse waves is usually written in the form

$$n^2 = 1 - \omega_{0e}^2/\omega^2$$

where $n = ck/\omega$ is the refractive index. As $\omega \to \infty$ the refractive index approaches unity. This conclusion refers to all substances and not only to plasma.

We shall take the collisions into account by a mere replacement $\omega_{0e} \to \omega_{0e}(1+i\nu_e/\omega)^{-1/2}$, where ν_e is the effective frequency of the electron collisions. Then

$$n^2 = \varepsilon = 1 - \frac{\omega_{0e}^2}{\omega(\omega + i\nu_e)}.$$

At $\omega \ll \nu_e$ this expression transforms into (3.23).

We shall accomplish the transition $\omega \to 0$ in (3.46) in two stages. First we make sure that

$$\varepsilon_{\alpha\beta} = \begin{pmatrix} \varepsilon_\perp & 0 & 0 \\ 0 & \varepsilon_\perp & 0 \\ 0 & 0 & \varepsilon_\| \end{pmatrix}$$

i.e. at low frequencies a collisionless plasma represents an anisotropic non-gyrotropic medium, similar to a single-axis crystal. Furthermore, we see that $\varepsilon_\perp = c^2/c_A^2$ as $\omega \to 0$ whereas $\varepsilon_\| \to -\infty$, where $c_A = B/(4\pi\rho)^{1/2}$ is the Alfvén velocity and $\rho = \sum mN$ is the plasma density. (Here, as well as in other places, the summation mark without the indication of an index signifies the summation over the types of particle.)

Let us turn to equation (1) in the previous exercise. Without limiting the generality we assume $n = (n_\perp, 0, n_\|)$ and write

$$M_{\alpha\beta} = \begin{pmatrix} n_\|^2 - \varepsilon_\perp & 0 & 0 \\ 0 & n^2 - \varepsilon_\perp & 0 \\ 0 & 0 & \varepsilon_\| \end{pmatrix}.$$

Now it is evident, that at $|\varepsilon_\|| \to \infty$ we have $E_\| \to 0$, i.e. the oscillations of the electric field are strictly transverse, not with respect to the wave vector $\boldsymbol{k}$, but with respect to the external magnetic field B. Equation (1) falls into two independent equations

$$(n_{\parallel}^2 - \varepsilon_{\perp})E_x = 0$$

$$(n^2 - \varepsilon_{\perp})E_y = 0.$$

The first of these describes the propagation of the Alfvén waves, and the second one corresponds to the magnetosonic waves. The respective dispersion laws have the form

$$n_1 = n_A/|\cos\theta| \qquad n_2 = n_A$$

where $n_A = c/c_A$ and θ is the angle between the vectors k and B. These dispersion laws coincide with the dispersion relations (2.22) and (2.24) that have been obtained for the case of ideal magnetohydrodynamics.

Now we have to find out what we have lost by finding the limiting transmissions before and not after the solution of the dispersion equation. We shall confine ourselves to discussing the case $\omega \to 0$.

At $\theta \neq 0$ we have $n_1 \neq n_2$, and at $\theta = 0$ it results in degeneration of the spectrum. But this should not be, since an external magnetic field is superimposed over the plasma. At $\theta = 0$ the general dispersion equation for the transverse waves falls into two different equations

$$n_{1,2}^2 = \varepsilon_{\perp} \pm g.$$

The waves have circular polarization with left (n_1) and right (n_2) rotation of the vector E (when viewed along B). Hence it follows, that the degeneration of the magnetohydrodynamic wave spectrum at $\theta = 0$ is associated with neglecting the gyrotropy. The limiting transition $\theta \to 0$ is impossible, if the transition $\omega \to 0$ has already been accomplished. The magnetohydrodynamic dispersion relations are valid only at $\theta^2 \gg 2g/\varepsilon_{\perp}$. This condition is termed the condition of quasi-perpendicular propagation. In this case we may ignore gyrotropy and consider the waves to have linear polarization. At $\theta^2 \ll 2g/\varepsilon_{\perp}$, quasi-longitudinal propagation, the gyrotropy effect is switched on and the waves acquire almost circular polarization. Degeneration does not exist at any θ.

Exercise 3.2.3.

Estimate the parameters of the gyrotropic layer, located in the lower ionosphere (height, thickness, integral conductivity).

Solution 3.2.3

In order to solve this problem it is more appropriate to use the conductivity tensor

$$\sigma_{\alpha\beta} = \begin{pmatrix} \sigma_{\perp} & \sigma_H & 0 \\ -\sigma_H & \sigma_{\perp} & 0 \\ 0 & 0 & \sigma_{\parallel} \end{pmatrix}$$

instead of $\varepsilon_{\alpha\beta}$. The components $\sigma_{\parallel}$ and $\sigma_{\perp}$ are termed longitudinal and transverse conductivities, and the component σ_H stands for the Hall conductivity. If the plasma consists of electrons, singly charged ions and neutral particles, then from (3.40), (3.46) and (3.47) it follows that

$$\sigma_{\parallel} = e^2 N / m_e v_e \tag{1}$$

for the longitudinal conductivity at $\omega \ll v_e$. Here we have taken into account that $m_e \ll m_i$, $v_i \ll v_e$ and $m_e v_e \ll m_i v_i$. These inequalities hold everywhere. As regards the condition $\omega \ll v_e$, it is accomplished in the ionosphere. In the magnetosphere an inverse inequality may be realized so then $\sigma_{\parallel} = i e^2 N / m_e \omega$. But here $\sigma_{\parallel}$ is so large that we can substitute $\sigma_{\parallel} \to i\infty$.

The transverse and the Hall conductivities equal

$$\sigma_{\perp} = e^2 N \left[\frac{1}{m_i} \frac{(v_i - i\omega)}{\Omega_i^2 + (v_i - i\omega)^2} + \frac{1}{m_e} \frac{(v_e - i\omega)}{\Omega_e^2 + (v_e - i\omega)^2} \right] \tag{2}$$

$$\sigma_H = e^2 N \left[\frac{1}{m_i} \frac{\Omega_i}{\Omega_i^2 + (v_i - i\omega)^2} - \frac{1}{m_e} \frac{\Omega_e}{\Omega_e^2 + (v_e - i\omega)^2} \right]. \tag{3}$$

The problem may be solved numerically, on constructing vertical profiles of the conductivities. We shall perform guiding estimates analytically, using the following consideration.

The gyrofrequencies $\Omega_{e,i}$ decrease with the height by powers, whereas the collision frequencies $v_{e,i}$, which are proportional to the neutral gas density, decrease exponentially. In the lower atmosphere $v_{e,i} \gg \Omega_{e,i}$. So, at some heights the profiles $\Omega_{e,i}(z)$ and $v_{e,i}(z)$ will have intersection points. Let us designate z_e and z_i as the heights, where $\Omega_e(z)$, $v_e(z)$ and $\Omega_i(z)$, $v_i(z)$ intersect respectively. We will show that $z_e < z_i$.

Let H be the scale of heights for the neutral gas. Then $v_{e,i}(z) \propto \exp(-z/H)$. We shall ignore the weak dependence of $\Omega_{e,i}$ on z at the ionospheric heights. Then

$$\Delta z = H \ln (m_i v_i / m_e v_e) \tag{4}$$

where $\Delta z = z_i - z_e$.

We now make a rough estimation of the ratio v_i / v_e. The collision frequency is of the order of $v \simeq v_T / l$, where $v_T \simeq \sqrt{T/m}$ is the thermal velocity, l is the average free path length and T is the temperature. If s is the collision cross section and N_n is the concentration of the scattered particles (neutral molecules), then $l \simeq 1 / N_n s$. The cross section of collisions $s \simeq a^2$, where a is the dimension of the molecule. Hence, it follows that $v_i / v_e \simeq (m_e / m_i)^{1/2}$.

Let us specify the estimation within the framework of a simple model of elastic collisions of round balls. The first point refers to the ion collision cross sections s_{in} with the neutral molecules. From the geometrical consideration it is clear that $s_{in} = 4 s_{en}$. The second point refers to the value v_T. In the ion–molecule collisions, instead of the ion thermal velocity, an average relative velocity of ion and molecule motions (which increases v_i in $\sqrt{2}$ times if $m_i = m_n$) should be taken into account. As a result $v_i/v_e \simeq 4(2m_e/m_i)^{1/2}$ and instead of (4) we may obtain

$$\Delta z = H \ln\left[4(2m_i/m_e)^{1/2}\right] \tag{5}$$

or $\Delta z \simeq 5.5H$, if the molecular weight of ions equals 28. The scale height $H = T/m_i g$, where g is the acceleration due to gravity. At a height of about one hundred kilometres $T \simeq 230$ K, $H \simeq 7$ km and $\Delta z \simeq 40$ km.

So, $\Delta z > 0$, i.e. $z_i > z_e$. Let us show that the value Δz is of the order of the thickness of the ionospheric gyrotropic layer.

Below the level z_e we have $v_{e,i} \gg \Omega_{e,i}$ and according to (1)–(3), $\sigma_\perp \simeq \sigma_\parallel \gg \sigma_H$. The medium may be considered isotropic and non-gyrotropic at $z < z_e$. The upper boundary of the isotropic layer is at a height of 70–80 km.

Above the level z_i we have $v_{e,i} \ll \Omega_{e,i}$ and respectively $\sigma_H \ll \sigma_\perp \ll \sigma_\parallel$. The ratio $\sigma_H/\sigma_\perp$ is as small as v_i/Ω_i and decreases rapidly with height. The medium is highly anisotropic, but not gyrotropic. Here we usually assume $\sigma_H = 0$, $\sigma_\parallel = \infty$, and instead of $\sigma_\perp$ we use

$$\varepsilon_\perp \simeq \varepsilon_A(1 + iv_i/\omega) \tag{6}$$

where $\varepsilon_A = c^2/c_A^2$ and c_A is the Alfvén velocity.

The gyrotropic layer is at $z_e < z < z_i$, i.e. at heights of 80–120 km. Here we have $v_e < \Omega_e$, $v_i > \Omega_i$. If these inequalities are enhanced, then from (2) and (3) it follows that

$$\sigma_\perp \simeq \frac{e^2 N}{m_i}\left(\frac{1}{v_i} + \frac{v_e}{\Omega_e\Omega_i}\right) \qquad \sigma_H \simeq -\frac{eNc}{B}. \tag{7}$$

In orders of value we have $\sigma_H/\sigma_\perp \simeq v_i/\Omega_i$ in the gyrotropic layer. The integral Hall conductivity is of the order of $\Sigma_H \simeq \sigma_H \Delta z$, where Δz is defined by (5). The typical values are: $-\Sigma_H \simeq 3\Sigma_\perp \simeq 3 \times 10^{13}$ cm s^{-1} in the daytime and $-\Sigma_H \simeq \Sigma_\perp \simeq 10^{11}$ cm s^{-1} at night.

Exercise 3.2.4.

Theme for discussion: atmosphere–magnetosphere coupling through the mediation of the ionosphere.

Comment 3.2.4.

This theme covers a broad spectrum of problems: charge, mass, energy, momentum, entropy and heat transfer mechanisms; interaction among neutral and ionized species in the ionosphere–atmosphere system; generation of infrasound owing to the motions of the auroral arcs; Joule heating of the upper atmosphere as a result of dissipation of the magnetospheric waves, etc. Certain problems are specific and very difficult, for example, the atmospheric oxygen transfer into the magnetosphere and, as a consequence, the modification of the ion cyclotron wave spectrum, resulting in a change in the lifetime of the energetic particles in the radiation belt.

As an example let us consider the generation of field-aligned current $j_\parallel$ in the magnetosphere by the atmospheric gravity waves in the long-wavelength limit.

For simplicity assume that the geomagnetic field is vertical and the magnetosphere is located immediately above the gyrotropic film (figure 3.3). It is known that the oscillating velocity v of a neutral gas is horizontal and parallel to the direction of propagation of long gravity waves. The ions are picked up by the motion of neutral molecules in the gyrotropic layer. This creates an external current $j_{\text{ext}} \simeq eNv$, which generates the magnetic field

$$b^+ \simeq c_A^{-1}(B \times v)\frac{\Sigma_H}{(\Sigma_\perp + \Sigma_w)}$$

above the layer.

Now, using the equation $j = (c/4\pi)\nabla \times b^+$ and the dispersion relation (2.36) we obtain

$$j_\parallel \simeq \left(\frac{m_i N}{4\pi}\right)^{1/2} \frac{i\omega v c \Sigma_H}{c_S(\Sigma_\perp + \Sigma_w)}.$$

3.3   The ionosphere as a resonator and waveguide

If the ionosphere behaves as a thin film with respect to the long waves, then with respect to the short waves it behaves as a resonator and a waveguide. In the ionospheric layers the Alfvén waves resonate in the Pc1 frequency range. Here they have the structure of standing waves along the geomagnetic field lines, which are almost vertical at high latitudes. Magnetosonic waves of the same frequency range are ducted along the ionospheric layers, covering long distances in the horizontal direction.

The gyrotropy of the ionosphere causes mutual transformation of magnetosonic and Alfvén waves, which considerably complicates the theory. Vertical inhomogeneity of the ionosphere leads to the same results if the direction of waves propagation deviates from the magnetic meridian. Nevertheless, realizing the fatal consequences of such an approach, we shall disregard the gyrotropy. At night (not during daytime) we may then try to take the gyrotropy into account using the perturbation method. The vertical inhomogeneity cannot be disregarded, since it defines the resonant and waveguide properties of the ionosphere. Therefore we shall consider the geomagnetic field to be vertical and then any direction of propagation will be meridional. (In reality, it takes place strictly only at the geomagnetic poles.) After that we rewrite equations (3.50) and (3.51) in the following form

$$\left(\frac{d^2}{dz^2} + k_0^2 \varepsilon_\perp\right) E_x = 0 \tag{3.77}$$

$$\left(\frac{d^2}{dz^2} + k_0^2 \varepsilon_\perp - k_\perp^2\right) E_y = 0. \tag{3.78}$$

Here we have taken into account the inequality $\varepsilon_\parallel \gg (k_\perp/k_0)^2$, which does not lead to any additional error.

Equation (3.77) describes the Alfvén waves. This equation is similar to the equation of the string oscillations. In order to determine the spectrum of oscillations it is necessary to specify the form of the function $\varepsilon_\perp(z)$ and set the boundary conditions. In the F-layer and in the magnetosphere $\nu_i \ll \Omega_i$, so that $\varepsilon_\perp = \varepsilon_A(1 + i\nu_i/\omega)$ at $\omega \ll \Omega_i$, where $\varepsilon_A = (c/c_A)^2$. Above the maximum of the F-layer we may neglect collisions and assume $\varepsilon_\perp \simeq \varepsilon_A$. In the lower layers the collisions play an important role, causing energy dissipation. For simplicity, instead of setting the profile $\nu_i(z)$, we shall suppose $\varepsilon_\perp = \varepsilon_A$ everywhere at $z > h$, and regard the dissipation using the boundary condition

$$\frac{1}{E_x}\frac{dE_x}{dz} = -\frac{4\pi i\omega}{c^2}\Sigma_\perp \tag{3.79}$$

which is given at the level $z = h$.

The profile $\varepsilon_A(z)$ approximately repeats the vertical profile of the plasma concentration $N(z)$, i.e. it has a maximum at a height of about 300 km. Above the maximum, ε_A decreases somewhat faster than $N(z)$ due to the rapid decrease of the average ion mass with height. Let us introduce the following model

$$\varepsilon_A = \varepsilon\left[\delta^2 + \exp\left(-2\frac{z-H}{\Lambda}\right)\right] \qquad z > H \tag{3.80}$$
$$\varepsilon_A = \varepsilon(1 + \delta^2) \qquad h < z < H \qquad \delta \ll 1.$$

Equations (3.79) and (3.80) have six parameters, which we can choose so that they imitate reality.

At $z > H$ equation (3.77) has the following solution

$$E_x = C_1 J_{i\nu}(\xi) + C_2 J_{-i\nu}(\xi) \tag{3.81}$$

where $C_{1,2}$ are constants, $J_{\pm i\nu}$ are Bessel functions, $\nu = \Lambda k_A \delta$, $\xi = \Lambda k_A \exp[-(z - H)/\Lambda]$ and $k_A = k_0 \varepsilon^{1/2}$. As $z \to \infty$ we have

$$J_{\pm i\nu} \sim \exp(\mp i\delta k_A z).$$

Therefore, at large altitudes the solution (3.81) is the superposition of two travelling waves: one of which propagates away from the Earth and the other towards the Earth. The Sommerfeld criterion for the radiation requires that there are no waves coming from infinity, i.e. it should be assumed that $C_1 = 0$.

At $h < z < H$ the solution of equation (3.77) has the form

$$E_x = C_3 e^{ik_A(z-H)} + C_4 e^{-ik_A(z-H)}. \tag{3.82}$$

At the boundary $z = H$, the function $E_x(z)$ and its derivation must be continuous and (3.79) applies. These three conditions and the radiation determine the spectrum of the Alfvén oscillations of the ionosphere.

Let us find the frequencies and the qualities of the resonances. It is clear that at fairly high frequencies there are no resonances, owing to the fact that the energy of oscillations freely escapes through the upper wall of the resonator. The frequency range of resonances is limited by the condition of the 'violation' of geometrical optics, which in this case has the form $\pi \delta k_A \Lambda \sim 1$. At $\varepsilon = 2.5 \times 10^5$, $\Lambda = 400$ km, $\delta^2 = 10^{-3}$, the resonant frequencies are below the frequency 2.5 Hz.

We now suppose that $(\pi/2)\delta k_A \Lambda \ll 1$. Then the resonances are expressed clearly and their parameters may be easily evaluated analytically. Let us designate $\omega_n = Re\omega_n - i\gamma_n$, where γ_n is the damping decrement, $\gamma_n > 0$. For the resonant frequencies and qualities we approximately obtain

$$Re\omega_n = \frac{c_A}{\Lambda + l} \left(\pi n + \frac{\pi}{4} - \frac{\varphi}{2} \right) \tag{3.83}$$

$$Q_n^{-1} = \frac{2\gamma_n}{Re\omega_n} = \frac{-\ln|R|}{\pi n + (\pi/4) - (\varphi/2)} + \frac{\Lambda}{\Lambda + 1} \pi \delta. \tag{3.84}$$

Here $l = H - h$; c_A is the Alfvén velocity in the maximum of the F-layer and $R = |R| \exp(i\varphi)$ is the coefficient of reflection from the lower wall of the resonator, defined by (3.69). At $\Sigma_\perp < \Sigma_w$ (night) we have $\varphi = 0$ and $n = 0, 1, \ldots$; then at $\Sigma_\perp > \Sigma_w$ (day) we have $\varphi = \pi$ and $n = 1, 2, \ldots$. Numerical estimates show that there exist two–three well-expressed resonances in the range 0.2–1 Hz. However, their exact positions and respective qualities should be defined using the numerical analysis of a more realistic ionospheric model.

Now we use the same model (3.80) in order to demonstrate the ducting of magnetosonic waves along the ionospheric layers. At $z > H$ (3.78) has the solution

$$E_y = C_1 J_p(\xi) + C_2 J_{-p}(\xi) \tag{3.85}$$

where $p = (k_\perp^2 k_A^{-2} - v^2)^{1/2}$ and the other notations are as before. If $z \gg H$ we have $|\xi| \ll 1$ and

$$J_{\pm p} \sim \exp[\pm(k_\perp^2 - k_A^2 \delta^2)^{1/2} z].$$

Evidently, at $k_\perp > k_A \delta$ there exists a solution which increases exponentially towards infinity. Nevertheless, it should be neglected as physically meaningless within the framework of this problem. Consequently, we put $C_1 = 0$. The exponentially decreasing solution $J_{-p}(\xi)$ corresponds to the waves trapped in the waveguide which are experiencing complete internal reflection from the upper half-space.

In the layer $h < z < H$ the solution has the form

$$E_y = C_3 e^{ik_z(z-H)} + C_4 e^{-ik_z(z-H)} \tag{3.86}$$

where $k_z = k_0 n_z$, $n_z = (\varepsilon - n_\perp^2)^{1/2}$ and $n_\perp = ck_\perp/\omega$. At $z = H$ the electric field E_y and its derivative dE_y/dz are continuous. At $z = h$ let us set the boundary condition

$$\frac{1}{E_y}\frac{dE_y}{dz} = -ik_z\frac{1 - R}{1 + R} \tag{3.87}$$

where R is the coefficient of reflection from the lower wall of the waveguide.

While accomplishing all these conditions the nontrivial solutions of the wave equation do not exist at any value of ω, but only at those that yield the dispersion relation

$$\frac{J'_{-p}(\Lambda k_A)}{J_{-p}(\Lambda k_A)} = i\frac{k_z}{k_A}\frac{e^{-ik_z l} - R e^{ik_z l}}{e^{-ik_z l} + R e^{ik_z l}}. \tag{3.88}$$

Here $J'_p(\xi) = dJ_p/d\xi$.

Finding the roots of the dispersion equation (3.88) is a fairly difficult problem. Usually it is solved numerically. As a result, critical frequencies of the waveguide modes, phase and group velocities, and coefficients of the waveguide attenuation are determined and the dependence of these characteristics of the waveguide on the state of the ionosphere (day, night, phase of the solar cycle, etc) are studied.

Sometimes for analytical estimations it is supposed that $\Lambda = 0$, i.e. the half-space $z > H$ is considered to be a homogeneous medium with the dielectric permeability $\varepsilon_\perp = \varepsilon\delta^2$. Then it is not difficult to take into account collisions of ions with neutral molecules in the waveguide, supposing $\varepsilon_\perp = \varepsilon(1 + i\nu_i/\omega)$, in the range of altitudes of $h < z < H$, where $\nu_i = $ constant. This is important, since, in general, it is only at night in the minimum of the solar cycle that dissipation appears below 150 km. At other time the area of the energy dissipation stretches from 100 to 400 km. The values ε, δ, R, l and ν_i are considered here as effective parameters that model the features of the real ionosphere.

The critical frequency, below which waveguide propagation is impossible, is of the order of $\omega_{min} \simeq c_A/l$, where $c_A = c\varepsilon^{-1/2}$. At these frequencies, considerably exceeding the critical frequency, the group velocity of the horizontal wave propagation is of the order of $v_g \simeq c_A$. The waveguide damping in the waveguide, owing to Joule losses, is of the order of $\mathrm{Im}\,k_\perp \simeq \nu_i/c_A$. These parameters change in the wide limits. The typical values of $f_{min} \simeq 0.3-0.5$ Hz, $v_g \simeq 300-700$ km s^{-1}, and both values during the daytime were lower than at night. Damping in the waveguide changes from units to dozens of dB/1000 km, and in the daytime damping is considerably higher than at night.

The indicated numbers refer to the discrete spectrum. Obviously, there is a continuous part of the spectrum connected with the propagation of so-called side waves in the half-space $z > H$. Side waves, if they exist, propagate in the upper medium at a velocity of several thousands of kilometres per second and reach the observation point earlier than the other waves.

Bibliography

For section 3.1

Global resonances of the Earth–ionosphere cavity were predicted by Schumann (1952a,b), who also showed that lightning discharges are an effective source for the resonance excitation. Balser and Wagner (1960) presented the experimental study of Schumann resonances. Wait (1962), Galejs (1972), Bliokh *et al* (1980), who considerably advanced the theory of the Schumann resonances, wrote monographs, in which the reader will find information on the origin and the fundamentals of the theory, as well as an analysis of the extensive literature on this subject. The reviews made by Galejs (1964), Polk (1969, 1982) and Ogawa *et al* (1969) may serve the same purpose. Fraser and Sentman (1991) gave a brief, but informative discussion of the problem of monitoring over the global lightning activity when observing the Schumann resonances. In this article the reader will find examples of Q-bursts, a description of the resonator modulations by x-rays of the solar flashes, examples of spectral line splitting and other interesting material.

The abundance and availability of the review and monographic literature on the problem of Schumann resonances forces us to confine ourselves to a few references.

When discussing the boundary problem (3.24) and (3.25) we took advantage of the article by Bliokh *et al* (1977). Within the framework of this problem the spectrum is $(2n + 1)$-fold degenerate. The observations give the multiplet structure of the spectrum, which testifies to the fact that the degeneration is removed, at least partly. Sentman (1987) observed splitting of the lines by 0.5 Hz in California. It was he who made a report on the preferentially clockwise (anticlockwise) rotation of the horizontal projection of the magnetic vector during daytime (night), which signifies the influence of the gyrotropy and spherical asymmetry of the ionosphere on the properties of the resonator. Rabinovich (1986) investigated the influence of spherical asymmetry within the framework of a relatively simple model. He took advantage of the fact, that the surface impedance of the ionosphere represents a small value ($|\zeta| \simeq 10^{-2}$) and considered it as a small parameter with which to construct a series of the perturbation theory. The article by Jones and Kemp (1971) is dedicated to the modelling of impulse sources that excite the resonator.

Lightning discharges excite not only the Schumann oscillations. For information on spherics, tweaks and whistlers see Helliwell (1965). The frequencies of the whistlers lie above the frequencies of the Schumann oscillations. On the other hand, Belyaev *et al* (1987) report on the excitation of oscillations in the frequency range 0.5–3 Hz, i.e. below the frequencies of the Schumann oscillations. According to those authors, the radiation of lightning discharges penetrates deep into the ionosphere and here it experiences resonant enhancement at the frequencies of the so-called ionospheric Alfvén resonator.

Nikolaenko and Rabinovich (1982) made a calculation of the resonant frequencies and qualities for the planets of the solar system (e.g. exercise 3.1.1). Guglielmi and Pokhotelov (1993a) estimated the photon mass by using the Schumann resonances.

For section 3.2

For the fundamentals of plasma electrodynamics see, for example, the monograph by Lifshitz and Pitaevsky (1979). The structure of the ionosphere is described in a number of manuals, for example, in the monograph by Risbeth and Garriot (1969).

The nontriviality of the problem of magnetohydrodynamic wave transmission from the magnetosphere to the Earth through the ionosphere and atmosphere was theoretically predicted by Dungey (1963). He expanded the field in the air by TM and TE fields and found the screening influence of the ionosphere on the field of the incident TM (Alfvén) wave. Nishida (1964) found the $\pi/2$ rotation of the Alfvén wave polarization when it was transmitting through the thin gyrotropic layer of the ionosphere.

The reflection and transmission properties of the ionosphere were analytically investigated by Inoue (1973), Hughes (1974), Hughes and Southwood (1976a,b), Belyaev and Polyakov (1980), Leonovich and Mazur (1991a) within the model of the gyrotropic absorbent layer. The results, obtained by these authors, were used when writing this section of the book. The reviews by Lanzerotti and Southwood (1979), Southwood and Hughes (1983) and the monograph by Nishida (1978) were also employed.

We have not touched upon the numerical analysis of the propagation of hydromagnetic waves in the ionospheric layers. The first results here belong to Francis and Karplus (1960), Karplus *et al* (1962), Prince and Bostick (1964), Greifinger and Greifinger (1965), Field and Greifinger (1965), Prikner (1968), Altman and Fijalkow (1969). The typical approach to the problem is as follows. A plane monochromatic wave is incident tangentially upon the horizontal stratified ionosphere. The geomagnetic field is considered to be homogeneous and tilted at some arbitrary angle to the horizon. It is necessary to determine the coefficients of reflection and transmission, as well as the vertical structure of the wave field. The most important result lies in finding ionospheric resonances, caused by the vertical inhomogeneity of ionization distribution. The calculation of wave-energy dissipation within the Pc1 frequency range, depending on the time of the day, the solar activity cycle, etc, are also of interest.

The dissipation of magnetohydrodynamic waves is displayed in the heating of the ionosphere (Dessler 1959a,b). According to Akasofu (1960) and Sorenson (1968), temperature rise of the ionosphere is insignificant if we choose a 'typical' value of the wave amplitude, say, 1γ at a frequency of 1 Hz. At the Earth's surface, stronger waves within this range are observed, but only rarely. However, waves with amplitudes up to 10^2 γ within the Pc1 frequency range were observed at ionospheric heights (see Chapter 7). Obviously, such high-amplitude waves may significantly heat the ionosphere.

The calculations of reflection and transmission parameters of the ionosphere are widely used when synthesizing the general picture of excitation and propagation of magnetohydrodynamic waves in the magnetosphere and interpreting the observations of wave fields on the ground and in space. Out of the many interesting works on the subject, we shall choose two. Newton *et al* (1978) made a calculation of the damping decrement γ of the Alfvén waves in the magnetosphere.

The mechanism of damping is due to Joule dissipation of the oscillation energy in the ionosphere. It was found that damping is lower in the daytime than at night ($\gamma \simeq 0.01\omega$ and $\gamma \simeq 0.1\omega$ respectively). Knowing this, the authors explain the morphological difference in geomagnetic pulsations over the daytime and at night (permanent Pc3-4 in the daytime and sporadic Pi2 at night). Undoubtedly, apart from this reason, the difference in the day and night pulsations is also caused by the asymmetry of the excitation conditions, which is connected, in the final analysis, with sharp asymmetry of the magnetosphere relative to the plane of the morning–evening meridian.

Another paper is dedicated to the experimental research of the influence of sunrise on the pulsations Pc4 (Itonaga *et al* 1981). At dawn, when along with the rotation of the Earth the observation point crosses the terminator and the ionosphere in the Zenith becomes subject to intensive solar UV radiation, the polarization of the pulsations undergoes notable modification. The effect is explained by the ionization of the E-layer in the morning, the increasing of the ratio $\Sigma_H / \Sigma_\perp$ and increasing of the coefficient of transformation (3.70).

Exercises 3.2.1 and 3.2.2 were composed using the review by Shafranov (1963).

We have not considered the problem of wave interaction between the neutral and ionized components in the upper atmosphere. Exercise 3.2.4 only compensates for this deficiency to a small extent. A discussion of the problem of atmosphere–ionosphere coupling and dynamics may be found in Blanc (1985), Liperovsky *et al* (1992), Petviashvili and Pokhotelov (1992).

For section 3.3

The ionosphere as a resonator for Alfvén waves was studied by Polyakov and Rapoport (1980). We gave an account of the main ideas of the resonator.

The analysis of magnetic signals from nuclear explosions in the upper atmosphere (Berthold *et al* 1960, Bomke *et al* 1960, Knox 1962, Kovach and Ben-Menahem 1966), and Pc1 pulsations (Tepley and Landshoff 1966, Manchester 1966, Greifinger and Greifinger 1968, 1973) has led to the idea of the existence of an ionospheric waveguide. Analytical calculation of the waveguide characteristics for the waveguide simplified models and numerical calculation with the use of more realistic models were carried out. In all cases the waveguide was considered to be horizontally homogeneous. The models differ from each other in the choice of the vertical profile of the complex dielectric permeability and the orientation of the external magnetic field. The model (3.80) and the dispersion equation (3.88) were taken from Greifinger and Greifinger (1968).

Fujita (1988) continued the detailed study of the waveguide attenuation, that was started by Greifinger and Greifinger (1968, 1973). He showed that the mechanism of Joule dissipation prevails over the mechanism of mode conversion of the magnetosonic waves into Alfvén waves when the transverse conductivity of the ionosphere exceeds the Hall conductivity, or is at least not much lower. If the Hall conductivity is appreciably higher than the transverse conductivity, then the mechanism of conversion is effective enough, and the frequency dependence of the waveguide weakening has a quasi-periodic form i.e. it is connected with the effective excitation of the standing Alfvén waves in the vertical direction. (The external magnetic field was chosen to be vertical.)

Fujita and Tamao (1988) investigated the excitation of the waveguide by a beam of Alfvén waves which are incident upon the ionosphere. The mechanism of excitation is connected with the mode conversion in the gyrotropic ionosphere.

Ovchinnikov (1991) developed an analytical approach to the theory of the ionospheric magnetohydrodynamic waveguide. In contrast to the earlier studies, Ovchinnikov (1991) employed a fairly complicated model of the vertical structure of the waveguide. In particular, above the maximum of the $F2$-layer he placed the so-called Epstein layer. Below this maximum the solution is constructed in the form of a quasi-static series by even powers of the wavenumber.

The idea of the waveguide is widely used to explain the observed properties of the Pc1 signals. In particular, it explains why signals with frequency $f < 0.4$ Hz are observed locally, and those with frequency $f > 0.4$ Hz are observed nearly simultaneously over vast areas of the Earth's surface (Troitskaya 1964, Manchester 1966). Wentworth *et al* (1966) observed Pc1 simultaneously within the range of latitudes from the subauroral zone to the equator. Other interesting examples of this kind may be found in the book by Nishida (1978), where information on the measurements of the Pc1 group velocity in the ionospheric waveguide (500–700 km s^{-1}) is also presented.

Chapter 4

The magnetosphere

The term magnetosphere applies to a cavity of a complex form in which the magnetic field of the Earth predominates. Outside the magnetosphere an interplanetary magnetic field of solar origin prevails. The solar wind flux flows over the magnetosphere along a surface which is termed the magnetopause. The distance from the centre of the Earth to the subsolar point of the magnetopause (the 'radius' of the magnetosphere) depends on the solar wind force and varies from 8 to $12R_E$ approximately, where R_E is the Earth's radius. In front of the magnetopause there is a bow shock. The distance from the magnetopause to the bow shock in the solar direction is approximately $3R_E$. The length of the geomagnetic tail stretching in the antisolar direction is fairly large, up to $1000R_E$.

At short distances from the Earth the geomagnetic field may be approximated by the magnetic dipole field. It should be noted that the dipole axis is inclined at an angle of approximately $11°$ from the axis of the Earth's rotation. At the periphery of the magnetosphere the geomagnetic field is strongly deformed due to the existence of currents flowing along the magnetopause and in the geomagnetic tail.

The specific distribution of the magnetic field, cold plasma and energetic particles that fill the magnetosphere creates an intricate mosaic of electrodynamic parameters. We shall single out the system of waveguides and resonators from this complicated picture and analyse their properties.

In this and the following chapters we shall consider the medium to be passive and the waves to be linear. In Chapter 6, without renouncing the linear approximation, we shall take into account the 'active fillings' in the magnetosphere, i.e. the presence of nonequilibrium distributions of energetic particles, shear flows and other structures, causing loss of stability of the magnetosphere and the self-excitation of waves. We shall defer the analysis of nonlinear effects until Chapter 7.

Such a gradual analysis of the picture of magnetospheric waves is methodically convenient. In some cases it may be justified by considering that the nonlinear effects are very small, that the deviations from equilibrium are also small and so on. However, we should admit that our knowledge of wave phenomena in the magnetosphere is incomplete especially when referring to real geomagnetic pulsations found from ground and satellite observations. A unifying theory of geomagnetic pulsations has not yet been constructed.

4.1   Waves in a cold plasma

We continue the study waves in a plasma, started in the preceding chapter. Our intention is to present a more complete picture of waves in the magnetoactive plasma[1]. The properties of the plasma will be specified by the use of the dielectric permeability tensor (3.46), i.e. we shall consider the plasma to be cold and collisionless. One more simplification will be assumed in this section, the plasma and the magnetic field will be considered to be homogeneous.

As a basis we take the analysis of the dispersion equation and its roots $\omega_\sigma(k)$, usually called dispersion relations or dispersion laws. Here the subscript σ numbers the normal modes, i.e. separate branches of the dispersion equation roots. (In cases where it does not lead to any misunderstanding, the subscript σ is omitted.)

The dispersion equation in its general form is presented in exercise 3.2.1. Having obtained the roots of the dispersion equation and respective polarization vectors, an arbitrary wave field in a homogeneous medium may be presented in the form of an expansion by normal waves. The application of the method of normal waves, based on the Fourier transformation and the transition into k-space, has a number of advantages over direct investigation of the dynamic equation in configurational space. The advantage of simplicity is not the most important. What is more essential is the fact that the language of dispersion laws is well adapted for penetration into the essence of natural wave phenomena. In addition, the spectral representation allows the unification of the description of wave processes of different origin, which significantly saves brainwork.

The dispersion relations allow us to describe the evolution of the wave packets and examine the stability of the initial state of the dynamic system. By using these relations the rays may be traced in a smoothly inhomogeneous medium. Nonlinear corrections to the dispersion relations permits the study of self-focusing and self-modulation phenomena of small but finite-amplitude waves.

Let us mention one more useful application of dispersion relations. We refer to the reduction of the initial dynamic equations, which are usually quite complicated in structure and sometimes contain redundant information. Proceeding to the spectral representation we may retain the leading terms in

---

[1] Such a term is used to describe a plasma immersed in an external magnetic field. The origin of the term is associated with the fact that the plasma in the magnetic field is an optically active (gyrotropic) medium.

the dispersion relations as they are the main points of interest for this specific problem. Simplified dynamic equations are synthesized using the replacements

$$\omega \to i\partial/\partial t \qquad k \to -i\nabla.$$

After these general remarks we proceed to the analysis of plane monochromatic waves in a cold plasma. The dispersion equation may be presented in the form

$$An^4 - Bn^2 + C = 0 \tag{4.1}$$

where

$$A = \varepsilon_\perp \sin^2\theta + \varepsilon_\| \cos^2\theta$$
$$B = (\varepsilon_\perp^2 - g^2)\sin^2\theta + \varepsilon_\perp\varepsilon_\|(1 + \cos^2\theta)$$
$$C = (\varepsilon_\perp^2 - g^2)\varepsilon_\|.$$

Here $n = ck/\omega$ is the refractive index; θ is the angle between the vectors k and B and $\varepsilon_\|$, $\varepsilon_\perp$ and g are the components of the dielectric permeability tensor (3.46). Instead of (4.1) the dispersion equation in the equivalent form may be used

$$\tan^2\theta = -\frac{\varepsilon_\|(n^2 - \mathcal{L})(n^2 - \mathcal{R})}{(\varepsilon_\perp n^2 - \mathcal{L}\mathcal{R})(n^2 - \varepsilon_\|)} \tag{4.2}$$

where

$$\mathcal{L} = \varepsilon_\perp + g \qquad \mathcal{R} = \varepsilon_\perp - g. \tag{4.3}$$

For a given k, equation (4.1) is of the tenth power with respect to ω. However, the frequency ω enters the dispersion equation in the form of even powers, similar to the case of ideal magnetohydrodynamics (see section 2.1). Therefore, only five essentially different branches of the dispersion equation are available. They are termed dispersion curves.

The dispersion curves may be conveniently presented in the form of the dependence of the square of the refractive index on the frequency at different values of θ. When solving (4.1) we find two roots $n_1^2(\omega, \theta)$ and $n_2^2(\omega, \theta)$. For example, at longitudinal propagation ($\theta = 0$) we have

$$n_{1,2}^2 = \varepsilon_\perp \pm g. \tag{4.4}$$

For one wave $n_1 = \sqrt{\mathcal{L}}$, while for the other $n_2 = \sqrt{\mathcal{R}}$. In the geophysical literature they are usually called $\mathcal{L}$- and $\mathcal{R}$-waves respectively. If the plasma consists of electrons and ions of one kind, then

$$n_{1,2}^2(\omega) = 1 - \frac{\omega_{0e}^2}{(\omega \pm \Omega_e)(\omega \mp \Omega_i)}. \tag{4.5}$$

Here we have taken into account that $\omega_{0i}^2\Omega_e = \omega_{0e}^2\Omega_i$ due to the quasineutrality of the plasma, and additionally we have neglected the term ω_{0i}^2 as compared with ω_{0e}^2, since $m_e \ll m_i$.

The dispersion curves for this case are indicated in figure 4.1 by solid lines in the region of transparency ($n^2 > 0$). They are extended into the nontransparent region by dashed lines. To the four branches, defined by (4.5), one more is added. It has the form of a vertical straight line, crossing the horizontal axis at the frequency $\omega = \omega_{0e}$. The need for this follows from the fact that at $\theta = 0$ the dispersion equation (4.1) is yielded at $\varepsilon_\parallel = 0$ at an arbitrary value of n.

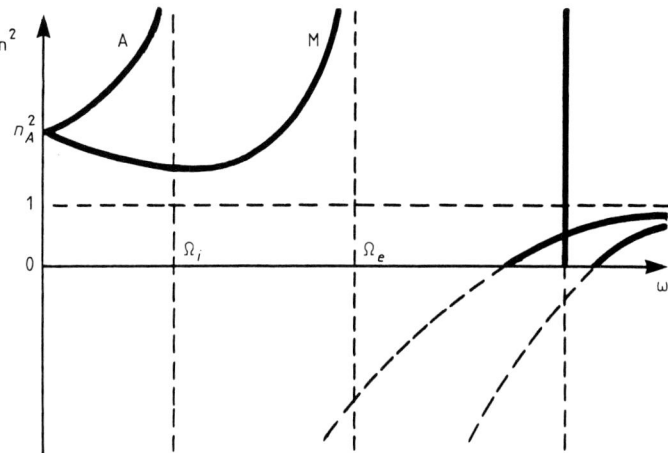

Figure 4.1. Dispersion curves for the longitudinal propagation. A—Alfvén branch, M–magnetosonic branch.

As $\omega \to \infty$ the two right branches of the dispersion curves asymptotically approach unity. As $\omega \to 0$ the two left branches approach the value $n_A^2 = 1 + \omega_{0i}^2/\Omega_i^2$. One of them is termed the Alfvén branch. It has a pole (resonance) at the ion gyrofrequency. The other one is termed the magnetosonic branch (resonance at the electron gyrofrequency). Other terms generally used are ion-cyclotron and electron-cyclotron branches. We note that usually $\omega_{0i} \gg \Omega_i$ and consequently $n_A \approx c/c_A$.

It follows from (4.5) that for longitudinal propagation the contribution of ions to the refractive index is small if $\omega \gg \Omega_i$. The value Ω_i varies over a wide range. Thus, for the protons $\Omega_p/2\pi$ equals approximately 900 Hz in the polar ionosphere and is of the order of 0.1 Hz in the interplanetary medium in front of the magnetosphere.

The dispersion curves for other directions of propagation may be studied similarly. With θ increasing from 0 to $\pi/2$ the position of singularity for the Alfvén branch does not change. However, the branch as a whole moves upwards. The intersection point of this branch with the vertical axis depends on θ in the following way: $n_1^2 = n_A^2/\cos^2\theta$ as $\omega \to 0$. The pole of the magnetosonic branch is shifted to the left: $n_2^2 \to \infty$ at $\omega = \Omega_e \cos\theta$, if the angle θ does not approach $\pi/2$ (to be more precise $\cos^2\theta \gg m_e/m_i$). If $\theta \to \pi/2$, then $n_2^2 \to \infty$

at the lower hybrid resonance frequency. This frequency equals $(\Omega_e \Omega_i)^{1/2}$ if $\omega_{0e}^2 \gg \Omega_e^2$. We also note that $n_2^2 = n_A^2$ as $\omega \to 0$ and at any θ.

It is clear that the behaviour of the dispersion curves is quite complicated even in the simplest model of the cold two-component plasma. This complicates the general analytical study of wave propagation. To facilitate the matter, different approximations are used. One such approximation lies in the assumption that ions are infinitely heavy and therefore motionless: they just create the background that neutralizes the charge of electrons. Such an approximation is valid at fairly high frequencies. Therefore, waves in which ions practically do not oscillate are called high-frequency waves. On the other hand, waves in which motion of ions plays an essential role are called low-frequency waves. The characteristic frequency, that separates low-frequency waves from high-frequency ones, is the frequency of the lower hybrid resonance $\sim (\Omega_e \Omega_i)^{1/2}$. However, in some cases the waves may be considered high-frequency even at frequencies considerably lower than $(\Omega_e \Omega_i)^{1/2}$. For example, at $\theta = 0$ the influence of ions may be ignored if $\omega \gg \Omega_i$, which has been mentioned above.

In addition to (4.4) and (4.5), let us present some generally used formulae, which determine the oscillation spectrum of the cold plasma. In the case when $\tan^2 \theta \ll \varepsilon_\parallel / \varepsilon_\perp$ and $\mathcal{L}\mathcal{R} \ll \varepsilon_\parallel \varepsilon_\perp$ we have

$$n_{1,2}^2 = \frac{\varepsilon_\perp}{\cos^2 \theta} \left[\frac{1 + \cos^2 \theta}{2} \pm \left(\frac{\sin^4 \theta}{4} + \frac{g^2}{\varepsilon_\perp^2} \cos^2 \theta \right)^{1/2} \right]. \tag{4.6}$$

If in addition to the mentioned inequalities the condition for quasi-longitudinal propagation is accomplished, i.e.

$$\frac{\sin^4 \theta}{4 \cos^2 \theta} \ll \frac{g^2}{\varepsilon_\perp^2} \tag{4.7}$$

then from (4.6) we obtain

$$n_{1,2}^2 = \tfrac{1}{4}[\mathcal{L}(1 \pm |\sec \theta|)^2 + \mathcal{R}(1 \mp |\sec \theta|)^2]. \tag{4.8}$$

For the strong inequality, which is the opposite of (4.7), one may speak of quasi-perpendicular propagation. But here, we present the formulae only for purely perpendicular propagation ($\theta = \pi/2$)

$$n_1^2 = \varepsilon_\parallel \qquad n_2^2 = \varepsilon_\perp - g^2/\varepsilon_\perp. \tag{4.9}$$

The given formulae for the refractive indices express the wave spectra $\omega(\mathbf{k})$ in an implicit form. The explicit form for the expressions for $\omega(\mathbf{k})$ in the general case cannot be found. Thus, approximate analytical expressions for $\omega(\mathbf{k})$ in different limiting cases are of particular importance.

At high frequencies ($\omega \gg \Omega_e$) in a relatively dense plasma we have

$$\omega = (\omega_{0e}^2 + c^2 k^2)^{1/2}. \tag{4.10}$$

At low frequencies ($\omega \ll \Omega_e$) from (4.6) it follows that

$$(\omega^2 - c_A^2 k_z^2)(\omega^2 - c_A^2 k^2) - \omega^2 k_z^2 k^2 c_A^4 / \Omega_i^2 = 0, \tag{4.11}$$

where the plasma is two-component, the z axis is directed along B, $k_z/k_\perp \gg \sqrt{m_e/m_i}$, $\omega_{0i}^2 \gg \Omega_i^2$. At $\omega \ll \Omega_i$ from (4.11) we obtain the dispersion relations

$$\omega = c_A k_z \tag{4.12}$$

$$\omega = c_A k \tag{4.13}$$

for the Alfvén and magnetosonic waves respectively. We note that (4.12) corresponds to quasi-perpendicular propagation, i.e.

$$\frac{\sin^2 \theta}{2 \cos \theta} \gg \frac{\omega}{\Omega_i}. \tag{4.14}$$

We have had the opportunity to ascertain that the range of the Alfvén waves is limited from above by the gyrofrequency of ions, i.e. $\omega < \Omega_i$. The magnetosonic waves may propagate at frequencies limited by electron gyrofrequency Ω_e at $k_\perp = 0$ and by lower hybrid frequency $\sim (\Omega_e \Omega_i)^{1/2}$ at $k_z = 0$.

If $\omega \gg \Omega_i$, $k_z/k_\perp > \sqrt{m_e/m_i}$ then the dispersion relations differ essentially from (4.13). When these inequalities are satisfied the waves of magnetosonic type are termed whistlers. From (4.11) we obtain the dispersion relation for the whistlers

$$\omega = \Omega_e \left[\frac{c}{\omega_{0e}} \right]^2 k_z k. \tag{4.15}$$

Equations (4.13) and (4.15) describe different fragments of a particular dispersion curve. They may be united in the form of the following interpolation formula

$$\omega = \alpha k (k_z^2 + k_{0i}^2)^{1/2} \tag{4.16}$$

where $\alpha = cB/4\pi eN$, and $k_{0i} = \omega_{0i}/c$, ω_{0i} is the ion plasma frequency (see exercise 4.1.1).

Finally, let us focus on the problem of the polarization of plane monochromatic waves in a cold magnetoactive plasma. A number of general interrelations between the vectors of the electromagnetic field follows immediately from Maxwell's equations:

(i) the vectors D, b and k are mutually perpendicular and form a right-oriented triplet of vectors;

(ii) the vectors E and b are also mutually perpendicular;

(iii) the vectors E, D and k are in one plane;
(iv) the relation $E \times D = b^2$ holds.

We introduce the Cartesian coordinate system (x, y, z) so that $B = (0, 0, B)$ and $k = (k \sin \theta, 0, k \cos \theta)$. We put $\omega > 0$ and consider the wave polarization with respect to the external magnetic field B. Let us present Maxwell's equations for the Fourier components in the form

$$M_{\alpha\beta} E_\beta = 0. \tag{4.17}$$

Here

$$M_{\alpha\beta} = n^2 \delta_{\alpha\beta} - n_\alpha n_\beta - \varepsilon_{\alpha\beta}$$

and $n = ck/\omega$. Suppose $\alpha = y$ in (4.17) and find the ratio of perpendicular components of the electric vector in the plane wave

$$\frac{E_y}{E_x} = \frac{ig}{\varepsilon_\perp - n^2}. \tag{4.18}$$

Similarly we find

$$\frac{E_z}{E_x} = \frac{n^2 \sin \theta \cos \theta}{n^2 \sin^2 \theta - \varepsilon_\parallel}. \tag{4.19}$$

At $\theta = 0$ the ratio of transversal components is defined from (4.18), where expression (4.4) should be substituted, i.e. $E_y/E_x = \mp i$. Hence we see that at $\theta = 0$ the transverse waves have circular polarization, and in the $\mathcal{L}$-wave (upper sign) the vector E has counter-clockwise rotation, and the case of the $\mathcal{R}$-wave (lower sign) corresponds to the clockwise rotation. The longitudinal wave at the frequency $\omega = \omega_{0e}$ is polarized along the external magnetic field.

At $\theta = \pi/2$ and $n = n_1$ we have $E \| B$. At $n = n_2$ we have $E_z = 0$ and $E_x/E_y = -ig/\varepsilon_\perp$. This wave is polarized elliptically, and the vector E rotates in the plane, perpendicular to the direction of the B. It should be noted, that there is a nonzero component E_x, parallel to the wave vector k. This is typical for waves in a magnetoactive plasma, i.e. in the general case, the waves are neither longitudinal nor transverse with respect to the vector k.

We also note that for quasi-parallel propagation at frequencies $\omega < \Omega_i$ the polarization is almost circular, and for quasi-perpendicular propagation it is almost linear (a strongly flattened ellipse).

Exercises

Exercise 4.1.1.

Construct the interpolation formula connecting the dispersion laws for magnetosonic waves (4.13) and whistlers (4.15).

Solution 4.1.1.

One of the calculating difficulties when solving the problems of geometric optics in an inhomogeneous magnetoactive plasma is the unwieldy form of the local dispersion relations. So different limiting cases that allow simple approximation of the dependence of ω on k are widely employed. Equation (4.15) may serve as an example for the whistlers. A simpler example is (4.13) for the fast magnetosonic wave. Both equations approximate different fragments of the same dispersion curve. The formula (4.13) is valid in a strong field region or at low frequencies ($\omega \ll \Omega_i$), and (4.15) corresponds to a weak field or high frequencies ($\omega \gg \Omega_i$). Furthermore, the applicability of (4.15) is limited by the inequality $\omega^2 \ll \Omega_e \Omega_i$ as well as by the condition that the angle θ between the external magnetic field and the wave vector does not approach $\pi/2$.

There is no simple expression for the dispersion relation in the intermediate region. Consequently, it is useful to construct the interpolation formula which becomes (4.13) and (4.15) under proper conditions.

Let the interpolation formula connect both limiting cases continuously and smoothly in the intermediate region. Also, let it be free of the parameters that do not appear in (4.13) and (4.15). This imposes restrictions on the plasma, to which the interpolation formula may be applied: the plasma should be cold and contain ions of one kind. Finally, we shall take into account the condition of analyticity of ω^2 as a function of k.

Let us present the formula for which we are searching in the form $\omega = kF(k)$. From (4.13) and (4.15) it follows that $F = c_A$ in the long-wavelength limit ($k^2 \ll k_{0i}^2$), and $F = \alpha k_\parallel$ in the short-wavelength limit ($k^2 \gg k_{0i}^2$). Here $\alpha = \Omega_i / k_{0i}^2$. Therefore, it is natural to assume $F = \alpha(k_{0i}^2 + k_\parallel^2)^{1/2}$. This choice seems to be reasonable, though the solution of this problem is not single-valued. Note that the simpler function $F = \alpha(k_{0i} + k_\parallel)$ does not yield the requirement of analyticity, and therefore it must be rejected.

The interpolation formula has the form

$$\omega = \alpha k (k_{0i}^2 + k_\parallel^2)^{1/2}.$$

At $k_\parallel^2 \ll k_{0i}^2$ it reduces to (4.13), and at $k_\parallel^2 \gg k_{0i}^2$ it corresponds to (4.15). Thus, we have one simple formula, that unites two essentially different types of dispersion. It qualitatively reproduces the course of the dispersion curve in a wide range of frequencies ($0 < \omega^2 \ll \Omega_e \Omega_i$) and angles of propagation ($0 \le \theta \le \pi/2$).

In a narrower range of frequencies ($\Omega_i^2 \ll \omega^2 \ll \Omega_e \Omega_i$) the interpolation formula may be derived from the dispersion equation (4.1). The polarization ratios $E_y/E_x = (i/k)(k_{0i}^2 + k_z^2)^{1/2}$ and $E_z/E_x = k_x k_z / k_{0e}^2$, where $k_{0e} = \omega_{0e}/c$, also follow from (4.16) and (4.17).

Exercise 4.1.2.

Find the refractive index for circularly polarized transverse waves, propagating along an external magnetic field in the proton–helium plasma.

Solution 4.1.2

The magnetospheric plasma contains several kinds of ions with different charge to mass ratios. We find oxygen, nitrogen, helium and hydrogen ions up to altitudes of 1–2 thousand kilometres. When moving away from the Earth, the concentration of heavy ions decreases rapidly. As for the He^+ ions, they are found up to altitudes of $\simeq 3 \times 10^4$ km with relative concentration $\xi = N[He^+]/N[H^+] \simeq 3 \times 10^{-3}$. It turns out that even a small admixture of these ions may essentially influence the propagation of waves in the Pc1 frequency range.

Suppose, a plasma consists of electrons and ions of two kinds (1 and 2). Taking into account the condition of quasineutrality $eN = e_1 N_1 + e_2 N_2$ the formula (4.4) at $\omega \ll \Omega_e$ may be presented in the form

$$n^2_{1,2} = \frac{\omega^2_{01}}{\Omega_1 (\Omega_1 \mp \omega)} \left\{ 1 + \left[\frac{1 \mp \omega/\Omega_1}{1 \mp \omega/\Omega_2} \right] \frac{N_2 m_2}{N_1 m_1} \right\} + 1.$$

In a proton–helium plasma under the additional condition $\xi \ll 1$ we have

$$n_{1,2} = \frac{n^2_A}{1 \mp \omega/\Omega_p} \left\{ 1 + \frac{3\xi}{1 \mp \omega/\Omega_{He^+}} \right\} + 1.$$

We note, that the presence of small admixture of He^+ ions strongly affects the behaviour of dispersion curves only in the immediate vicinity of the gyrofrequency Ω_{He^+}.

Exercise 4.1.3.

Using the result of exercise 4.1.2, we would like to show the distribution n^2_1 along the marked line of the geomagnetic field and find the coefficient of wave transition through the nontransparency (or opaqueness) bands, appearing due to the presence of He^+ ions in the plasma.

Solution 4.1.3.

Figure 4.2 presents the change of the square of the refractive index for waves radiated at the latitude $63.5°$ and propagating along the field line. l/R_E is the distance along the field line in units of the radius of the Earth, counted from the equator plane. The geomagnetic field is assumed to be dipolar; the electron concentration decreases according to the law $N = 1.25 \times 10^4 (r/R_E)^{-3}$ and the relative concentration of helium ions is $\xi = 0.01$. The curves in the top picture correspond to the case when the wave gyrofrequency ω is smaller than the gyrofrequency of helium ions $\Omega_{He^+}^{min}$ in the peak of the field line: $\omega < \Omega_{He^+}^{min} \simeq 6$ rad s^{-1}. The curves in the bottom picture refer to the case when the wave frequency is larger than the gyrofrequency of helium ions, but smaller than the gyrofrequency of protons: $\Omega_{He^+}^{min} < \omega < \Omega_p^{min}$. In the second case there are two opaqueness bands situated symmetrically relative to the equator plane.

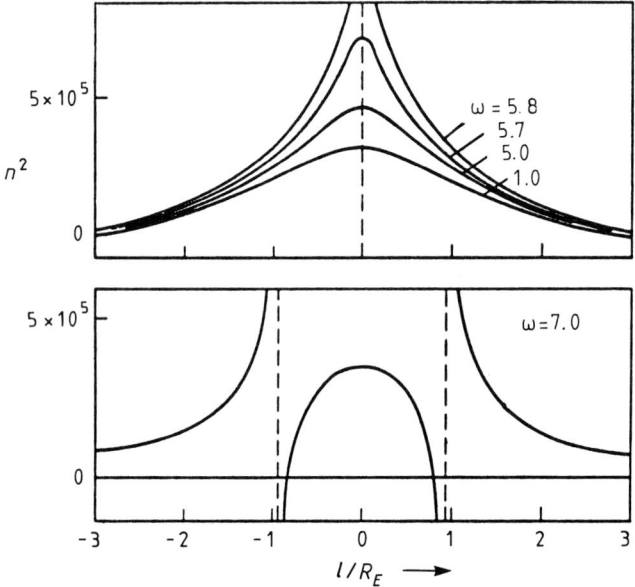

Figure 4.2. The square of the refractive index for the $\mathcal{L}$-waves in the magnetosphere. The coordinate l is counted from the equatorial plane along the field line which crosses the Earth's surface at latitude 63.5°. The frequencies ω are given in s^{-1}.

The amplitude coefficient of the wave transition through these two opaqueness bands may be estimated by the Badden formula

$$D = e^{-\pi\beta}.$$

Here $\beta = (\omega/c_A) |l' - l''|$ is the width of opaqueness band in the units of the Alfvén wavelength, l' is the pole coordinate of n_1^2, and l'' is the zero coordinate of n_1^2. At $n_1^2 \gg 1$, $\Omega_{He^+}(l') = \omega$ and $\Omega_{He^+}(l'') = \omega/(1 + 3\xi)$.

Figure 4.3 shows the dependence of the attenuation coefficient D on the relative concentration of helium ions ξ, when a wave with frequency $\omega > \Omega_{He^+}^{min}$ propagates along the field line of the geomagnetic field crossing the Earth's surface at latitude 63.5°.

We point out that all the above refers to the ideal case of strictly longitudinal propagation ($\theta = 0$). Even at small, but nonzero values of θ the picture of

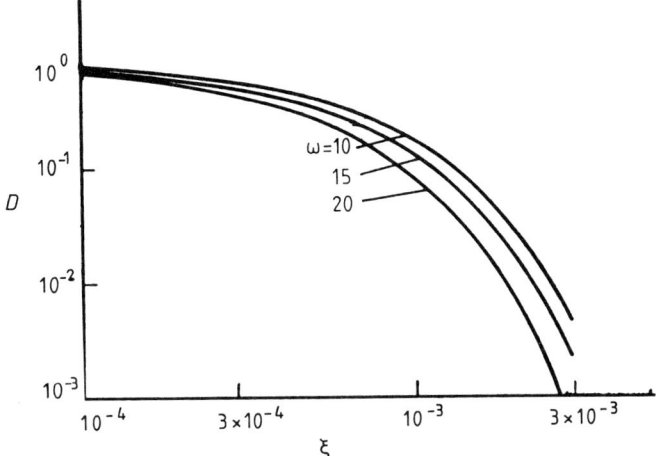

Figure 4.3. The attenuation coefficient of the $\mathcal{L}$–waves in the magnetosphere on the trace between the conjugate points in the case of two nontransparency bands (see figure 4.2, lower panel).

propagation changes quantitatively, since mutual transformation of $\mathcal{L}$- and $\mathcal{R}$-waves appears. The problem has not yet been investigated thoroughly enough.

4.2   Ray theory

The ray theory of wave propagation is based on the local dispersion law which is understood as the Hamilton–Jacobi equation. It is called the eikonal equation in wave theory. The characteristics of the eikonal equation satisfy Hamilton canonical equations and are the rays, i.e. trajectories along which the energy of perturbations propagates, if the fairly stringent conditions for the applicability of the ray approximation are held in the vicinity of each point of the family of rays.

We restrict the present analysis to monochromatic waves. This means that the perturbations depend on time as $\exp(-i\omega t)$. The medium is assumed to be inhomogeneous, but stationary. In other words, $\omega(k, x)$ depends on the coordinate x but not on the time. Then the wave vector k varies in such a way along the ray that $\omega(k(t), x(t)) = $ constant. The ray equations have the form

$$\frac{\mathrm{d}x}{\mathrm{d}t} = \frac{\partial\omega}{\partial k} \qquad \frac{\mathrm{d}k}{\mathrm{d}t} = -\frac{\partial\omega}{\partial x}. \tag{4.20}$$

For illustrative purposes we consider the ray theory of MHD waves, i.e. geometric optics for MHD waves, or more precisely, geometric magnetohydrodynamics.   Evidently the goals and methods of geometric magnetohydrodynamics and geometric optics (acoustics, seismics, etc) coincide; the only distinction is the subject matter which is determined by the particular form of the dispersion law and, of course, the applications.

Let us choose the simplest dispersion relations for the Alfvén waves (4.12) and (4.13) for magnetosonic waves. Formally, (4.13) does not differ from the dispersion law for acoustic waves or for light in an isotropic medium. Consequently, all the well-known results from optics and acoustics concerning ray tracing can be applied to the case in hand without any changes. For example, the shape of the rays is determined by the equation

$$\mathrm{d}1/\mathrm{d}l = n^{-1}[\nabla n - 1(1 \cdot \nabla n)] \tag{4.21}$$

where 1 is a unit vector tangential to the ray trace, $\mathrm{d}l$ is the element of the ray trace length and $n = ck/\omega$ is the refractive index, with $n = c/c_A$ in the case of (4.13). From (4.21), and noting (4.13), it follows that

$$R = -c_A(N \cdot \nabla c_A)^{-1}$$

where R is the radius of the ray curvature, and the unit vector N is the principal normal to the ray. This means that the magnetosonic ray bends towards decreasing Alfvén velocity, as a light ray bends towards increasing refractive index.

One might say that, formally, the relation (4.13) has nothing which can be considered specifically 'magnetohydrodynamic'. On the contrary, the dispersion law for the Alfvén waves (4.12) is extremely specific. It follows from (4.12), that the group velocity $v_g = \partial\omega/\partial k$ is always parallel or antiparallel to B. This result means that the rays of Alfvén waves coincide with the magnetic field lines, i.e. the shape of the rays is determined completely once we specify the field $B(x)$. All that remains is to calculate the refraction, the variation of k along the ray, by means of the second equation in (4.20). This problem can be solved in quadratures (see exercise 4.2.1).

The simplicity of this description, however, has been achieved at the cost of a far-ranging idealization. As a result, for example, we conclude from (4.12) that there are no simple caustics for Alfvén waves. Let us instead assume that the simple caustic exists. Near it two rays will pass through each point of the illuminated part of space: one ray which has already touched the caustic and one which has not yet done so. But this situation would be impossible, since magnetic field lines cannot intersect each other.

Let us discuss another view, which corresponds to a different formulation of the problem. Suppose, there is a point source of Alfvén waves. All the rays, emerging from this source coincide with one and the same magnetic field line. This line is a caustic (the envelope of a family of rays). There are no other rays, besides the caustic one, i.e. the entire space, except the field line, which passes through the source, is immersed in the caustic shadow.

This unusual picture is unacceptable for many reasons and, in particular, because it does not possess structural stability. The account of any weak dependence of the wave frequency on the transverse component of the wave vector completely changes the character of propagation, i.e. leads to the deviation of the ray trace from the field line and the partial restoration of the normal structure of caustics.

In a cold plasma a weak dependence of ω on $k_\perp$ (transverse dispersion) arises for Alfvén waves owing to the gyrotropy of the medium and/or the electron inertia. The gyrotropic effect is predominant at small values $k_\perp$, and the inertial effect at large values. In a hot plasma a dependence of ω on $k_\perp$ also arises because of spatial dispersion, which is manifested in this case as a finite ion Larmor radius effect.

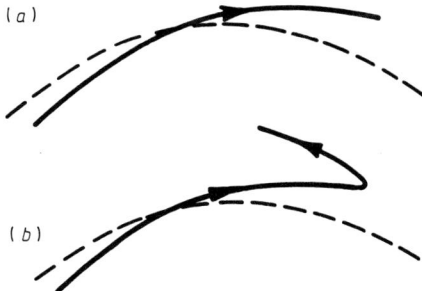

Figure 4.4. Ray traces for the (*a*) Alfvén wave and (*b*) whistling atmospherics. The dashed line corresponds to the external magnetic field lines.

The most important result of accounting for transverse dispersion is the regularization of the structure of caustics. The corresponding deviation of rays partly deprives the Alfvén waves of the peculiarity under consideration. However, there remains something no less interesting. The Alfvén waves are obviously the only type of waves, the rays of which never turn back (figure 4.4(*a*)). In other words, the Alfvén caustics do not touch surfaces orthogonal to the magnetic field lines, at any point. This property of 'never rolling up' appears to be very stable. It disappears neither when gyrotropy is taken into account, nor at other modifications of the dispersion law[2].

On the face of it, whistling atmospherics possess the same property. According to Storey's dispersion law (4.15) the direction of their group velocity deviates from the direction of the external magnetic field by not more than 19°. However, in reality (4.15) should be replaced by (4.13) in the limit $k_\parallel \to 0$. As a result, the ray of a whistler can turn back, as shown in figure 4.4(*b*), in contrast to the ray of an Alfvén wave.

The dispersion laws (4.12) and (4.13) correspond to quasi-perpendicular propagation and all the MHD rays mentioned above refer to this case. For quasi-

[2] This refers to conducting gases and liquids. In conducting solid bodies the Alfvén ray may turn back (see exercise 1.1.1).

longitudinal propagation the problem of the turning of the ray does not arise since turning is forbidden by the condition of applicability of this approximation. The following statement may seem somewhat unexpected: Alfvén and magnetosonic waves have identical rays under identical conditions. A still stronger statement is as follows. The equation describing the shape of the MHD rays for quasi-longitudinal propagation is reduced to the ray equation for waves in an isotropic medium (see section 4.4).

The general character of ray traces is frequently judged by the expression for the group velocity, without solving the system of equations (4.20). In an explicit form the dependence $\omega(k)$, necessary for the calculation of group velocity using the formula $v_g = \partial\omega/\partial k$, is known only in some limiting cases. So it is useful to know the expression for v_g through the refractive index.

Let us write the dispersion equation for the given type of waves in the form

$$M(\omega, k) = ck - \omega n(\omega, k) = 0.$$

Then

$$\frac{\partial \omega}{\partial k} = -\frac{\partial M}{\partial k} \Big/ \frac{\partial M}{\partial \omega}$$

and

$$v_g = \frac{c}{n_g}\left(\frac{k}{k} - \frac{\omega}{c}\frac{\partial n}{\partial k}\right) \tag{4.22}$$

where

$$n_g = \frac{\partial \omega n}{\partial \omega}.$$

In a cold plasma the refractive index depends on $\gamma = |\cos\theta|$, but does not depend on $|k|$. Therefore, (4.22) may be rewritten in the form

$$v_g = \frac{c}{n_g}\left[\frac{k}{k} - \frac{1}{n}\frac{\partial n}{\partial \gamma}\left(\frac{B}{B} - \gamma\frac{k}{k}\right)\right]. \tag{4.23}$$

We see that v_g lies in the same plane as the vectors k and B. The modulus of group velocity is

$$|v_g| = \frac{c}{n_g}\left[1 + \left(\frac{1}{n}\frac{\partial n}{\partial \gamma}\right)^2(1 - \gamma^2)\right]^{1/2}$$

and the projection v_g upon the direction k is

$$v_{g,k} = \frac{c}{n_g}.$$

We also present the expressions for the longitudinal and transverse components of v_g with respect to the external magnetic field

$$v_{g\parallel} = (c/n_g)\left[\gamma - (1/n)\partial n/\partial\gamma\left(1 - \gamma^2\right)\right]$$
$$v_{g\perp} = (c/n_g)\left[1 + (\gamma/n)\partial n/\partial\gamma\right]\left(1 - \gamma^2\right)^{1/2}. \tag{4.24}$$

If $v_g \| k$ we say that the waves propagate isotropically. This takes place, for example, in a vacuum where $n = 1$ and according to (4.23) $v_g = c(k/k)$. Magnetosonic waves, for which $n = n_A$, propagate isotropically in a cold plasma. The Alfvén waves possess extremely strong anisotropy. For them $n = n_A \gamma^{-1}$ and in accordance with (4.24), $v_{g\perp} = 0$. For the whistlers we have

$$n = \omega_{oe}(\omega \Omega_e \gamma)^{-1/2}.$$

Substituting this into (4.24) we find that the angle between v_g and B does not exceed $\tan^{-1}(2\sqrt{2})^{-1} \approx 19°$ (Storey theorem). Therefore, the energy of whistlers is guided by the external magnetic field although it is not as efficient as the energy of Alfvén waves.

At $\theta \neq 0$ and at small but finite ratio (ω/Ω_i) the group velocity of waves of the Alfvén branch deviates from the direction of the external magnetic field. However, within the whole range $0 < \omega < \Omega_i$ the angle between v_g and B does not exceed $\sim 12°$.

Similarly the waves of the magnetosonic branch at small but finite value (ω/Ω_i) possess some anisotropy. However, even at the frequencies $\omega \sim \Omega_i$ the anisotropy of propagation in a cold plasma is rather weak. In any case the angle between v_g and k does not exceed $30°$.

Exercises

Exercise 4.2.1.

Calculate the refraction of the Alfvén waves in an inhomogeneous medium.

Solution 4.2.1.

The traces of the Alfvén waves with the dispersion law $\omega = c_A k_\|$ are known. They are the field lines of the external magnetic field. We can choose an arbitrary trace and calculate the refraction, i.e. change of orientation of k along the trace.

Let us introduce the coordinate system x, y, l, where l is the distance along the field line, the x axis is directed along the main normal, and the y axis—along the binormal. If there is no twisting of the field lines, then nonzero components of the metric tensor are

$$g_{11} = g_{22} = 1 \qquad g_{33} = (1 - \kappa x)^2$$

where $\kappa(l)$ is the curvative of the axial line.

The magnetic field in the small vicinity of the chosen field line has two components $B = (B_1, 0, B_3)$. In order to seek them we use the equations $\nabla \cdot B = 0$ and $\nabla \times B = 0$: $B_1 = -B_0'(l)x$, $B_3 = B_0(l)$, where $B_0' = \mathrm{d}B_0/\mathrm{d}l$. Below we shall need the formulae connecting contravariant (index at the top), covariant (superscript) and physical (subscript) vector components

$$a_i = \sqrt{g_{ii}}a_{(i)} = g_{ii}a^i.$$

(Here the tensor g_{ij} is assumed diagonal.) In particular, for $\boldsymbol{B}$ we have

$$B_1 = B^1 = B_{(1)} \qquad B_3 = B_0(l) \qquad B^3 = B_0(l)[1 - \kappa(l)x]^{-2}$$

$$B_{(3)} = B_0(l)[1 - \kappa(l)x]^{-1} \qquad |\boldsymbol{B}| = |B_i B^i|^{1/2} = B_0(l)[1 + \kappa(l)x].$$

It should be noted that summing up by the index $i = 1, 2, 3$ is implicit in the expression $B_i B^i$, as is assumed in such cases.

To solve the problem we also specify the distribution of the medium density in the vicinity of the ray. Suppose $\rho = \rho_0(l)[1 + \alpha(l)x]$, where $\alpha = \partial \ln \rho / \partial x$ is the steepness of change of ρ across the magnetic shell, containing the trace.

Let us use the dispersion relation $\omega = c_A^i k_i$, where $c_A^i = B^i / \sqrt{4\pi\rho}$ and write the Hamilton canonical equation in the form

$$dk_x/dt = -k_l \partial (c_A/\sqrt{g_{33}})/\partial x$$
$$dk_l/dt = -k_l \partial (c_A/\sqrt{g_{33}})/\partial l$$
$$dl/dt = c_A/\sqrt{g_{33}}$$

where $c_A = |\boldsymbol{B}|/\sqrt{4\pi\rho}$. Here for simplicity we assume $k_y = 0$ (meridianal propagation). Integrating the first equation we get

$$k_x(l) = k_x(0) - k_l(0) \int\limits_0^l \frac{c_A(0)}{c_A(l)} (2\kappa - \alpha/2) \, dl.$$

Integrating the second equation gives

$$k_l(l) = [c_A(0)/c_A(l)]k_l(0)$$

which certainly could have been foreseen, since $\omega(\boldsymbol{k}(t), \boldsymbol{x}(t) = $ constant along the wave trace in the stationary medium. The third equation gives the velocity of the signal movement along the field line.

Exercise 4.2.2.

Find the amplitude variation of Alfvén wave along the ray trace.

Solution 4.2.2.

The geometric magnetohydrodynamics considers the waves as locally plane. From this standpoint the vectors $\boldsymbol{E}$ and $\boldsymbol{B}$ in the Alfvén wave at every point are directed in the way shown in figure 2.1. Therefore the Poynting vector $\boldsymbol{S} = (c/4\pi)\boldsymbol{E} \times \boldsymbol{b}$ is parallel (or antiparallel) to the vector $\boldsymbol{B}$ everywhere.

Next, in any plane wave the vectors $\boldsymbol{E}$, $\boldsymbol{D}$ and $\boldsymbol{b}$ are connected by the relations

$$\boldsymbol{E} \cdot \boldsymbol{D} = b^2 \qquad \boldsymbol{D} = n^2 \boldsymbol{E} - n(\boldsymbol{n} \cdot \boldsymbol{E})$$

where $n = ck/\omega$. Hence, there follows the relation $E = (c_A/c)b$ for the Alfvén wave. Therefore, $\boldsymbol{S} \propto \boldsymbol{B}(b^2/\sqrt{\rho})$. Under stationary conditions $\nabla \cdot \boldsymbol{S} = 0$, hence it follows that

$$b \propto \rho^{1/4}$$

if we take into account that $\nabla \cdot \boldsymbol{B} = 0$.

Thus, the amplitude of magnetic field oscillations changes along the ray so that $b\rho^{-1/4} = \text{constant}$. The amplitude of electric field oscillations changes so, that $\rho^{1/4}E/B = \text{constant}$.

Exercise 4.2.3.

Evaluate the deviation of the Alfvén wave rays.

Solution 4.2.3.

It is necessary first of all to generalize the dispersion law (4.12) so that ω should depend on $k_\perp$. In a cold plasma ($\beta \ll m_e/m_i$), in the case of quasi-perpendicular propagation we have

$$\omega = c_A k_\|(1 + \mu k_\perp^2)^{-1/2}$$

where $\mu = k_{0e}^{-2}$ and $k_{0e} = \omega_{0e}/c$. In a relatively hot plasma ($\beta \gg m_e/m_i$)

$$\omega = c_A k_\|(1 + \mu k_\perp^2/2) \tag{1}$$

where now $\mu \approx r_i^2$, $(k_\perp r_i)^2 \ll 1$ and r_i is the ion gyroradius. In both cases we shall use the dispersion law in the form of (1) assuming $k_\perp^2 \ll k_{0e}^2$ and substituting the corresponding value of μ into (1).

Let the field lines of the external magnetic field be plane curves. For simplicity we shall consider the rays lying in these planes. Let us introduce the coordinate system (s, x, y) so, that $\boldsymbol{B} = (B^s, 0, 0)$. Then the Hamiltonian will have the form (1) if we substitute $k_\| \to k_s$, $k_\perp \to k_x$, $c_A \to \tilde{c}_A = c_A/h_s$, $\mu \to \tilde{\mu} = \mu/h_x$, where h_s and h_x are the Lame coefficients. From (4.19) we obtain the equations

$$dx/ds = (\omega\tilde{\mu}/\tilde{c}_A)k_x \qquad dk_x/ds = -(\omega/\tilde{c}_A^2)\partial\tilde{c}_A/\partial x$$

that determine the shape of the rays.

Instead of solving the Cauchy problem for this system of differential equations we may seek the solution of the integral equation

$$\xi(s) = \xi_0 + \int_0^s H(s', \xi(s'))\, ds'.$$

Here for shorter presentation we have introduced the following notations

$$\xi = \{x; k_x\}$$
$$H = \{k_x F; G\}$$
$$F = \tilde{\mu}\omega/\tilde{c}_A$$
$$G = -(\omega/\tilde{c}_A^2)\partial\tilde{c}_A/\partial x.$$

Let us take advantage of the small parameter μ and construct the solution using the method of Peakar successive iterations. We choose ξ_0 as the zero approximation. The consequent iterations are determined by the recurrent relations

$$\xi_n(s) = \xi_0 + \int_0^s H(s', \xi_{n-1}(s'))\,ds'.$$

In the first approximation ($n = 1$) we obtain the formulae that describe the refraction

$$k_x(s) = k_x(0) + \int_0^s G(0, s')\,ds'$$

and the deviation of the ray from the field line

$$x(s) = k_x(0)\int_0^s F(0, s')\,ds' + \int_0^s ds'\, F(0, s') \int_0^{s'} G(0, s'')\,ds''.$$

Let us estimate the deviation in the case when the parameters of the medium do not depend on s

$$x(s) = \mu_0 k_s[k_x(0)s + k_s s^2(\kappa - \alpha/4)].$$

Here κ is the curvature of the field line, $\alpha = \partial \ln N/\partial x$ and $\mu_0 = \mu(0, 0)$.

Exercise 4.2.4.

Determine the trace of a whistler in the axisymmetric magnetosphere using the interpolation formula (4.16).

Solution 4.2.4.

The equations of geometric optics with the dispersion law (4.16) have the form

$$d\boldsymbol{x}/dt = (1/\omega)[(c_A^2 + v^2)\boldsymbol{k} + vk^2\boldsymbol{\alpha}]$$

$$d\boldsymbol{k}/dt = -(1/2\omega)\nabla[(c_A^2 + v^2)k^2]$$

where $\alpha = \alpha(B/B)$, $v = \alpha \cdot k$. In the axisymmetric magnetosphere the azimuth φ is a cyclic coordinate and consequently, $k_\varphi = $ constant. This allows us to lower the order of equations. For simplicity suppose $k_\varphi = 0$, i.e. we shall consider the traces lying in the plane of the geomagnetic meridian.

Further lowering of the order may be done by means of the integral $\omega(x(t)$, $k(t)) = $ constant, which exists if the magnetosphere is stationary. As a result we get a set of second-order canonical equations

$$dk_s/dx = \partial H/\partial s \qquad ds/dx = -\partial H/\partial k_s. \qquad (1)$$

Here

$$H = [\omega^2/(a + bk_s^2) - fk_s^2]^{1/2}$$

$$a = c_A^2/g_{xx} \qquad b = \alpha^2/g_{xx}g_{ss} \qquad f = g_{xx}/g_{ss}$$

where g_{ik} are the components of the metric tensor. The orthogonal coordinates s, x in the meridianal plane of the magnetosphere are chosen so, that $B = (B^s, 0, 0)$.

Equations (1) are equivalent to the equations of an open mechanical system with the Hamiltonian $H(k_s, s; x)$, that performs one-dimensional motion. In general, rigorous solutions of (5) do not exist. Approximate solutions may be found using the method of perturbations. The problem where H does not depend on x is considered to be unperturbed.

Let us investigate the general character of the traces and calculate the adiabatic invariant within the framework of the following model of the medium: $a = a_0(x) + a_2(x)s^2/2$, $b = b_0(x)$. This corresponds to the equatorial regions of the dipole magnetosphere with the distribution of the plasma along the field lines $N(s) \propto B(s)$. Note that $a_2 = 9a_0$. Let, for simplicity, the inequalities $h^2/f \gg k_s^2(0) \gg a_0/b_0$, where $h = H(k_s, s; x)$ and $k_s(0) = \omega/h\sqrt{b_0}$. Here h is analogous to the energy of the mechanical system. Then the motion along s is finite and the integral

$$I = \frac{1}{2\pi} \oint k_s \, ds$$

taken along the trace at constant values of x and h is

$$I = \tfrac{1}{2}k_s(0)s_m \qquad (2)$$

where $s_m = \sqrt{2/a_2}(\omega/h)$. If I is the adiabatic invariant, then the spatial quasi-period of motion equals $X = -2\pi \, \partial I/\partial h$, or taking into account (2) we have

$$X = (4\pi/\omega)I^{3/2}(2b_0a_2)^{1/4}.$$

The condition of adiabaticity has the form $X \ll \min(X_a, X_b)$ where X_a, X_b are the typical scale variations of a and b along x. The trace oscillates along s between the turning points $s = \pm s_m$ and slowly moves along x. The period and amplitude of oscillations depend on x in the following way

$$X \propto [a_0(x)b_0(x)]^{1/4} \qquad s_m \propto [b_0(x)/a_0(x)]^{1/4}.$$

If we take into account that $a_0 \propto B_0^2/N_0$ and $b_0 \propto B_0^2/N_0^2$, then $X \propto B_0/N_0^{3/4}$ and $s_m \propto N_0^{-1/4}$, where $B_0(x)$ and $N_0(x)$ are the magnetic field and plasma concentration in the equatorial plane.

4.3   A transverse waveguide under the arch of plasmasphere

Let us apply the ray theory to describe the waveguides in the magnetosphere.

The refractive waveguide, which directs the magnetosonic waves across the field lines of the geomagnetic field, is shown in figure 4.5. If we ignore axial asymmetry of the magnetosphere, we can say that the waveguide axis coincides with the equator of the magnetic shell under the plasmapause. Here $c_A = B/(4\pi\rho)^{1/2}$ increases with distance from the axis in any direction: northwards, southwards and earthwards because of the increase in B, and away from the Earth because of the sharp decrease of ρ in the plasmapause[3]. According to (4.20) the magnetosonic rays bend in the direction of decreasing c_A which leads to their channelling along the waveguide axis.

Strictly speaking, these considerations are valid only at frequencies much lower than the gyrofrequencies of ions on the waveguide axis. The gyrofrequency is approximately 8.7 Hz for the protons in the case shown in figure 4.5. But actually the waveguide also exists at higher frequencies, up to the frequency of the lower hybrid resonance Ω_{LH} (approximately 370 Hz for the case shown in figure 4.5).

The waveguide structure is much more complex in the frequency range $\Omega_i < \omega \leq \Omega_{LH}$, than in the MHD range. First of all, this is so due to the frequency dependence of the refractive index. Figure 4.6 shows the vertical profiles $n^2(\omega)$ for the case of strictly transverse propagation ($\theta = \pi/2$). Here L is the distance from the Earth's centre to the equator of the magnetic shell (McIllwain parameter). We can see a strong frequency dependence of the structure of the waveguide, which still exists for the plane rays ($\theta = \pi/2$) at $L \lesssim 4$. At $L \gtrsim 4$ anti-waveguide propagation takes place.

The second factor that causes complications in the waveguide structure at frequencies $\omega > \Omega_i$ is associated with the local anisotropy of the wave propagation. It cannot be ignored, since the deviation of the ray from the equator plane leads to the oscillations of the angle θ about $\theta = \pi/2$. Complicated equations, describing the angular dependence of the refractive index in this frequency range, are awkward for the analytical research of the ray traces. So we shall employ the interpolation formula (4.16).

Let us introduce orthogonal coordinates (s, x, y) near the plasmapause whose coordinate lines $x =$ constant, $y =$ constant coincide with the magnetic

---

[3] If we ignore the thickness of the plasmapause, we shall obtain something like a whispering gallery. Note, that a similar wave duct exists in the ionosphere at nearly the maximum of the F2 layer. The refractive index decreases here when moving away from the waveguide axis northwards or southwards owing to the increase of the magnetic field, and upwards and downwards because of the decrease of plasma density.

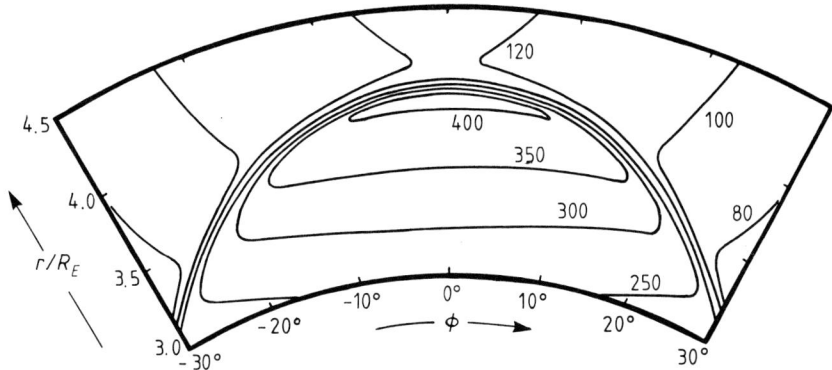

Figure 4.5. The transverse waveguide cross section under the arch of plasmasphere. The isolines of the refractive index $n = c/c_A$ in the plane of geomagnetic meridian are presented. Plasmapause is located at the magnetic shell with the parameter $L = 4$; r is the distance from the Earth's centre, R_E is the Earth's radius and ϕ is the geomagnetic latitude.

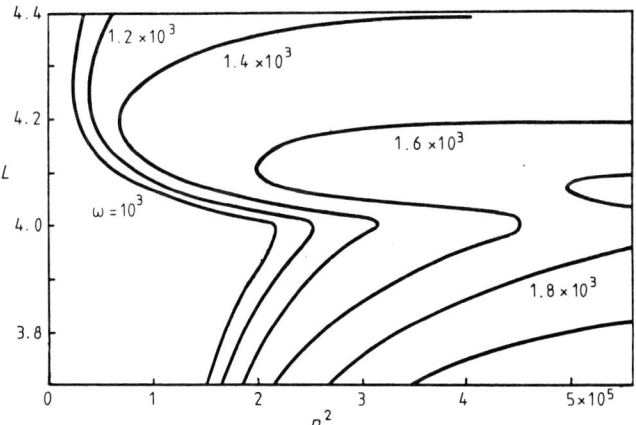

Figure 4.6. Vertical profiles of the square of refractive index for the strictly perpendicular propagation near the lower hybrid frequency. The frequencies ω are given in units s^{-1}.

field lines. Let the y axis be directed along and the x axis across the magnetic shells. Choose y as a course variable. Then from (4.16), (4.20) by means of the first integral $\omega(k(t), x(t))$ we obtain the canonical equations

$$\frac{dp}{dy} = \frac{\partial H}{\partial q} \qquad \frac{dq}{dy} = -\frac{\partial H}{\partial p} \qquad (4.25)$$

with the Hamiltonian

$$H = [\omega^2(a + bk_s^2)^{-1} - fk_s^2 - hk_x^2]^{1/2}$$

where $p = (k_s, k_x)$, $q = (s, x)$, $a = c_A^2/g_y$, $b = \alpha^2/g_s g_y$, $f = g_y/g_s$, $h = g_y/g_x$ and g_i are the nonzero (diagonal) components of the metric tensor.

In the following simplification we ignore the weak y dependence of H, and proceed to the 'action-angle' variables. Then, after choosing the model of the medium (i.e. the dependence of a, b, f and h on s and x) we can then analyse (4.25) by the methods of the theory of autonomous dynamic systems.

We restrict the present discussion to pointing out the character of the only singular point of the system which is found by equating the right-hand sides of (4.25) to zero. In the dipole magnetosphere the coordinates of the singular point are $k_s = 0$, $k_x = 0$, $s = 0$, $x = x_0$, where x_0 is determined as the solution of the equation

$$N^{-1}\partial N/\partial x = 8/L_p R_E. \tag{4.26}$$

Here s is reckoned from the equatorial plane, x is reckoned from the plasmapause (earthwards), and L_p is the McIllwain parameter of the plasmapause. The singular point will be attractive if the following condition holds

$$\partial^2 N/\partial x^2 < 0. \tag{4.27}$$

The conditions (4.26) and (4.27) hold near the plasmapause (closer to the Earth). We also note that the movement of the ray along x and s is double periodic. The ratio of the spatial periods is of the order of $\Omega_i \Delta L/\omega L$, where ΔL is the characteristic scale of the variation of Alfvén velocity across the magnetic shells in the vicinity of the waveguide axis. Since $\Delta L < L_p$, the ray rapidly oscillates along the altitude and slowly along the latitude at $\omega \gg \Omega_i$. The velocity of propagation along the longitude approaches the Alfvén velocity over the waveguide axis, $c_A \simeq 700$ km s^{-1}.

Let us explain our interest in the toroidal waveguide that exists in the equatorial vicinity of the plasmapause. The problem is the following. The instability of the radiation belt of the Earth usually has a convective character. Therefore, wave generation arises, for example, in the cases when the wave packet may cross repeatedly the region of interaction with the energetic particles. Such opportunity for Alfvén waves is realized owing to the magnetic focusing and the reflection of waves from the ionosphere at opposite ends of the magnetic field tube.

Magnetosonic waves do not experience magnetic focusing. Their trajectories are quite complicated curves. As a rule these waves rapidly leave the region of resonant interaction with the energetic particles. However, the waves trapped in the transverse waveguide may have interaction with the resonant particles over a long period of time, and if the distribution of particles is unstable the wave amplitude will increase.

4.4 Longitudinal waveguides

We proceed to the description of longitudinal waveguides without leaving the framework of the ray considerations.

It is known that space plasma is characterized by a stratified and fibrous structure with strata and fibres stretched along the magnetic field lines. In the magnetosphere this structure forms a system of refractive waveguides, along which electromagnetic waves propagate from one hemisphere to the other. In fact, the fibrous structure of magnetospheric plasma was discovered in the course of whistler observations. The strata and fibres are capable of channelling even short radio waves, thus creating the effect of the 'world echo'.

For MHD waves the idea of the longitudinal waveguide works only in the case when the wavelength is much shorter than the length of the geomagnetic field line. This condition holds at the frequencies in the Pc1 frequency range (0.2–5 Hz) in the central and peripheral regions of the magnetosphere.

The theory of longitudinal waveguides faces difficult problems. Even the existence of longitudinal waveguides as stable channels where the wave packet oscillates repeatedly between the reflection points from the ionosphere has not been proved purely theoretically. It is true, that there is strong experimental evidence on that score, and we shall now consider this evidence.

4.4.1 Pearls at conjugate points

The term 'pearls' applies to quasi-monochromatic signals in the Pc1 frequency range. The oscillogram actually resembles a string of pearls (figure 4.7). Sometimes successive signals are clearly separated in time. However, more often the signals crawl over each other, creating a complicated picture of beats. These may be easily analysed by means of a sonograph—the device used to construct a dynamic spectrum of oscillations. A typical sonogram of a series of pearls is shown in figure 4.8. The frequency is measured by the vertical axis, the time by the horizontal axis, and the spectral density is proportional to the blackness of the record. Analysis of the dynamic spectra proves that the pearls represent a periodic succession of signals. In most cases the carrier frequency increases in time within the limits of each signal. These signals are termed hydromagnetic whistlers.

Here are the main parameters of pearls. The carrier frequency of oscillations varies from one case to another in the limits $f \simeq 0.2$–3 Hz, the period of signal repetition is $\tau \simeq 50$–300 s and the bandwidth is $\Delta f \simeq 0.05$–0.3 Hz. On average $f \simeq 0.8$ Hz, $\Delta f \simeq 0.1$ Hz and $\tau \simeq 120$ s. There is a fairly stable relation $\tau f \simeq 10^2$ between f and τ. The characteristic growth of the carrier frequency within one signal is $df/dt \simeq 0.1$ Hz min^{-1}. The amplitude of oscillations is of the order of 10–100 nT at mid-latitudes.

Pearls are a comparatively rare phenomenon. They appear sporadically and last for about half an hour on average. However, sometimes a series of pearls

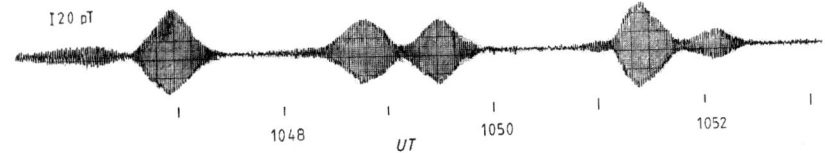

Figure 4.7. Oscillogram of 'pearls' registered at Sogra Observatory (Arkhangelsk region) on 17 November 1968. The evolution of pearls is similar to the interaction of a pair of solitons.

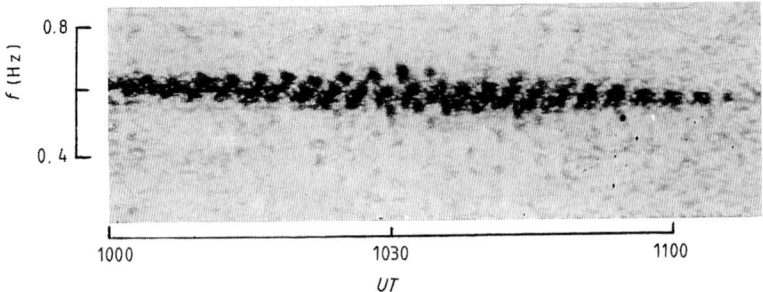

Figure 4.8. Dynamic spectrum of pearls (Sogra Observatory, 21 February 1967). The wave packet passes about 50 times between the conjugated points.

is excited continuously for many hours. We observe a complex daily latitudinal variation of the probability of appearance of pearls. The maximum probablility of appearance falls in the morning and at midday at mid- and high latitudes respectively.

The very fact of the periodicity of the signal repetition gives us reasons to suppose that a series of pearls represent a succession of echo-signals, oscillating in one of the longitudinal waveguides and reflecting from the ends of this waveguide in the opposite hemispheres. When reflecting from the ends the signal excites the ionospheric MHD waveguide. As a result the pearls may be observed at distances of thousands of kilometres from the ionospheric source region.

We shall supplement this scenario with a hypothesis on the mechanism of excitation of the longitudinal waveguide given in section 6.3, and here we raise a question: how can we test experimentally the concept of a series of pearls as being a succession of echo signals? The answer suggests itself: two observatories should be situated in the opposite hemispheres symmetrically relative to the ends of the hypothetical waveguide. In other words the observatories should be located at magnetic conjugate points, i.e. at the points of intersection of the geomagnetic field line with the Earth's surface[4]. Then,

---

[4] It is unlikely that this very field line will be the axis of the longitudinal waveguide. However, in case such a waveguide exists somewhere in the vicinity of this line, the signals can be observed

within the framework of the given scenario we expect that the discrete elements of the dynamic spectrum of pearls should be observed at the conjugate points in turn. In terms of waveforms it means that the envelope of the oscillation amplitude at the conjugate points must be anticorrelated.

Such an experiment was carried out at conjugate points Sogra (a village in Arkhangelsk region) and Kerguelen (an island in the Indian ocean). The result of observations proved expectations. The hydromagnetic whistlers appeared in turn at one moment in Sogra, and at another at Kerguelen and vice versa. The time of signal propagation from one hemisphere to the other was about half the period of repetition of τ. The observations at other pairs of conjugate points prove this conclusion and testify in favour of the hypothesis of the waveguide propagation of pearls in the magnetosphere.

From time to time we observe simultaneously several series of pearls, which differ in carrier frequency and period of signal repetition. The anticorrelation of the envelope of the oscillation amplitude at conjugate points takes place for each of these series. This proves the existence of several longitudinal waveguides in the magnetosphere at least when the multiband pearls were observed.

4.4.2 Longitudinal channelling of $\mathcal{L}$- and $\mathcal{R}$-waves

Occasionally one hears the opinion that Alfvén waves are ducted by magnetic fields 'better' than magnetosonic waves. In the case of quasi-longitudinal propagation this idea is not correct. It turns out that the conditions for longitudinal ducting are identical for both types of waves. Moreover, the equations which describe the shape of the rays of both magnetosonic and Alfvén waves, reduce to the ray equation (4.21) for the waves in an isotropic medium.

In light of the above comments the usual argument, which associates pulsations of Pc1 in the magnetosphere with packets of Alfvén waves, seems anachronistic. The observed propagation of Pc1 between the conjugate points in channels stretched along the field lines of the geomagnetic field does not eliminate the formation of Pc1 by waves of the magnetosonic type at certain segments of the trajectory.

So, in the quasi-longitudinal approximation, the rays of Alfvén waves have no distinguishing features. They do not differ from the magnetosonic rays or, in general, from the rays of any waves in an isotropic medium. This conclusion is of methodical importance, since it allows us to make direct use of the known results from the geometric optics of isotropic media.

Instead of Hamilton's equations (4.20) it is convenient to use the eikonal equation

$$(\nabla\varphi)^2 = n^2. \tag{4.28}$$

Here φ is eikonal and n is the refractive index found from the local dispersion equation and is dependent, in general, on the coordinate x, the instantaneous

owing to the action of the ionospheric waveguide.

frequency $\omega = -\partial\varphi/\partial t$ and the local wave vector $k = (\omega/c)\nabla\varphi$.

Under quasi-longitudinal propagation the refractive indices of the Alfvén and magnetoacoustic waves are determined by (4.8). We recall that in the given approximation we use the terms $\mathcal{L}$- and $\mathcal{R}$-wave, that reflect the character of polarization of these two types of wave.

Let us consider the plasma to contain electrons and ions of one sort. Then

$$\varepsilon_\perp = \frac{\omega_{0i}^2}{\Omega_i^2 - \omega^2} \qquad g = \frac{\omega}{\Omega_i}\varepsilon_\perp. \tag{4.29}$$

Here we have taken into account that $\omega_{0i} \gg \Omega_i$ in the magnetosphere. The applicability of (4.8) with regard (4.29) is limited by the inequalities

$$\sin^4\theta \ll \left(\frac{2\omega}{\Omega_i}\right)^2 \cos^2\theta \qquad \omega^2 \ll \Omega_e\Omega_i. \tag{4.30}$$

If $\omega^2 \ll \Omega_i^2$, then it follows from (4.30) that $\theta^2 \ll 1$, and (4.8) may be written as

$$n^2 = (\varepsilon_\perp \pm g)\left(1 + \frac{\theta^2}{2}\right). \tag{4.31}$$

This expression also holds over a wider range of frequencies if the condition $\theta^2 \ll 1$ is assumed to be independent.

Now we take into account that $\theta \approx k_\perp/k_\parallel$, designate $\varepsilon \pm g = n_\pm^2$ and rewrite (4.31) in the following way

$$\frac{c^2}{\omega^2}\left(k_\parallel^2 + \frac{k_\perp^2}{2}\right) = n_\pm^2. \tag{4.32}$$

Let us make the scale transformation $x_\perp \to x_\perp/\sqrt{2}$, $k_\perp \to \sqrt{2}k_\perp$, where $x_\perp$ are the coordinates on the surfaces orthogonal to the field lines of the external magnetic field. In place of (4.31) we then have

$$n = n_\pm(\omega, x). \tag{4.33}$$

In this form the refractive index, as in the case of an isotropic medium, depends on coordinate x and the frequency ω, and not on the orientation of the wavefront.

The conditions of applicability (4.33) contain one of the following two inequalities: either $\omega^2 \ll \Omega_i^2$ or $\theta^2 \ll 1$. If the former occurs, then $n_+ \approx n_-$, and the Alfvén rays coincide with the magnetosonic with corresponding accuracy.

This approach makes possible a consistent and concise description of a fairly wide range of phenomena connected with the propagation of MHD waves in longitudinal waveguides in terms of the geometric optics of isotropic media. Essentially, we may immediately use nearly all the results of this well-advanced theory by identifying n with n_+ or n_-, by making a scaling transformation, and by following the conditions for the applicability of (4.33). Let us illustrate this by analysing paraxial rays in the refractive waveguide.

Keeping in mind (4.33), the shape of rays, i.e. the spatial projections of the characteristics of equation (4.28), is described by equation (4.21). According to the meaning of the evaluation of (4.33), the direction of rays must deviate only slightly from the direction of the magnetic field lines. Let us denote one of them to be the axis. Assuming s to be the unit vector along the axis, let us superimpose the following supplementary condition: we consider not only the deviation of direction but also the perpendicular shift of the ray relative to the axis to be small. Let, for simplicity, the axis and rays be plane curves. Then $dx = (1 - s) \cdot N ds$, where ds is the element of the axis length, and dx is the ray shift over the route of ds. It is known that $ds/ds = N/R$, where R is the radius of the axial line curvature. Hence, taking into account (4.21) we obtain

$$\frac{d^2 x}{ds^2} = \frac{1}{n} \frac{\partial n}{\partial x} - \frac{1}{R}. \tag{4.34}$$

Here it was assumed that $dl \approx ds$ and $1 \cdot s \approx 1$, i.e. linearization was carried out over dx/ds. The direction x was chosen to be along the principal normal to the axial line. Turning back to the real scale ($x \rightarrow \sqrt{2}x$), we at last obtain the equation that describes the deviation of rays from the geomagnetic field line

$$\frac{d^2 x}{ds^2} = -\frac{1}{2} \frac{\partial U}{\partial x} \qquad U = \kappa(s)x - \ln n_{\pm}(x, s). \tag{4.35}$$

Here $\kappa = R^{-1}$. The only distinctive feature of paraxial MHD rays is the potential $U(x, s)$ being half that in the isotropic case.

Equation (4.35) may be deduced directly from the canonical equations of ray theory. This allows us to specify the conditions of applicability (4.35)

$$\kappa x \ll 1 \qquad \theta x (d \ln \kappa / ds) \ll 1$$

$$\theta \partial n / \partial s \ll \partial n / \partial x \qquad (dx/ds)^2 \ll 1$$

where $\theta = 2(dx/ds)$. The second and the third of these inequalities will be yielded in the case in which the medium is almost stratified, i.e. if the potential is only slowly changing along the axial line

$$\partial U / \partial s \ll \partial U / \partial x. \tag{4.36}$$

Equation (4.35) coincides in its form with, for example, the equation of motion of a point mass in a field of external forces and the equation of the ray theory of radio-wave propagation, etc. This may be used to apply a detailed method of adiabatic invariant to construct a general picture of MHD rays in an almost stratified medium relatively easily, without solving equations (4.20).

Under condition (4.36) there exists an adiabatic invariant

$$I = 4 \int_{x_1}^{x_2} \sqrt{E - U} \, dx. \tag{4.37}$$

Here the turning points $x_{1,2}$ are found from the equation $U(x_{1,2}, s) = E(s)$. It is supposed that $E > U$ in between, i.e. there exists a waveguide (channel, duct). The integration is carried out at fixed s. The dependence $E(s, I_0)$ may be found from the asymptotic conservation of the adiabatic invariant $I(E, s) = I_0$. This dependence is used to determine the most important characteristics of propagation: the ray oscillation step along s, absorption (enhancement), group delay, the conditions for capture in a channel and conditions for escape from a channel, etc. Of course, these and other problems may be solved without any reference to the similar results of the geometric optics of isotropic media, but this is hardly advisable.

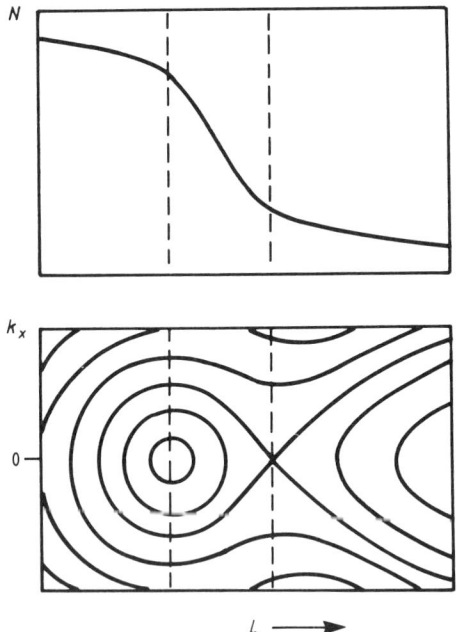

Figure 4.9. The distribution of plasma density in the vicinity of the plasmasphere (upper panel) and a phase portrait of the longitudinal waveguide (lower panel).

The approach, presented here, has not yet been realized over the whole volume. It has only ascertained that there exists at least a short segment of the waveguide at the plasmapause that is orthogonal to the equatorial plane. Figure 4.9 helps us understand its structure. The L-profile of plasma concentration N is shown in the upper panel. N decreases at the plasmapause by one–two orders over the interval of $\Delta L \simeq 0.1$–0.5 when moving away from the Earth. The phase portrait of the system (4.35) is presented at the bottom. This picture is qualitatively described by the position of singular points and separatrices. The singular points (of centre and saddle type) are located at the straight line $k_x = 0$. Generally speaking, the position of the points on the axis L depends on the frequency ω and the type of wave. If $\omega \ll \Omega_i$, then the coordinates of the singular points are determined as the roots of the equation

$$\frac{R_E}{R} + \frac{1}{4N}\frac{\partial N}{\partial L} = 0. \tag{4.38}$$

The saddle point is located somewhat further from the Earth than the centre. The channel for $\mathcal{L}$- and $\mathcal{R}$-waves is situated around the centre inside the closed section of the separatrix.

Exercise

Exercise 4.4.1.

Find the frequency dependence of coefficients of equations like (4.38).

Solution 4.4.1.

In the critical points of the potential (4.35) we have $\partial U/\partial x = 0$, or

$$\kappa - \partial \ln n_\pm/dx = 0.$$

Taking into account (4.29) we have

$$n_\pm = \frac{\omega_{0i}}{\Omega_i}\left(1 \mp \frac{\omega}{\Omega_i}\right)^{-1/2}.$$

Now we take into account that $\omega_{0i} \propto N^{1/2}$, $\Omega_i \propto (1 + \kappa x)$ and $\kappa|x| \ll 1$. Then instead of (4.38) we have

$$\frac{R_E}{\mathcal{R}_\pm(\omega)} + \frac{1}{4N}\frac{\partial N}{\partial L} = 0$$

where $\mathcal{R}_\pm(\omega) = [4(\Omega_i \mp \omega)/(4\Omega_i \mp 3\omega)]R$. We have taken into account that L increases and x decreases when moving away from the Earth. If $\omega \ll \Omega_i$, then $\mathcal{R}_\pm(\omega) \simeq R(1 \mp \omega/4\Omega_i)$.

We see that the effect of the frequency dispersion $n_\pm(\omega)$ may be presented as the frequency dependence of the effective radius of curvature $\mathcal{R}_\pm(\omega)$ of the geomagnetic field axial line. For the $\mathcal{L}$-wave, $\mathcal{R}_+(\omega) < R$, and for the $\mathcal{R}$-wave, $\mathcal{R}_-(\omega) > R$.

4.5   Oscillations of magnetic shells

In the Pc3–5 range (2–100 mHz) the lengths of MHD waves are comparable to the dimensions of the magnetosphere, so geometric optics is generally inapplicable. However, there is an approach along which the eikonal depends on two and not on three coordinates, as was assumed above. In other words, the ray pattern is used along two spatial directions, while the wave (mode) field structure is retained along the third. In this way underwater acoustics deals with horizontal rays and vertical modes. For reasons, which will become clear below, we will speak here in terms of transverse rays and longitudinal modes.

4.5.1   The equation of toroidal oscillations

Let us examine the problem of the spectrum of Alfvén eigen oscillations of magnetic shells. We use the Maxwell equations

$$\nabla \times \boldsymbol{E} = \mathrm{i}\left(\frac{\omega}{c}\right)\boldsymbol{b} \qquad \nabla \times \boldsymbol{b} = -\mathrm{i}\left(\frac{\omega}{c}\right)\hat{\varepsilon}\boldsymbol{E} \qquad (4.39)$$

with the dielectric permeability tensor of the form $\hat{\varepsilon} = \mathrm{diag}(\varepsilon_\parallel, \varepsilon_\perp, \varepsilon_\perp)$.

We introduce the orthogonal curvilinear coordinates (x^1, x^2, x^3) defined in such a way that the anisotropy axis coincides at each point with the tangent to the x^1 coordinate line. For simplicity let $\hat{\varepsilon}$ be independent of x^3. Then from (4.39) we may single out poloidal and toroidal modes. For the toroidal mode

$$\boldsymbol{b} = (0, 0, b_3) \qquad \boldsymbol{E} = (E_1, E_2, 0)$$

from (4.39) we have

$$\frac{\partial}{\partial x^1}\frac{g_{22}}{\varepsilon_\perp g^{1/2}}\frac{\partial b_3}{\partial x^1} + \frac{\partial}{\partial x^2}\frac{g_{11}}{\varepsilon_\parallel g^{1/2}}\frac{\partial b_3}{\partial x^2} + \frac{\omega^2}{c^2}\frac{g^{1/2}}{g_{33}}b_3 = 0 \qquad (4.40)$$

where g_{ik} is the diagonal metric tensor and $g = \det g_{ik}$. The electric field is found by the magnetic field in the following way

$$\begin{aligned}
E_1 &= (\mathrm{i}c/\omega\varepsilon_\parallel)(g_{11}/g^{1/2})\partial b_3/\partial x^2 \\
E_2 &= -(\mathrm{i}c/\omega\varepsilon_\perp)(g_{22}/g^{1/2})\partial b_3/\partial x^1.
\end{aligned} \qquad (4.41)$$

When deducing (4.40) we took advantage of the well-known formula of tensor analysis

$$(\nabla \times \boldsymbol{E})^\alpha = \frac{e^{\alpha\beta\gamma}}{2\sqrt{g}}\left(\frac{\partial F_{,\gamma}}{\partial x^\beta} - \frac{\partial E_\beta}{\partial x^\gamma}\right).$$

Here $e^{\alpha\beta\gamma}$ changes sign with the permutation of any two indices, and $e^{123} = e_{123} = 1$. Moreover, we used the relationships (see also exercise 4.2.1)

$$\varepsilon_{11} = g_{11}\varepsilon_\parallel \qquad \varepsilon_{22} = g_{22}\varepsilon_\perp \qquad \varepsilon_{33} = g_{33}\varepsilon_\perp.$$

In a cold plasma in the low-frequency limit ($\omega \ll \Omega_i$), we have $\varepsilon_\perp = \omega_{0i}^2/\Omega_i^2$, $\varepsilon_\parallel = -\omega_{0e}^2/\omega^2$, and correspondingly (4.40) becomes

$$\frac{\partial}{\partial s}E\frac{\partial\Psi}{\partial s} + \omega^2\left(F\Psi - \frac{\partial}{\partial x}G\frac{\partial\Psi}{\partial x}\right) = 0 \qquad (4.42)$$

where $s = x^1$, $x = x^2$ and $\Psi = b_3$. The coefficients of equation (4.42) depend on x and s and are

$$E = c_A^2 g_{22}/g^{1/2} \qquad F = g_{33}^{1/2} \qquad G = g_{11}/k_{0e}^2 g^{1/2}$$

where $k_{0e} = \omega_{0e}/c$.

In (4.41) the transverse dispersion appears due to electron inertia. In the limit $k_{0e} \rightarrow \infty$ equation (4.42) transforms to the well-known Dungey equation.

Equation (4.42) has an interesting feature: the small parameter is found only in the transverse operator. In geometrical optics terms, the motion of the ray is fast along s and slow along x. This fact shows us a way to solve (4.42). Since the small parameter appears in a nonuniform way in the higher derivatives, the transition to the short-wave asymptotic here is generally not accompanied by the transition to the high-frequency limit. There arises a possibility to study low-frequency oscillations by using the computational advantages of the short-wave approximation.

For greater clarity instead of (4.42) we consider the equation

$$\frac{\partial}{\partial s} \frac{E}{\omega^2} \frac{\partial \Psi}{\partial s} + \mu \frac{\partial^2 \Psi}{\partial x^2} + \Psi = 0. \tag{4.43}$$

It retains all the basic features and all the complexity of equation (4.42), but is more convenient for the analysis. Besides, (4.43) allows us to take account of the thermal motion of particles at the qualitative level. For this it should be assumed that $\mu = -1/k_{0e}^2$ in a cold and $\mu \approx r_i^2$ in a hot plasma and it should be considered for uniformity that $k_\perp^2 |\mu| \ll 1$ in both cases (here r_i is the gyroradius of the thermal ions).

Let us supplement (4.43) with a simplified boundary condition at the ionosphere

$$\left. \frac{\partial \Psi}{\partial s} \right|_{s=\pm s_0} = 0 \tag{4.44}$$

where $s = s_0$ and $s = -s_0$ are the coordinates of the ionosphere in the northern and southern hemispheres. One more restriction, now completely determining the spectrum, is the absence of exponential field growth at $x \rightarrow \pm\infty$.

4.5.2 Fast and slow subsystems

Wishing to retain the mode structure of the field along s and thus not to pass to the high-frequency limit, we will attempt to use a version of perturbation theory to find the eigenfrequencies. This version is based on splitting of the system into fast and slow subsystems.

We first find the solution of the Dungey problem

$$-\frac{\partial}{\partial s} E(s, x) \frac{\partial \Phi_n}{\partial s} = \lambda_n \Phi_n \qquad \left. \frac{\partial \Phi_n}{\partial s} \right|_{s=\pm s_0} = 0 \tag{4.45}$$

where $\Phi_n(s, x)$ and $\lambda_n(x)$ are the eigenfunctions and eigenvalues of the longitudinal operator, which depend parametrically on the slow variable x. The solution of the initial problems (4.43) and (4.44) is sought as an expansion in Φ_n

$$\Psi(s, x) = \sum_n a_n(x)\Phi_n(s, x).$$

Substituting this expression into (4.43) and using (4.45) we obtain

$$\sum_n a_n \left(1 - \frac{\lambda_n}{\omega^2}\right) \Phi_n = -\mu \sum_n \frac{\partial^2}{\partial x^2} a_n \Phi_n.$$

Multiplying this equation by Φ_m^*, integrating over s and applying the condition of orthonormalization

$$\int_{-s_0}^{s} \Phi_m^* \Phi_n \mathrm{d}s = \delta_{mn}$$

we get

$$\frac{\mathrm{d}^2 a_m}{\mathrm{d}x^2} + \frac{1}{\mu}\left(1 - \frac{\lambda_m}{\omega^2}\right) a_m = -2\sum_n \frac{\mathrm{d}a_n}{\mathrm{d}x} \int_{-s_0}^{s_0} \Phi_m^* \frac{\partial \Phi_n}{\partial x} \mathrm{d}s - \sum_n a_n \int_{-s_0}^{s_0} \Phi_m^* \frac{\partial^2 \Phi_n}{\partial x^2} \mathrm{d}s.$$

Considering the dependence of Φ on x to be weaker than $a(x)$, we employ the method of successive approximations. The equation of the zero approximation for the coefficients of expansion Ψ by Φ_n is

$$\mu \frac{\mathrm{d}^2 a_n}{\mathrm{d}x^2} + \left(1 - \frac{\lambda_n(x)}{\omega^2}\right) a_n = 0.$$

This is an equation of the Schrödinger type. We find its solution in the WKB approximation and define the spectrum from the quantization condition

$$\int_{x_1}^{x_2} \frac{\mathrm{d}x}{\mu^{1/2}} \left(1 - \frac{\lambda_n(x)}{\omega^2}\right)^{1/2} = \pi \left(\nu + \tfrac{1}{2}\right). \tag{4.46}$$

The turning points $x_{1,2}$ are found from the vanishing of the expression in the radical; between the turning points, this expression must be positive. It follows that the Alfvén eigen oscillations exist and that in this case they have a discrete spectrum $\omega_{n\nu}$, $n = 1, 2 \ldots$, $\nu = 0, 1, \ldots$ only if we have $\mu \neq 0$, and the Dungey spectrum λ_n has its maximum ($\mu < 0$) or its minimum ($\mu > 0$) as a function of x.

If $\mu = 0$, then there are no eigen oscillations. In this context sometimes it is asserted that eigen oscillations with a continuous spectrum exist, but that assertion contradicts the conception of eigen oscillations as they are, their frequency being determined by the system itself, not by the external agent. Clearly, this entire situation is related in a definite way to the pathological behaviour of Alfvén rays at $\mu = 0$, as discussed in section 4.2.

4.5.3   Transverse rays and longitudinal modes

The procedure to obtain (4.46) is not quite justified since the series of successive approximations is not proved. The conclusion described may be considered as a guiding argument, and the correctness of (4.46) at sufficiently small μ may be considered as a hypothesis.

Now let us outline the deduction of (4.46), using the concept of rays and modes in a nearly stratified medium.

The medium, the properties of which change rapidly with the change of any one coordinate, is termed nearly stratified. It may seem that in our case this would be the coordinate x, as the plasma easily spreads out along the magnetic field lines of the magnetic field. But as a matter of fact it is coordinate s. A peculiar turn of the stratification direction by $\pi/2$ is connected with the influence of the small parameter μ in (4.43). All this becomes obvious if instead of (s, x) we introduce new coordinates (s, χ) in which the distances along x are stretched: $\chi = x/|\mu|^{1/2}$. Then the Alfvén velocity $c_A(s, |\mu|^{1/2}\chi)$ will depend on the transverse coordinate only via the combination $|\mu|^{1/2}\chi$, and the small parameter in front of the transverse operator in (4.43) will disappear.

On establishing this fact we may construct an asymptotic theory for transverse rays and longitudinal modes, reproducing an almost analogous construction to that found in underwater acoustics. In the zero approximation we get an eikonal equation, from which (4.46) follows.

4.5.4   The Dungey spectrum

In order to determine the discrete spectrum of the toroidal oscillations from the condition (4.46), the Dungey spectrum $\lambda_n(x)$ should first be found, i.e. the problem (4.45) should be solved. It will be more convenient to reformulate this problem choosing E_2 as a dependent variable and not b_3. Then the Dungey equation will take the form

$$\frac{\mathrm{d}}{\mathrm{d}x^1}\left(\frac{g_{33}}{\sqrt{g}}\frac{\mathrm{d}E_2}{\mathrm{d}x^1}\right) + \left(\frac{\omega}{c_A}\right)^2 \frac{\sqrt{g}}{g_{22}}E_2 = 0. \tag{4.47}$$

Here we used the relationship

$$b_3 = \frac{c}{i\omega}\frac{g_{33}}{\sqrt{g}}\frac{\mathrm{d}E_2}{\mathrm{d}x^1}$$

which follows from (4.39) with taking into account the polarization of toroidal oscillations in the Dungey approximation.

Let us choose the dipole approximation of the geomagnetic field

$$B_r = -\frac{2M}{r^3}\cos\vartheta \qquad B_\vartheta = -\frac{M}{r^3}\sin\vartheta \tag{4.48}$$

where M is the magnetic moment of the Earth, other notation is given in figure 4.10. The equation of the field line has the form

$$r = r_0 \sin^2 \vartheta. \qquad (4.49)$$

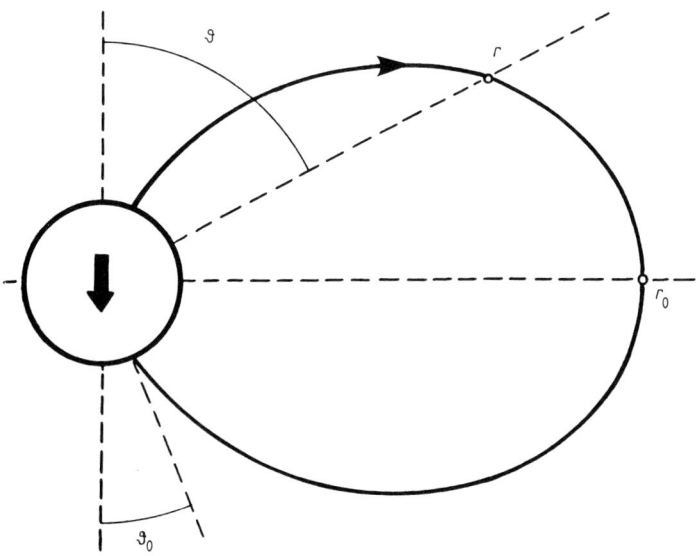

Figure 4.10. The geomagnetic field line.

Let us imagine a magnetic shell with the parameter $L = r_0/R_E$, formed by the rotation of this line around the axis of the geomagnetic dipole. We shall consider the axially symmetric oscillations of this shell. It is clear, that distribution of the plasma in the magnetosphere is taken to be axially symmetric, i.e. ρ depends on the distance r, co-latitude ϑ, but does not depend on the longitude φ.

Under the dipole approximation the coordinates (x^1, x^2, x^3) are related to the spherical coordinates (r, ϑ, φ) in the following way:

$$x^1 = \cos \vartheta / r^2 \qquad x^2 = \sin^2 \vartheta / r \qquad x^3 = \varphi.$$

The metric tensor components have the form

$$g_{11} = \left(\frac{M}{B}\right)^2 \qquad g_{22} = \left(\frac{M}{Br \sin \vartheta}\right)^2 \qquad g_{33} = (r \sin \vartheta)^2$$

where $B = (B_r^2 + B_\vartheta^2)^{1/2}$.

Using the given formulae and passing to a new independent variable $z = \cos \vartheta$ in (4.47) we get the equation of toroidal oscillations in the form

$$\hat{L}\Psi = \Lambda\Gamma\Psi \qquad (4.50)$$

where

$$\hat{L} = -\frac{d^2}{dz^2} \qquad \Gamma(z) = \frac{\rho(z)}{\rho(0)}(1 - z^2)^6$$

$$\Lambda = \frac{4\pi\rho(0)\omega^2 R_E^8}{M^2(1 - z_0^2)^8} \qquad z_0 = \left(1 - \frac{1}{L}\right)^{1/2}.$$

Obviously, $z_0 = \cos\vartheta_0$, where ϑ_0 is the co-latitude of intersection of the oscillating magnetic shell with the Earth's surface. It is also clear that $\rho(0)$ is the plasma density at the equator of this magnetic shell. We designate the component E_2 via Ψ, which will certainly cause no misunderstanding (see (4.42)). The idealized boundary conditions on the ionosphere now have the form

$$\Psi(\pm z_0) = 0. \tag{4.51}$$

If $\rho \propto (1 - z^2)^{-6}$, then $\Gamma = 1$, and we shall easily determine eigenfunctions and eigenvalues of (4.50) and (4.51)

$$\Psi_n = z_0^{-1/2}\sin[\Lambda_n^{1/2}(z - z_0)] \tag{4.52}$$

$$\Lambda_n = (\pi n/2z_0)^2 \qquad n = 1, 2, \ldots. \tag{4.53}$$

There are some more functions $\rho(z)$ for which the problem has rigorous solutions. However, in practice, in cases of interest we have to use the methods of approximate solution (see exercises 4.5.1 and 4.5.2). Numerical methods to calculate the spectrum are more often used.

The general dependence of the oscillation frequency ω_n on the parameter L may be understood if we pay attention to the weak dependence of Λ_n on L at sufficiently large L. Neglecting this latter dependence we obtain

$$\omega_n(L) \propto L^{-4}[\rho(0, L)]^{-1/2} \tag{4.54}$$

where $\rho(0, L)$ is the plasma density at the equator of the shell with the parameter L.

Figure 4.11, borrowed from the paper by Yumoto and Saito (1983) with kind permission of the authors, shows the dependence of ω_1, ω_2 and ω_3 on L. The characteristic bending of the curves is caused by the transition from a dense plasma to a rarefied one when crossing the plasmapause ($L=4$ for the given case). We suppose that $\mu > 0$ in the vicinity of the plasmapause, and hence the discrete spectrum $\omega_{n\nu}$ of eigen oscillations is evidently concentrated close to the minima of curves on the shell $L = 4$ (see (4.46)).

All this is certainly interesting and noteworthy, as it clears up the structure of Alfvén oscillations; however, there arises the question of whether the discrete spectrum $\omega_{n\nu}$ exists in the real magnetosphere. To put it simply: can we observe the fine structure of the spectrum of Alfvén oscillations in the experiment, i.e. splitting of the harmonics with the fixed number n into a series of spectral lines

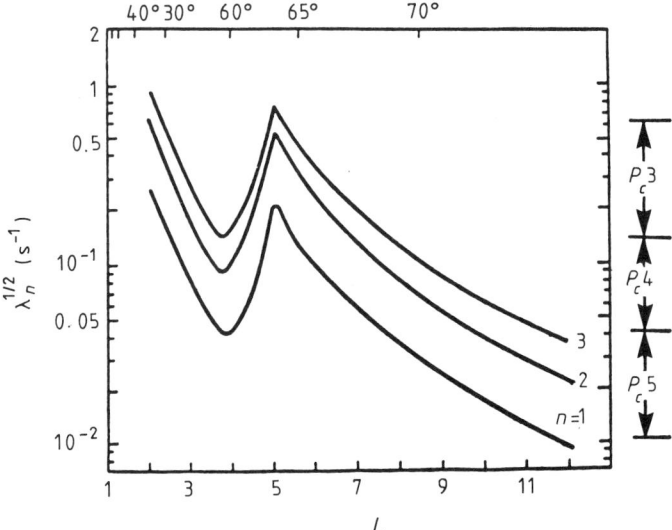

Figure 4.11. The dependence of the Dungey spectrum on the parameter of the magnetic shell L. The upper panel indicates the latitudes where the magnetic shells cross the Earth's surface. Here, with some modifications, the plot from the paper by Yumoto and Saito (1983) is presented with the kind permission of the authors.

numbered by the index ν? Unfortunately, and for the most prosaic reason, the answer should be negative.

Let us evaluate the shift $\delta\omega = \omega_{\nu+1} - \omega_{\nu}$ between the adjacent components of the multiplet. Here, for shorter presentation, we do not indicate the fixed index n. Using the formula (2) in exercise 4.5.4 we find

$$\frac{\delta\omega}{\omega} = \left(\frac{\mu a_2}{2a_0}\right)^{1/2}.$$

If we compare this splitting with the natural broadening $\Delta\omega$ of the harmonic n which arises owing to the absorption of the oscillation energy in the ionosphere, we find out that almost always $\Delta\omega \gg \delta\omega$, probably with the exception of the most extraordinary conditions in the magnetosphere. Consequently, because of the low quality of the magnetospheric resonator the fine lines of the multiplet cannot be distinguished against the background of the general broadening of the level of n.

Exercises

Exercise 4.5.1.

Select a plasma distribution along the field lines, when the Dungey equation (4.50) has exact solutions.

Solution 4.5.1.

Let us assume

$$\rho(z) = \rho(0)(1 - z^2)^{-6} F(z)$$

with $F(0) = 1$. If $F = (1 + az)$, then the Bessel functions will be the solutions of equation (4.50); at $F = (1 - az)^{-4}$ the formulae of the WKB approximation will be the exact solutions. Both these distributions are asymmetric relative to the equatorial plane. In the case of a symmetric distribution of the form $F = (1 + az^2)^{-2}$, the solutions may be found in terms of trigonometrical functions. Taking into account (4.51), the eigenvalues are

$$\Lambda_n = \left(\frac{\pi n}{2z_0}\right)^2 \left(\frac{\sqrt{a}z_0}{\tan^{-1}\sqrt{a}z_0}\right)^2 - a \tag{1}$$

where $a \geq 0$, $n = 1, 2, \ldots$ It should be noted that the spectrum ω_n is not equidistant at $a \neq 0$. At $a = 0$ the formula (1) evidently coincides with (4.53) and describes the equidistant spectrum, but it is an exception. One should realize that in general the Alfvén oscillations of the magnetosphere have a non-equidistant spectrum. It is also evident that the value of non-equidistance decreases rapidly with the growth of the harmonic number.

Exercise 4.5.2.

Obtain the Dungey spectrum (*a*) using the WKB approximation; (*b*) using the Ritz method.

Solution 4.5.2

(*a*) At $\Lambda \to \infty$ equation (4.50) has asymptotic solutions

$$\Psi \propto \exp\left\{\pm i\Lambda^{1/2} \int\limits_0^z [\Gamma(z)]^{1/2} \, dz\right\}. \tag{1}$$

They are universal in the sense that their validity in the asymptotic region does not depend on the real form of the function $\Gamma(z)$. Composing a linear combination of the solutions (1) and imposing boundary conditions (4.51) we find the spectrum in the WKB approximation

$$\Lambda_n^{1/2} = \frac{\pi n}{2z_0} \left\{\frac{1}{z_0} \int\limits_0^{z_0} [\Gamma(z)]^{1/2} \, dz\right\}^{-1} \tag{2}$$

where, according to the condition of applicability of the method, the integer n should be sufficiently high.

However, if $\rho \propto (1 - z^2)^{-6}$, then $\Gamma = 1$, and the spectrum (2) is exact at any values of n (see (4.53)). Let us consider a more realistic distribution $\rho \propto (1 - z^2)^{-4}$. Concerning (2) we have

$$\Lambda_n^{1/2} = \left(\frac{\pi n}{2z_0}\right)\left(1 - \frac{z_0^2}{3}\right)^{-1}.$$

Let the McIllwain parameter for the oscillating shell be $L = 5$, i.e. $z_0 = 0.894$. Then $\Lambda_n^{1/2} = 2.39n$. If we compare this with the result of the numerical estimation of $\Lambda_n^{1/2}$, then at $n \geq 3$ the error of the WKB approximation will constitute not more than 10%. The error increases significantly at $n = 1, 2$.

(b) In accordance with the Ritz method instead of the unknown wavefunction $\Psi(z)$ we should choose a relatively simple function $u(z, \alpha)$, which depends on the parameter α and obeys boundary conditions. The estimate of Λ_1 from the above is obtained by varying (selecting) the parameter α

$$\Lambda_1 = \min\frac{\int\limits_{-z_0}^{z_0} (du/dz)^2\,dz}{\int\limits_{-z_0}^{z_0} u^2\Gamma\,dz}.$$

Usually we get a satisfactory result even when the test function resembles the rigorous eigenfunction in general outline only. Take, for example, the triangle function $u(z) = z_0 - |z|$, $z \leq z_0$. It does not have a selecting parameter at all. At $\Gamma = 1$ simple computations give $\Lambda_1^{1/2} = 1.73/z_0$. This estimate differs from the accurate value of $\Lambda_1^{1/2} = 1.57/z_0$ by only 10%.

Exercise 4.5.3.

Estimate the polarization splitting of the Dungey spectrum (a) using the perturbation method, (b) using the WKB method.

Solution 4.5.3

(a) Let us introduce curvilinear coordinates so that $B = (B_1, 0, 0)$. Formally we proceed to the limit $\varepsilon_{\|} \to \infty$ in (4.39) and prove that $E = (0, E_2, E_3)$ in this case. One-dimensional wave equations for E_2 and E_3 are not connected to each other if

$$\partial E_2/\partial x^3 = \partial E_3/\partial x^2. \tag{1}$$

This condition may be obeyed in the axially symmetric magnetosphere, when $\partial B/\partial\varphi = 0$, $\partial\rho/\partial\varphi = 0$, where φ is the geomagnetic longitude. In this case the general solution for E may be presented in the form of a superposition of azimuthal harmonics $\propto \exp(im\varphi)$, where $m = 0, \pm 1, \pm 2, \ldots$. For the toroidal oscillations ($m = 0$) condition (1) will be yielded if we choose the polarization $E = (0, E_2, 0)$, which brings us to equation (4.47).

Now suppose, that there exist one-dimensional oscillations with the polarization $E = (0, 0, E_3)$. We shall have to proceed to the limit of an infinitely large azimuthal number, since according to (1)

$$E_2 = -\frac{i}{m} \frac{\partial E_3}{\partial x^2} \rightarrow 0.$$

as $m \rightarrow \infty$. Let us abstract from the complicated problems arising here and consider the equation for E_3 formally

$$\frac{d}{dx^1} \left(\frac{g_{22}}{\sqrt{g}} \frac{dE_3}{dx^1} \right) + \left(\frac{\omega}{c_A} \right)^2 \frac{\sqrt{g}}{g_{33}} E_3 = 0. \tag{2}$$

Compare (2) and (4.47). These equations describe the Alfvén waves with two independent polarizations. If the field lines of B had been straight ($g_{\alpha\beta} = \delta_{\alpha\beta}$) then (2) would have differed from (4.47) in nothing but the polarization. However, the curvature of the geomagnetic field lines causes more radical differences between (2) and (4.47).

Let us compare the oscillation spectra of E_2 and E_3 in the dipole magnetosphere. For E_2 we have (4.50) and (4.51). On making some simple transformations we get for E_3

$$(\hat{L} + \hat{U})\bar{\Psi} = \bar{\Lambda}\Gamma\bar{\Psi} \qquad \bar{\Psi}(\pm z_0) = 0 \tag{3}$$

where $\hat{U} = [6z/(1 + 3z^2)]d/dz$. Here the bar above the symbols signifies that the respective values are referred to the oscillations of E_3. Let us estimate the difference $\Delta_n = \bar{\Lambda}_n - \Lambda_n$ using the method of perturbations, and considering $\hat{U}$ as a small correction to the operator $\hat{L}$. The regular procedure of the theory of perturbations leads to the series

$$\Delta_n = U_{nn} + \sum_m{}' \frac{U_{mn}^2}{\Lambda_n - \Lambda_m}$$

where the prime means that $m \neq n$ when summing up is being carried out. Matrix elements are

$$U_{mn} = \int_{-z_0}^{z_0} \Psi_n \hat{U} \Psi_m \, dz.$$

If eigenfunctions and eigenvalues of the unperturbed problem are known, then we use them to find the value of the polarization splitting of the spectrum $\delta\omega_n/\omega_n \simeq \Delta_n/2\Lambda_n$. For example, at $\Gamma = 1$ in the Born approximation we have

$$\Delta_n = -\frac{6}{z_0} \int_0^{z_0} \frac{1 - 3z^2}{(1 + 3z^2)^2} \sin^2\left[\left(\frac{\pi n}{2z_0} \right) (z - z_0) \right] dz.$$

At $z_0 \lesssim 1$ the numerical estimate of the integral gives $\delta\omega_1/\omega_1 \simeq -0.23$.

(*b*) At $n \gg 1$ the oscillation spectrum E_2 is found approximately as in the previous exercise. In order to obtain the oscillation spectrum E_3 in the same approximation we bring equation (3) to the form

$$u'' + [\bar{\Lambda}\Gamma - P]u = 0$$

by substituting $u = \sqrt{v}\,\bar{\Psi}$, where $P = (v''/2v) - (v'/2v)^2$, $v = (1 + 3z^2)^{-1}$ and the prime stands for differentiation by z. The boundary conditions do not alter. Since there are no turning points inside the interval $(-z_0, z_0)$, the spectrum is determined by the formula

$$\int_{-z_0}^{z_0} \left[\bar{\Lambda}_n \Gamma(z) - P(z)\right]^{1/2} dz = \pi n.$$

As a result we find

$$\frac{\delta\omega_n}{\omega_n} = -\frac{6}{\pi^2 n^2} \left[\int_0^{z_0} \sqrt{\Gamma(z)}dz\right]\left[\int_0^{z_0} \frac{(1 - 6z^2)}{(1 + 3z^2)^2}\frac{dz}{\sqrt{\Gamma(z)}}\right].$$

If $\Gamma = 1$, $z_0 \lesssim 1$, then $\delta\omega_n/\omega_n \simeq -0.2/n^2$.

So, the frequency of the E_2-oscillations is somewhat higher than that of the E_3-oscillations, and splitting of the spectrum decreases rapidly with the growth of the harmonic number.

Exercise 4.5.4.

Find the discrete spectrum of Alfvén oscillations by the method of separating the variables.

Solution 4.5.4.

Assume the coefficients of equation (4.42) are independent of s. Then the variables s and x are separated: $\Psi(s, x) = \varphi(s)\phi(x)$. The separation constant may be labelled by $-k_\parallel^2$ and then instead of (4.42) we have

$$\frac{d^2\varphi}{ds^2} + k_\parallel^2\varphi = 0$$

$$\frac{d}{dx}\left(G\frac{d\phi}{dx}\right) + \left[\left(\frac{k_\parallel}{\omega}\right)^2 E - F\right]\phi = 0.$$

Now we integrate the equation for φ and knowing (4.44) we find

$$k_\parallel = \pi n/2s_0 \qquad n = 1, 2 \ldots.$$

The solutions of the equation for ϕ are sought by means of the WKB method. The discrete spectrum, if it exists, is determined by the condition

$$\int_{x_1}^{x_2} \left[\left(\frac{k_\parallel}{\omega} \right)^2 E - F \right]^{1/2} \frac{dx}{\sqrt{G}} = \pi \left(v + \tfrac{1}{2} \right)$$

which is analogous to the condition (4.46).

If the field lines of the external magnetic field are straight, then $F = 1$ and $E = c_A^2(x)$. In this case it is necessary that the profile $c_A^2(x)$ should have a hump for the discrete spectrum to exist. If the plasma is hot, then G should be substituted for $\mu \simeq r_i^2$, and then a cavity in the profile of $c_A^2(x)$ will be a necessary condition for spectrum discretion. It should also be noted that when the field lines of the external magnetic field are curved, nonmonotonic behaviour of the function of $c_A(x)$ is not obligatory for the discrete spectrum to exist.

In the vicinity of the maximum or minimum of $c_A^2(x)$ we use the parabolic approximation

$$c_A^2(x) = a_0 + a_2 x^2/2. \tag{1}$$

Then we consider $\mu a_2 > 0$, with $\mu = -k_{0e}^{-2}$ at $\beta \ll m_e/m_i$ and $\mu \simeq r_i^2$ at $\beta \gg m_e/m_i$. Then

$$\omega_{nv} = \frac{\pi n}{2 s_0} c_A(0) \left[1 + \sqrt{\frac{\mu a_2}{2 a_0}} \left(v + \tfrac{1}{2} \right) \right]. \tag{2}$$

Suppose that (1) is valid at $|x| < x_m$, and at $|x| \geq x_m$ the Alfvén velocity is equal to the constant value of $c_A(x_m)$. In this case the necessary condition for the existence of spectrum (2) has the form $|x_{1,2}| < x_m$ or

$$x_m (\Delta a / \mu a_0)^{1/2} > 2v + 1$$

where $\Delta a = c_A^2(x_m) - c_A^2(0)$.

4.6 Resonance in the caustic shadow

A priori there are no eigen oscillations in the regions of the magnetosphere, where the Dungey spectrum monotonically depends on L, but there exist the so-called Alfvén resonances, i.e. non-eigen oscillations of the magnetic shells. The resonances are excited by magnetosonic waves, penetrating into the magnetosphere from the interplanetary medium, by surface waves, propagating along the magnetopause, and also by sources inside the magnetosphere.

The notion of Alfvén resonances was introduced by Hasegawa and Chen and independently by Southwood in the early 1970s. It is often used for the interpretation of geoelectromagnetic waves. At the same time, this notion cannot be judged to be simple one. It touches conceptual questions in the theory of wave propagation. Not surprisingly, widespread use of the notion of field line resonances sometimes leads to inaccuracies and misunderstandings. Errors are usually caused by treating the resonances as eigen oscillations of the magnetosphere.

This fact stimulates the search for methodical tools capable of clearing up the picture of the field-line resonances. We shall draw an analogy with Försterling's problem of the oblique incidence of an electromagnetic wave on a slab of an isotropic dielectric. Analysis of common and distinctive features of the two problems allows us to eliminate any traces of vagueness in the problem of the field-line resonances. This analogy has a heuristic value since Försterling's problem is rich in physical content and has been studied thoroughly.

4.6.1 Formulation of the MHD problem

Let us consider MHD waves in a plane-stratified medium in order to understand the essence of the Alfvén resonance. We shall use the following initial equations

$$\nabla \times E = i(\omega/c)b$$
$$\nabla \times b = (4\pi/c)j$$
$$j_\perp = -i(\omega/4\pi)\varepsilon_\perp E_\perp \qquad (4.55)$$
$$E_\parallel = 0$$
$$\varepsilon_\perp = (c/c_A)^2$$

which may be deduced from (3.34) using (3.46) at $g \to 0$, $\varepsilon_\parallel \to \infty$, $\varepsilon_\perp \gg 1$. They describe MHD waves of small amplitude in a cold collisionless plasma.

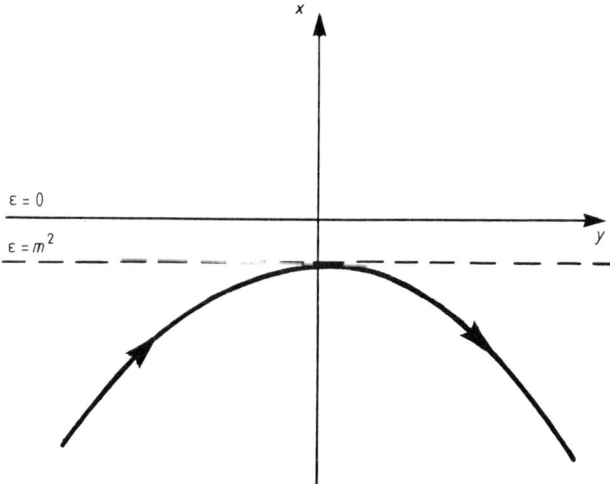

Figure 4.12. The incidence of a magnetosonic wave on an inhomogeneous plasma layer.

In a homogeneous plasma ($c_A = $ constant) equations (4.55) describe normal waves of two types: Alfvén and magnetosonic. It is known from the general theory of wave propagation, that normal waves in an inhomogeneous medium are close to normal waves in a homogeneous medium only in the region of applicability of the ray approximation. However, in this case there are no normal Alfvén waves in an inhomogeneous medium at all. Therefore, we shall do the following in order to retain the analogy with the usual ideas on Alfvén and magnetosonic waves.

We introduce the Cartesian coordinates as shown in figure 4.12. The z axis is perpendicular to the plane of the page. The field lines of the external magnetic field are straight and parallel to the z axis. The plasma is inhomogeneous along x, and $c_A(x)$ is a monotonic function, which smoothly converts into a constant as $x \to -\infty$. This ensures the applicability of the ray approximation at negative values of x, the modulus of which is sufficiently large, at least for magnetosonic waves. (The rays of the Alfvén waves have undesirable behaviour even in a homogeneous medium, see section 4.2).

Let us demonstrate that although there are no eigen Alfvén oscillations, there is a possibility of specific forced oscillation of the field lines, the frequency of which is given by the external force, and the localization along x is determined by the position of the Alfvén resonance.

We specify the incident field in the form of a magnetosonic wave, propagating upwards. Figure 4.12 shows a projection of the ray onto the x, y plane for the case where $c_A(x)$ is a growing function. We choose the following dependence on y and z in the incident wave: $\exp(ik_\parallel z + imy)$. Owing to the homogeneity of the medium along y and z, the total field will have the same dependence. Taking all this into account we rewrite equations (4.55) eliminating j

$$\varepsilon_\perp E_x = \frac{c}{\omega}(k_\parallel b_y - m b_z) \tag{4.56}$$

$$\varepsilon_\perp E_y = -\frac{c}{\omega}\left(k_\parallel b_x + i\frac{\partial b_z}{\partial x}\right) \tag{4.57}$$

$$b_x = -\frac{ck_\parallel}{\omega}E_y \tag{4.58}$$

$$b_y = \frac{ck_\parallel}{\omega}E_x \tag{4.59}$$

$$b_z = -\frac{c}{\omega}\left(mE_x + i\frac{\partial E_y}{\partial x}\right). \tag{4.60}$$

Substituting (4.59) and (4.60) into (4.56), and also (4.58), (4.60) into (4.57), we obtain the equation for the components of the electric vector

$$(\varepsilon - m^2)E_x = im\frac{\partial E_y}{\partial x} \tag{4.61}$$

$$\varepsilon E_y + \frac{\partial^2 E_y}{\partial x^2} = im\frac{\partial E_y}{\partial x} \tag{4.62}$$

where $\varepsilon = (\omega/c)^2 \varepsilon_\perp - k_\|^2$.

Let us introduce the explicit dependence of the field on x

$$E_y = \varphi(x) \exp(ik_\| z + imy). \tag{4.63}$$

From (4.61) and (4.62) we obtain

$$\frac{d}{dx}\left(\frac{\varepsilon}{\varepsilon - m^2}\right)\frac{d\varphi}{dx} + \varepsilon\varphi = 0. \tag{4.64}$$

Equation (4.64) is the basis for the analysis of local resonances of the magnetic field lines. For us, however, it will be more convenient to proceed from the equation for b_z, and not for E_y. The equation for b_z is simpler than for E_y, but even this is not the main point. Excluding all the components, except b_z, from (4.56)–(4.60) we obtain an equation that coincides with the analogous equation from Försterling's problem.

We introduce the explicit dependence of b_z on x

$$b_z = \psi(x) \exp(ik_\| z + imy). \tag{4.65}$$

The equation for ψ has the form

$$\varepsilon\frac{d}{dx}\left(\frac{1}{\varepsilon}\frac{d\psi}{dx}\right) + (\varepsilon - m^2)\psi = 0. \tag{4.66}$$

Transverse components of the magnetic field are expressed in terms of the longitudinal component according to the formulae

$$b_x = i\frac{k_\|}{\varepsilon}\frac{\partial b_z}{\partial x} \qquad b_y = -\frac{k_\| m}{\varepsilon}b_z. \tag{4.67}$$

4.6.2 Analogy with the Försterling problem

Propagation of electromagnetic waves in an isotropic dielectric is described by the equations

$$\nabla \times E = i(\omega/c)H \qquad \nabla \times H = -i(\omega/c)\bar\varepsilon E. \tag{4.68}$$

Here $\bar\varepsilon$ is the dielectric permeability, the bar over symbol is used to distinguish between $\bar\varepsilon$ in equation (4.68) and ε in equations (4.61)–(4.67). In contrast to the previous problem here we used H to mark the magnetic field of the wave. In both cases the magnetic permeability of the medium is assumed to be equal to unity, and thus it cannot lead to misunderstandings.

Let the dielectric permeability $\bar\varepsilon$ depend only on x, and the wave field only on x and y. We shall consider H-waves, i.e. waves with the electric vector lying in the plane of incidence (this plane is evidently formed by the x, y plane). The polarization of the H-wave has the form: $E = (E_x, E_y, 0)$, $H = (0, 0, H_z)$.

The wave equation for $H_z = \Psi(x)\exp(i\kappa y)$ formally coincides with equation (4.66).

Suppose that $\bar{\varepsilon}$ decreases with increasing x. The incident wave reflects from the point x_1, in which $\bar{\varepsilon}(x_1) = c^2\kappa^2/\omega^2$. Below this point $(x < x_1)$ the field oscillates owing to the interference of the incident and reflected waves. Above the point x_1 the field damps exponentially. The Försterling phenomenon appears in cases where $\bar{\varepsilon}$ passes through zero with increasing x.

At the point $\bar{\varepsilon} = 0$ against the background of general damping of the field there is a specific feature analogous to the 'field-line resonances' in magnetohydrodynamics.

If we now turn back to the magnetohydrodynamic problem, then we have to reconstruct almost literally the solution of Försterling's problem, described in detail in the literature. Instead we shall make a few remarks, and then avail ourselves of the mentioned analogy, in order to find the dependence of resonant phenomena on the parameters of the medium and the incident fields without solving the respective magnetohydromagnetic problem.

Both problems have not only formal, but also physical generality. In both cases there is spatial structure of the field with a turning point separating the region of accessibility and inaccessibility for the real rays, and the resonance in the forbidden region. In both cases there is an energy flux behind the turning point directed towards the resonant point and finite dissipation in the resonance at infinitely small nonconservatism. There are also some other general features.

Now, a few words on the distinctions between these two problems. There is no dependence of the field on z in the Försterling problem, and the magnetic field has only one nonzero component H_z. The dependence on z cannot be removed from the MHD problem without losing the effect, and in general the magnetic field has three nonzero components.

4.6.3 Resonance structure

In the region $\varepsilon(x) > m^2$ the field is described geometro-optically and has the form of the superposition of the incident and reflected magnetosonic waves. The reflecting surface $\varepsilon(x) = m^2$ is a caustic. In the caustic shadow $(\varepsilon(x) < m^2)$ the field usually decreases exponentially with distance from the caustic. In this case a singularity arises at $\varepsilon(x) = 0$, against the background of an overall decrease. This singularity is the Alfvén resonance.

In the vicinity of the resonance, let $\varepsilon = -ax$, $a > 0$. The coordinate of the reflection point is $x_0 = -m^2/a$. Then at $x \to +0$

$$b_x = -ib_z(k_\parallel m^2/a)\ln(mx) \qquad b_y = b_z(k_\parallel m/ax)$$

$$E_x = b_z(\omega m/acx) \qquad E_y = ib_z(\omega m^2/ac)\ln(mx)$$

(4.69)

where b_z is taken at $x = 0$. If $x \to -0$, then in (4.69) a substitution $\ln x \to \ln|x| - i\pi$ should be made in accordance with the limiting absorption principle.

Dissipative properties of the medium may be taken into account by introducing an effective parameter Δ : $x \rightarrow x - i\Delta$. Then the field takes a finite value in the resonance ($x = 0$). The parameter Δ signifies the distance over which $|b_y|^2$ reduces by half with distance from the resonance.

Absorption of the energy of the incident wave takes place in the resonance at any arbitrary small nonconservatism of the medium. This phenomenon is well studied within the framework of Försterling's problem, and so we only have to make some redesignations in order to obtain immediately a time-average of the flow of energy to the resonance at $x < 0$

$$S_x = \frac{\omega m^2}{8a} |\psi(0)|^2.$$

All the arriving energy is absorbed in the resonance since $S_x = 0$ at $x > 0$. The rate of energy dissipation depends on the parameters of the incident wave and on the conservative parameters of the medium only (the Gildenburg effect).

Owing to unavoidable absorption in the resonance, the amplitude of the reflected magnetosonic wave is always smaller than the amplitude of the incident one. Therefore, the field below the reflection point ($x < x_o$) does not have the structure of a standing wave. The relative intensity of running wave vitally depends on the angle of incidence.

The next argument was taken almost entirely from Ginzburg's monograph (1971), taking advantage of the analogy mentioned above.

Let ω and $k_{\parallel}$ be fixed, while m varies[5]. At $m = 0$ the singularity at the point $\varepsilon = 0$ disappears. This is obvious in the Försterling problem, since $m = 0$ corresponds to the normal incidence on the slab. In the MHD problem, $m = 0$ corresponds to the 'meridianal' propagation when there is no coupling between the Alfvén and magnetosonic waves. (We recall that gyrotropy of the medium is not taken into account here, see for comparison equations (3.50) and (3.51).) As m increases the field singularity also disappears sooner or later. The reasons are the increase in the distance between the reflection point $\varepsilon = m^2$ and the resonance point $\varepsilon = 0$, by exponential weakening of the field beyond the reflection point and the presence of absorption—even if extremely slight—in a real system. It is thus natural to expect that the resonance would be manifested most vividly at a certain intermediate value of m, not very large, but also not very small.

To seek the m dependence of the Alfvén resonance we do not have to solve the magnetohydrodynamic problem. An analogous solution is found within the framework of Försterling's problem, and only necessary redesignations have to be made.

The Hasegawa–Chen–Southwood theory is one of the most out-standing achievements of magnetospheric physics. It has a direct and important

[5] If the problem is reformulated in terms of magnetospheric physics, then m becomes an azimuthal number.

application to experiment, namely for the ground and satellite observations of Pc3-5 pulsations or to hydromagnetic diagnostics of the magnetosphere.

Exercise

Exercise 4.6.1

Estimate the photon mass by the observational data of the Alfvén oscillations of the magnetosphere.

Solution 4.6.1.

Let us introduce the photon mass m_γ into the MHD equations like we introduced it into the Maxwell equations when solving exercise 3.1.3. Then, instead of $\omega = c_A k_\parallel$, we shall have the following dispersion law for the Alfvén waves

$$\omega = c_A (k_\parallel^2 + \mu^2)^{1/2}$$

where $\mu = m_\gamma c / \hbar$ and $\hbar$ is the Planck constant. So, at $m_\gamma \neq 0$ the spectrum of the Alfvén waves has a low-frequency cut-off $\omega_{min} = c_A \mu$. Hence it follows that if we observe Alfvén oscillations with frequency ω, then $m_\gamma < \hbar \omega / c_A c$. According to the observations of the Alfvén resonances of the magnetosphere $m_\gamma < 2.4 \times 10^{-48} g$. Information on the galactic and intergalactic magnetic field imposes still stricter restrictions on the photon mass: $m_\gamma < 3 \times 10^{-60}$ g. For the present moment it is unclear how we could improve this estimate.

Bibliography

The term 'magnetosphere' was proposed by Gold (1959). Clear ideas on the magnetosphere of the Earth were suggested by Chapman and Ferraro in 1931. The authors arrived at a conclusion on the existence of the magnetic cavity in the solar plasma flow based on an analysis of magnetic storms that start suddenly (see, for example, Chapman and Bartels 1940). The sudden commencements (SCs) of magnetic storms themselves as a result of the interaction between the magnetosphere and collisionless shock waves, were interpreted by Gold also by analysing ground observations. Ground observations still play an essential role in the study of the magnetosphere. However, since the early 1960s satellite observations have been the major source of new information. The modern view on the structure and dynamics of the magnetosphere is presented by Akasofu and Chapman (1972) and Nishida (1978).

For section 4.1

Extensive literature is dedicated to plasma waves. The book by Stix (1962) has not lost its timeliness and is useful for studying the waves in a cold plasma.

The monograph by Akhiezer *et al* (1974) contains a systematic presentation of the material and a good bibliography. The monograph by Ginzburg (1971) is the most helpful for geophysical applications. The review by Shafranov (1963) was used when composing this section.

The concept of plasma was introduced by Langmuir in the early 1920s since when longitudinal oscillations of plasma at frequency $\omega \simeq \omega_{oe}$ have been called Langmuir oscillations. The deduction of the formulae for the refractive indices of high-frequency waves in a plasma is connected with names of Appleton, Hartree and Eckersley. Equation (4.15), which Storey (1953) applied successfully when investigating whistlers, was formulated by Eckersley. The fundamental dispersion laws (4.12) and (4.13) in the low-frequency limit were discovered by Alfvén.

Booker and Dyce (1965) proposed and studied in detail a rather useful approximation (4.6). The interpretation formula (4.16) was constructed by Guglielmi (1982).

Exercise 4.1.2 was constructed following an article by Gintzburg (1963). The information on the concentration of He^+ ions in the magnetosphere was borrowed from the article by Brinton *et al* (1968).

For section 4.2

The ray-trace equation (4.19) may be deduced heuristically, presenting the wave field in the form of locally plane waves (Landau and Lifshitz 1984, Synge 1960), but they may also be deduced as a zero (relative to the wavelength) approximation within the framework of the theory of asymptotic solutions (Guillemin and Sternberg 1977, Kravtzov and Orlov 1980). In this case the rays are considered as some framework for the 'wave drapery' [6].

The basis of the geometric MHD as short-wave asymptotics of linearized MHD equations were laid by Weinberg (1962). However, the powerful technique of the asymptotic theory (the method of reference equations, Maslov's canonical operator, etc) are seldom used when solving MHD problems. For the sake of justice it should be added that the real demand for this fairly complicated theory has only appeared recently, owing to the improvement of experimental technique. Concerted observations of geomagnetic pulsations by means of radars, satellites and using dense networks of magnetometers have revealed fine details of the spatio-temporal structure of the wave field, which cannot be understood in the framework of the ray-tracing concept.

The ray-tracing theory simplifies the propagation problem (Booker 1962, Rao and Booker 1963). However, the main aim of proceeding to the ray-tracing approximation will not be achieved if we fail to integrate the corresponding equations. In most cases of interest this can only be done by means of a computer. Of the many articles dedicated to the numerical simulation of MHD

---

[6] Here we have paraphrased the well known Keller's aphorism on the rays, which play the role of a skeleton for the 'wave flesh'.

ray tracing in the magnetosphere, we point out the articles by Sugiura (1965) and Kitamura and Jacobs (1968).

In the literature on the analytical research of simplified models, the uniqueness of geometric MHD is sometimes exaggerated. For example, the equations of paraxial MHD ray tracing in the refractive waveguide are deduced from first principles instead of reformulating a corresponding problem in geometric optics. On the other hand, the uniqueness of MHD waves is often underestimated. For example, in geometro-optic interpretations of observations the pathological character of caustics of the Alfvén waves is ignored. Critical remarks of this kind may be found in the review by Guglielmi (1985b). Obviously, they are appropriate, since the visual ray-tracing concept dominates explicitly or implicitly when posing and analysing observations of the MHD waves in the near-Earth plasma.

We do our best to be impartial in our historical commentary. Nevertheless, there may be some bias as it inevitably depends on our knowledge and (it cannot be excluded) on unintentional prejudice. As far as we know, Walen was the first to discover the change of amplitude of the Alfvén wave along the ray trace according to the law $b \propto \rho^{1/4}$ in the mid-1940s (see exercise 4.2.2).

For section 4.3

Guglielmi (1970a) considered the magnetosonic channel existing under the arch of the plasmasphere. In our opinion the most interesting feature of this waveguide is the self-excitation of waves at high harmonics of the ion gyrofrequency (see bibliography to section 6.2).

Buldyrev *et al* (1973) also discussed magnetosonic waves in the transverse waveguide, but from a different standpoint. In the opinion of these authors the properties of Pc3 pulsations are explained by the existence of the waveguide. In particular, if the waveguide is considered to be a toroidal resonator, then, according to Buldyrev *et al* (1973), we may interpret the dominating period in the spectrum of Pc3. A contrary opinion stating that Pc3 is caused by pulsations of extramagnetospheric origin is presented in section 6.5.

For section 4.4

Pearl pulsations were discovered by Harang (1936) and Sucksdorff (1936). Benioff (1960) and Troitskaya (1961, 1964) reviewed the interest in them in the 1960s (see also Hayashi *et al* 1962, Jacobs and Jolley 1962, Heacock 1963, Heacock and Hessler 1965). Lokken *et al* (1963) and Yanagihara (1963) detected a 180° phase shift in the Pc1 envelope at the conjugate points. The systematic study of Pc1 in the Sogra–Kerguelen conjugate points was undertaken by Gendrin and Troitskaya (1965). The observations at the conjugate points keep on supplying valuable information on the Pc1 propagation conditions (e.g. Ishizu *et al* 1981). Wentworth (1964) detected the Pc1 activization after a magnetic storm; Heacock and Hessler (1965) detected the appearance of Pc1 1–2 min after SCs. The works by Campbell and Stiltner (1965), Matveeva and Troitskaya (1965), Tepley (1965), Kenney and Knaflich (1967), Fraser (1968), Fukunishi *et al* (1981), Kawamura *et al* (1981), Kuwashima *et al* (1981) are dedicated to the problems of general morphology of pearl pulsations. Heacock (1970) compiled an atlas of pearl pulsations, cursory acquaintance with which provides mental pabulum. Matveeva *et al* (1968), Frazer–Smith (1970), Mursula *et al* (1991) analysed long series of observations and gave a reliable proof of strong negative correlation between the annual Pc1 activity and annual sunspot number. The search for the end of a longitudinal waveguide was undertaken by Heacock (1971) using the data on the amplitude and the polarization of Pc1 registered in Alaska and in Finland. In particular, in the course of this research a record amplitude of Pc1 on the Earth's surface was marked: on 8 February 1969 in Nurmijärvi ($L = 3.3$) oscillations with an amplitude of 1.6 nT were registered. With the help of the observations at the network of stations in Canada, Hayashi *et al* (1981) discovered that the radius of the end of the longitudinal waveguide does not exceed 300 km. Kikuchi and Taylor (1972) used the information on the plasma in the magnetosphere, obtained from the *OGO-3* satellite, and arrived at the conclusion that the presence of irregularities in the structure of the magnetopause favours the longitudinal canalization of Pc1. Simultaneous observations of Pc1 in the magnetosphere (*GEOS-2* satellite) and on the ground testify to the considerable influence of He^+ ions on the condition of Pc1 penetration from the magnetosphere to the Earth (Perraut *et al* 1984, Ludlow *et al* 1989). Mursula *et al* (1994) have observed Pc1 above the ionosphere on board the *Freja* satellite. The authors inferred that observations give evidence for a plasmapause-connected source region of Pc1.

Stratification and fibring of the space plasma are specially noted by Alfvén and Arrhenius (1975). The influence of the layers and fibres of the magnetospheric plasma on whistler propagation was revealed by Smith *et al* (1960), Smith (1961) and Carpenter (1966). The role of longitudinal waveguides in the Pc1 propagation was mentioned by Tverskoi (1967) and independently by Kitamura and Jacobs (1968). Attempts to construct the theory of longitudinal waveguides in the magnetosphere were made by Dmitrienko and Mazur (1985) and Guglielmi (1985b).

In addition to material presented in section 4.4 we point out that the geomagnetic tail is a specific waveguide (Patel 1968b, Potapov 1973). Along the tail, as if along a tube, the waves propagate towards and away from the Earth, creating a complex mosaic of electromagnetic perturbations in the polar caps and along the whole length of the tail.

For section 4.5

Dungey (1954) was the first to present the equation of oscillations of the dipole magnetosphere in explicit relation to the problem of interpretation of geomagnetic pulsations (see also Dungey 1958, Dungey and Southwood 1970, 1975). Under the direct influence of the Dungey's ideas there followed a 20 year period of experimental and theoretical studies of MHD oscillations of the magnetosphere, which was crowned with the creation of the theory of resonant oscillations of magnetic shells in 1974 (see bibliography to section 4.6).

The review by Lanzerotti and Southwood (1979) contains important information on the search for proof of the existence of standing Alfvén waves in the magnetosphere. Sugiura (1961) and Sugiura and Wilson (1964) discovered the expected correlation of long-period pulsations in conjugate points on the Earth's surface. Cummings *et al* (1969) observed transverse quasi-sinusoidal oscillations at the geosynchronous orbit. The observation was compared with the calculation of the Dungey spectrum at $m = 0$ and $m \to \infty$. The result of the calculation agrees with the observation and testifies to the polarization splitting of the spectrum (see exercise 4.5.3). Later the observations at the geostationary orbit and in the conjugate points were supplemented with the study of the spatio-temporal distribution of pulsations with the application of fairly dense networks and chains of magnetometers (Samson *et al* 1971). This study proved the idea of the existence of the standing Alfvén waves in the magnetosphere and stimulated the development of the theory (Lanzerotti and Fukunishi 1974).

Obayashi (1958), Watanabe (1959, 1961), MacDonald (1961), Westphal and Jacobs (1962), Siebert (1964), Carovillano and McClay (1965), Carovillano *et al* (1966), Kovach and Ben-Menachen (1966), Radoski and Carovillano (1966), Carovillano and Radoski (1967), Radoski (1967), Radoski and McClay (1967), Hruska (1968), Cummings *et al* (1969), Infeld (1969), Kahalas (1969), Guglielmi (1970b,c), Troitskaya and Guglielmi (1970), Orr and Matthew (1971), Radoski (1971), Guglielmi and Troitskaya (1973), Orr (1973), Krylov and Fedorov (1976), Singer *et al* (1981), Yumoto and Saito (1983) investigated the spectrum of toroidal and poloidal oscillations of the magnetosphere numerically and analytically. Special attention was paid to the dependence of the period of oscillations of magnetic shells on the latitude, since this dependence can be easily verified by experiment.

Guglielmi and Polyakov (1983) considered the problem of a discrete spectrum of toroidal oscillations (equation (4.42)). They borrowed the idea on the transverse rays and longitudinal modes from Burridge and Weinberg (1977). On the tensor operations in the curvilinear coordinates see, for example, Schrödinger (1950), Dirac (1975), or Landau and Lifshitz (1988b); on the WKB approximation and on the theory of perturbations see Fermi (1960), or Landau and Lifshitz (1963).

We have set aside the interesting problem of oscillations of the geomagnetic tail. The works by Patel (1968a), Siscoe (1969), McKenzie (1970, 1971), Ershkovich and Nusinov (1972), Ershkovich *et al* (1972) are dedicated to it.

For section 4.6

Paraphrasing von Neumann (1961) we may affirm that the criterion of success in geophysics is measured by the number of natural phenomena, previously thought to be heterogeneous and complicated, that can be understood by means of a simple and elegant theory. The theory of magnetospheric resonances, created by Chen and Hasegawa (1974), and Southwood (1974a) undoubtedly yields this criterion. In the reviews by Hasegawa and Chen (1974), Southwood (1974b), Lanzerotti and Southwood (1979), Rostoker (1979), Hughes (1983) and in the monograph by Nishida (1978) the reader will find more useful information on the theory in addition to this text.

Chen and Hasegawa (1974), and Southwood (1974a) were the first to remove the contradiction between the continuity of the Dungey spectra and the idea of the eigen oscillations of the magnetic shells. This idea was widely used in the literature, although it is evident from the physical point of view, that it is not correct to speak of eigen oscillations of the continuous medium if the period of oscillations changes from one point to another.

The theory predicts a fairly definite character of oscillation polarization on the ground and in space, and it was used by Lanzerotti *et al* (1981) in experimental research of polarization distribution of Pc3 at low latitudes. Previously Hasegawa and Lanzerotti (1978) demonstrated a 90° turn of the main axis of the polarization ellipse between the equator of the resonant magnetic shell and the Earth's surface. Hughes *et al* (1977, 1978) investigated the structure of the magnetospheric resonances by means of observation of Pc3-4 at several geostationary satellites simultaneously. Hillebrand *et al* (1982) studied the resonance structure of the 'giant pulsations' (a variety of Pc4) using the ATS-6 satellite and a ground-based magnetometer station network. The observations at GEOS-2 satellite definitely testify to Pc5 pulsations being resonant oscillations of the magnetic shells (Junginger *et al* 1984). The resonant origin of Pc5 was proved independently by the data on fluctuations of the electric field in the ionosphere registered by the STARE radar system (Walker *et al* 1979, Walker and Greenwald 1981). Samson *et al* (1992) observed the intensification of field-aligned currents in the region of resonant oscillations using a radar system and the Canadian magnetometer network. Takahashi and Anderson (1992) observed the resonant oscillations outside as well as inside the plasmasphere (AMPTE CCE satellite). The authors obtained the plasma density profile with a specific 'knee' on the plasmapause using the observed L-dependence of oscillation frequency. We shall conclude this list, which is far from being complete, with the works by Baumjohann and Glaßmeier (1984), and Lanzerotti and Medford (1984), where certain difficulties that arise when interpreting Pi2 are discussed.

The theory was developed along several. First of all dissipative, screening and gyrotropic properties of the ionosphere were taken into account (for example, see the review by Southwood and Hughes (1983), mentioned above). Then Hasegawa *et al* (1983) make a generalization, concerning the spectral structure of the source. They showed, that under the influence of the broadband source the magnetic shells resonate at frequencies, which changed continuously with the change of the L parameter in some interval ΔL, which is larger the broader the band of the source. Southwood and Kivelson (1990) used a so-called 'box-model' for modelling the oscillatory reaction of the magnetosphere to a sudden perturbation of the magnetopause. The effect of longitudinal inhomogeneity of the magnetospheric resonator was investigated by Southwood and Kivelson (1986), Chen and Cowley (1989), and Mond *et al* (1990). Influence of azimuthal inhomogeneity was investigated by Schulze-Berge *et al* (1992). A series of research works by Krylov and Fedorov (1976), Krylov *et al* (1979, 1981), Lifshitz (1980), are dedicated to the mathematical aspects of the theory of magnetospheric resonances. In particular, using the qualitative methods of investigating spectral problems, rigorous local formulae for the frequencies of the continuous spectrum with regard to the inhomogeneity of the plasma and curvature of the field lines were obtained. It follows from these formulae that two series of frequencies $\omega_n^{\pm}$ are related to each field line, where $\omega_n^+ (\omega_n^-)$ corresponds to the polarization of oscillations along the direction of the first (second) principal curvature of the surface, which is orthogonal to the field lines. Leonovich and Mazur (1991b) analytically investigated the distribution of the magnetic field over the Earth's surface when the standing Alfvén waves are excited in the magnetosphere. The cases of excitation by monochromatic, impulsive and stochastic sources are considered separately.

Jekulin in the early 1930s, Försterling in the late 1940s, and Denisov in the mid-1950s investigated the phenomenon of resonance of the wave field under oblique incidence of the H-wave on the layer of inhomogeneous isotropic plasma. Gershman *et al* (1957) and Ginzburg (1971) presented works on this aspect systematically. Gildenburg (1964) detected a universal effect of wave absorption in the resonance within the framework of the Försterling problem. Guglielmi and Potapov noticed the analogy between the Hasegawa–Chen–Southwood problem and the Fösterling problem (see references in the review by Guglielmi 1985b). Kivelson and Southwood (1985) also emphasize the similarity of these two problems.

Gintzburg (1964, 1974) deduced the dispersion law for the Alfvén waves with the finite photon mass taken into account. He was also the first to improve the estimate of m_γ, obtained by Schrödinger (1943), based on the satellite information on the geomagnetic field. Lanzerotti (1974) obtained the restriction $m_\gamma < 2.4 \times 10^{-48}$ g, which we reconstructed in the exercise 4.6.1 (see also Patel 1965, Whilliams and Park 1971, Chibisov 1976). Guglielmi and Pokhotelov (1993b) noted the latent contradiction in the paper by Lanzerotti (1974) and suggested their own variant of the estimation of m_γ using the spectrum of magnetospheric oscillations (see also Guglielmi and Pokhotelov 1993a).

Chapter 5

Modulation

In the preceding chapters we frequently used the notion of the monochromatic wave. It was an idealization, convenient and useful in the mathematical theory. Monochromatic waves are absent in the world around us. However, there are modulated waves. We shall not try to find an exhaustive definition but will imagine a modulated wave as 'nearly' monochromatic with parameters which (amplitude, phase, frequency) vary in space and in time smoothly and slowly. This will be enough in many cases, although it is necessary to understand that quasi-monochromatic waves do not exhaust the class of modulated waves. For example, we may speak about the modulation single impulse if its width and height vary smoothly during propagation.

The teaching of modulation is boundless and poses an inexhaustible source of problems. Here we shall present only some fragments. They all refer to the theory of wave propagation in linear dispersive media. There exist specific modulation effects in nonlinear dispersive media. One of the most essential effects of this kind lies in the fact that shocks and solitary waves appear as a result of quasi-monochromatic wave self-modulation. We shall consider this in Chapter 7.

5.1 Wave packets

5.1.1 Group velocity

Let us consider a wave packet $\psi(x, t)$ in a homogeneous medium, occupying a limited volume of space at each given moment of time. We present it in the form of superposition of plane waves

$$\psi(x, t) = \int \psi_k e^{i k \cdot x - i \omega_k t} dk. \tag{5.1}$$

Here $\psi(x, t)$ is one of the wave field components and $\omega_k = \omega(k)$ is the dispersion relation for waves of the type under consideration. If at the initial moment of time the field is described by the function

$$\psi(x, 0) = e^{ik_0 \cdot x} \varphi(x) \tag{5.2}$$

then in (5.1) the spectral density has the form

$$\psi_k = (2\pi)^{-3} \int \varphi(x) e^{-i(k-k_0) \cdot x} \, dx. \tag{5.3}$$

Suppose, the function $\varphi(x)$ decreases to zero rapidly but smoothly outside the region into which a large number of wavelengths go, and inside this region it is almost constant[1]. This means that (5.2) represents a group of plane waves with close wave vectors, i.e. ψ_k has a sharp maximum at $k = k_0$ and decreases rapidly with growth of the difference $|k - k_0|$. Let us apply the expansion $\omega(k)$ into a series in the vicinity of k_0 with an accuracy of up to the first two terms. Then

$$\omega(k) = \omega(k_0) + \left.\frac{\partial \omega}{\partial k}\right|_{k=k_0} (k - k_0). \tag{5.4}$$

Let us now substitute (5.3) and (5.4) into (5.1) and after integrating we get

$$\psi(x, t) = e^{ik_0 \cdot x - i\omega(k_0)t} \varphi\left(x - \left.\frac{\partial \omega}{\partial k}\right|_{k=k_0} t\right). \tag{5.5}$$

It is evident that the packet travels as a whole in time, without changing its form, at the velocity

$$v_g = \frac{\partial \omega}{\partial k}. \tag{5.6}$$

In the previous chapter we considered the group velocity v_g as a vector, tangent to the ray trace. Now we see that v_g may also be understood as the propagation velocity of the modulation.

The third interpretation, worth regarding, is that v_g is the velocity of the wave energy transfer. This seems almost obvious, as according to (5.5) the group velocity equals the transition velocity of the amplitude maximum of the wave packet. However, in reality this interpretation of the group velocity, as well as the other two, makes sense only when a series of conditions are yielded. We point them out for the case of electromagnetic waves in a plasma.

The conservation of energy law has the form

$$\partial W / \partial t + \nabla \cdot S = 0 \tag{5.7}$$

where

$$W = \frac{\partial \omega^2 \varepsilon_{\alpha\beta}}{\partial \omega^2} \frac{E_\alpha^* E_\beta}{8\pi} \tag{5.8}$$

---

[1] If $\varphi(x) = $ constant everywhere, i.e. there is one plane wave, then $\psi_k \propto \delta(k - k_0)$.

is the mean value averaged over time energy density of the electromagnetic field, and

$$S = v_g W \tag{5.9}$$

is the time-average density of the energy flux. Equation (5.7) may be derived from the Poynting theorem

$$(1/8\pi)\partial(E^2 + b^2)/\partial t + E \cdot j(E) + (c/4\pi)\nabla \cdot (E \times b) = 0 \tag{5.10}$$

which in its turn follows immediately from the Maxwell equations (3.32). The relation (5.10) is exact, while the equation of energy transfer (5.7) is approximate, i.e. is meaningless if the respective conditions of applicability are not specified.

The necessity of a linear relation between j and E is quite obvious. Then it is supposed, that the connection between j and E is local or in other words $\varepsilon_{\alpha\beta}(\omega)$ does not depend on the wave vector k, i.e. there is no spatial dispersion. In addition, the tensor $\varepsilon_{\alpha\beta}$ is considered to be Hermitian ($\varepsilon_{\alpha\beta} = \varepsilon_{\beta\alpha}^*$), i.e. we neglect dissipative processes. Finally, two more conditions correspond to the structure of the wave field.

The first one lies in the fact that the wave packet having the form

$$E(x, t) = E_0(x, t)\exp(ik_0 \cdot x - i\omega_{k_0} t) \tag{5.11}$$

consists of normal modes of the same type, i.e. of waves with an identical dispersion law $\omega_k = \omega(k)$ and identical polarization structure.

The second condition is much more rigid: the wave packet (5.11) must be quasi-monochromatic. This means that the complex amplitude $E_0(x, t)$ varies in time and space slowly and smoothly as compared with $\exp(ik_0 x - i\omega_{k_0} t)$.

We have to recall that the complex form of representation (5.11) is convenient and admissible, if the values are subject only to linear operations. At the end of calculations, the real part of the corresponding value should be singled out. While deriving the transport equation from the Poynting theorem, averaging of square combinations of the field components over the period $2\pi/\omega$ of high-frequency field oscillations is carried out. Instead of singling out the real part every time, the averaging rule, presented in section 3.1, was employed.

5.1.2   Dispersive broadening of the wave packet

Let us study the one-dimensional wave packet. This means that the wave vectors of all the spectral components are parallel to each other. Let us take into account one more expansion term in (5.4)

$$\omega(k) = \omega_0 + v_g \Delta k + (v_g'/2)(\Delta k)^2. \tag{5.12}$$

Here $\omega_0 = \omega(k_0)$, $v_g = \partial\omega/\partial k|_{k=k_0}$, $v_g' = \partial^2\omega/\partial k^2|_{k=k_0}$ and $\Delta k = k - k_0$. It should be noted, that ω_0 corresponds to the carrier frequency of the wave packet. The frequency ω_0 should not be confused with the Langmuir frequency ω_{0e} which is sometimes in other contexts also labelled ω_0.

Now suppose that the envelope of the signal amplitude (5.2) at $t = 0$ has the form of a Gaussian function:

$$\varphi(x) = \varphi_0 \exp[-(x/x_0)^2].$$

Then the spectral density (5.3) also has the form of the Gaussian curve

$$\psi_k = \psi_0 \exp\left[-\tfrac{1}{4}x_0^2 (k - k_0)^2\right]$$

where $\psi_0 = \varphi_0 x_0/2\sqrt{\pi}$. Substituting this into (5.1) and using the expansion (5.12) we obtain

$$\psi(x, t) = \psi_0 \exp(ik_0 x - i\omega_0 t)\left[\frac{\pi}{\alpha + i\beta t}\right]^{1/2} \exp\left[-\frac{(x - v_g t)^2}{4(\alpha + i\beta t)}\right]. \quad (5.13)$$

Here

$$\alpha = x_0^2/4 \qquad \beta = v_g'/2.$$

The envelope of the packet has the form

$$|\psi(x, t)| = \psi_0 \left(\frac{\pi^2}{\alpha^2 + \beta^2 t^2}\right)^{1/4} \exp\left[-\frac{\alpha(x - v_g t)^2}{4(\alpha^2 + \beta^2 t^2)}\right]. \quad (5.14)$$

The maximum of the packet amplitude travels at velocity v_g and decreases in time as $\propto (\alpha^2 + \beta^2 t^2)^{-1/4}$. The width of the packet increases in time as $\propto (\alpha^2 + \beta^2 t^2)^{1/2}$. The total wave energy is conserved, i.e.

$$\int_{-\infty}^{\infty} |\psi|^2 \, dx = \text{constant}. \quad (5.15)$$

The conservation law (5.15) as well as (5.7) is a consequence of the conservativity of the medium. The assumption of the absence of dissipation in the medium is formally expressed in the fact that the frequency ω is real when k is real. In terms of the dielectric permeability tensor this means that the tensor $\varepsilon_{\alpha\beta}$ is Hermitian.

The maximum of the wave-packet amplitude described by (5.13) propagates at the velocity group velocity v_g. However, at this point the similarity with (5.5) comes to an end. In contrast to (5.5) the wave packet (5.13) spreads out in time.

The deformation of the modulation is associated with the frequency dependence of the group velocity and by no means with the special choice of the Gaussian form of modulation at the initial moment of time. We made such a choice in order to demonstrate the effect of broadening of the packet by a fairly simple example.

The waves for which $\partial^2\omega/\partial k^2 \neq 0$ are called dispersive[2]. Sometimes one may speak of dispersive media in this connection. In the case of electromagnetic waves the medium is dispersive if the dielectric permeability tensor $\varepsilon_{\alpha\beta}$ depends on frequency.

We have encountered dispersive and nondispersive waves more than once. For example, body elastic waves in solid matter do not exhibit dispersion (Chapter 1). Dispersion is also absent in the case of Rayleigh waves, travelling along the surface of a homogeneous elastic semispace. On the other hand gravitational and capillary waves over a water surface belong to the class of dispersive waves (Chapter 2).

Electromagnetic waves in a vacuum do not possess dispersion. For them $\varepsilon = 1$, the dispersion law has the form $\omega = ck$ and the group velocity v_g equals the velocity of light. At the same time the vacuum gap between the parallel conducting plates represents the dispersive 'medium'. (Such a model was used in Chapter 3 for the modelling of local properties of the Earth–ionosphere waveguide.)

There is no contradiction or paradox at this point. Certainly dielectric permeability still equals unity in the vacuum gap, but the dispersion law cannot be presented in the form $\omega = ck$, since, due to the influence of the conducting plates, plane waves are no longer normal waves. The normal waves (modes) of our waveguide are TE and TM modes and the so-called cable wave or TEM mode.

The dispersion law for TE and TM modes has the form

$$\omega = c(\kappa^2 + k^2)^{1/2} \tag{5.16}$$

where $k = |\boldsymbol{k}|$ and $\boldsymbol{k}$ is a two-dimensional wave vector, parallel to the plates; $\kappa = \pi s/h$, $s = 1, 2, \ldots$. The plates, the distance between which is denoted h, is supposed to be ideally conducting. Consequently the group velocity is

$$v_g = \partial\omega/\partial k = ck/(\kappa^2 + k^2)^{1/2}$$

and its derivative is

$$v_g' = \partial^2\omega/\partial k^2 = c\kappa^2/(\kappa^2 + k^2)^{3/2}.$$

Thus the vacuum gap between the two conducting planes behaves as the medium does with respect to TE and TM modes. On the other hand, with respect to the TEM mode such a gap does not differ from the vacuum, i.e. the dispersion of the TEM signals is absent. We encountered this phenomenon when analysing oscillations in the Earth–ionosphere cavity.

It should be noted that in a dispersive medium having waveguide type dispersion (5.16) the following relation between phase and group velocities holds

$$v_g v_{\text{ph}} = c^2. \tag{5.17}$$

---

[2] In the three-dimensional case, dispersive waves are those for which $\det(\partial^2\omega/\partial k_i \partial k_j) \neq 0$.

It is obvious that a similar relation is also valid in any nondispersive medium.

Plasma represents the medium where one may find the most complete set of dispersion laws (e.g. Chapter 4). Rare are the cases when the waves may be considered nondispersive. For example, dispersion may sometimes be neglected in the high-frequency limit (radio-wave range) and in the low-frequency limit (MHD range). However, dispersive distortions of the signal modulation tend to accumulate and therefore at relatively long traces dispersion should be taken into account.

At high frequencies using (4.1) one may obtain a dispersion law of the waveguide type

$$\omega = (\omega_{0e}^2 + c^2 k^2)^{1/2} \tag{5.18}$$

where ω_{0e} is the plasma (Langmuir) frequency. Apparently, relation (5.17) holds.

Within the MHD range, the dispersion of Alfvén and magnetosonic waves in a cold plasma is small if the ratio (ω/Ω_i) is small. In a number of cases the dispersion may be neglected. However, in other cases the dispersion should be taken into account. It may be done easily under purely longitudinal propagation $(\theta = 0)$.

At $\theta = 0$ the waves of Alfvén and magnetosonic branches are termed $\mathcal{L}$- and $\mathcal{R}$-waves, which has been mentioned above. From (4.5) and (4.23) we get

$$v_g = c_A \left(1 - \frac{\omega}{\Omega_i}\right)^{3/2} \left(1 - \frac{\omega}{2\Omega_i}\right)^{-1} \tag{5.19}$$

for the $\mathcal{L}$-waves and

$$v_g = c_A \left[\left(1 + \frac{\omega}{\Omega_i}\right)\left(1 - \frac{\omega}{\Omega_e}\right)\right]^{3/2} \left(1 + \frac{\omega}{2\Omega_i}\right)^{-1} \tag{5.20}$$

for the $\mathcal{R}$-waves. Here it is assumed that $n^2 \gg 1$ and it is taken into account that the transparency region for the $\mathcal{L}$-waves is limited by the frequency Ω_i, and for the $\mathcal{R}$-waves by the frequency Ω_e.

If $\omega \ll \Omega_i$, then with help of (5.19) and (5.20) we may present the explicit dependence of v_g on k

$$v_g = c_A(1 \mp k/k_{0i}) \tag{5.21}$$

and calculate the parameter

$$v_g' = \mp c_A/k_{0i} \tag{5.22}$$

which characterizes the wave-packet broadening. Here $k_{0i} = \omega_{0i}/c$ and the upper (lower) sign refers to the $\mathcal{L}(\mathcal{R})$-waves. At $v_g' < 0$ the dispersion is sometimes called normal or negative, and at $v_g' > 0$ anomalous or positive. However, in connection with dispersion the terms 'normal' and 'anomalous' are also used to signify other phenomena, so we shall not employ them in order to avoid misunderstanding.

According to (5.14) the packet width doubles within the time interval $t = \sqrt{3}\alpha/|\beta|$. Taking into account (5.21) and (5.22) it follows that the distortion of the MHD wave packet may not be neglected along the trace of propagation, the length of which exceeds $x_0^2 k_{0i}$. We note that, irrespective of the sign of v_g', the packet width always grows during propagation.

While the packet width grows in ordinary space, the width of the spectrum in k-space decreases. The uncertainty relation always takes place

$$\Delta x \Delta k \geq \tfrac{1}{2} \tag{5.23}$$

where Δx and Δk are the corresponding mean-square deviations. For the Gaussian packet (5.13), (5.23) becomes an equality.

The analysis of the wave-packet evolution in the dissipative medium is much more complicated. The presence of dissipation is formally manifested by the fact that the frequency $\omega(k)$ becomes complex at real values of k. The dielectric permeability tensor ceases being Hermitian. The energy conservation law in the form of (5.7) or (5.15) is no longer valid. However, if the dissipation is not strong the notions of group velocity and wave energy density may be retained. For this the real part of frequency should be substituted into (5.6), and the Hermitian part of the dielectric permeability tensor should be used in (5.8). The transport equation (5.7) may also be 'improved' if substituted in the form

$$\partial W / \partial t + \nabla \cdot (v_g W) = 2\gamma W \tag{5.24}$$

where $\gamma = \mathrm{Im}\,\omega$, $|\gamma| \ll \mathrm{Re}\,\omega$.

At $\gamma < 0$ this value is termed decrement, and at $\gamma > 0$ it is termed growth rate, or increment. The explicit expression for γ via the anti-Hermitian part of the dielectric permeability tensor will be given later. Here we just note that at $\gamma > 0$ the medium is unstable or, as it is sometimes termed, it possesses negative dissipation. In an unstable medium the wave packet undergoes a complicated and interesting evolution.

Exercise

Exercise 5.1.1.

Consider the evolution of a one-dimensional wave packet in an unstable plasma.

Solution 5.1.1.

If at the initial moment $t = 0$ there is perturbation

$$E(x, 0) = \int E_k e^{ikx}\, dk$$

then in the following moments of time ($t > 0$) the perturbation is a superposition of normal waves

$$E(x, t) = \int E_k \exp\{\gamma(k)t + i[kx - \omega(k)t]\}\, dk$$

where the complex frequency $\omega + i\gamma$ is connected with the wavenumber k by some dispersion relation.

Let increment $\gamma(k)$ have a maximum at $k = k_m$. Expanding $\gamma(k)$ and $\omega(k)$ by powers $(k - k_m)$

$$\omega = \omega_m + v_g(k - k_m) + \alpha(k - k_m)^2$$
$$\gamma = \gamma_m - \delta(k - k_m)^2$$

where

$$\omega_m = \omega(k_m) \qquad \gamma_m = \gamma(k_m)$$
$$\alpha = \partial^2\omega/2\partial k^2|_{k=k_m} \qquad \delta = -\partial^2\gamma/2\partial k^2|_{k=k_m}.$$

Since at large times the packet evolution is weakly dependent on the initial state, we assume for simplicity that

$$E_k = E_0 \exp\left[-\tfrac{1}{4}x_0^2(k - k_0)^2\right].$$

Then

$$E(x,t) = E_0 \left[\frac{\pi}{(\delta + i\alpha)t + x_0^2/4}\right]^{1/2}$$
$$\times \exp\left\{i(k_m x - \omega_m t) + \gamma_m t - \frac{[x - v_g t + ix_0^2(k_m - k_0)/2]^2}{4(\delta + i\alpha)t + x_0^2}\right.$$
$$\left. -\frac{x_0^2}{4}(k_m - k_0)^2\right\}.$$

Hence it follows that at

$$v_g^2 < 4\gamma_m \frac{\alpha^2 + \delta^2}{\delta}$$

the wave amplitude increases without limit at $t \to \infty$ in every point of space. Such instability is called absolute. At the inverse inequality the amplitude is limited in every fixed point of space, but increases at the point $x = v_g t$. This instability is termed convective.

5.2 Impulses

5.2.1 The method of stationary phases

According to the definition the modulation is a slow and smooth variation of the wave parameters. The wave packet was defined so that it yields the condition of slowness and smoothness. An arbitrary impulse may not yield this condition initially. However, with time an arbitrary impulse signal transforms into a modulated wave during propagation in a dispersive medium.

To illustrate this, we shall consider one-dimensional waves propagating in the positive direction along the x axis in a homogeneous infinite medium. Let the dispersion law $\omega(k)$ be known and an arbitrary perturbation $\psi(x, 0)$ be given at the moment $t = 0$. It is required to find the form of the wave at $t > 0$.

The solution by the Fourier method proceeds as follows. We define the spectral components of the initial impulse

$$\psi_k(0) = \frac{1}{2\pi} \int_{-\infty}^{\infty} \psi(x, 0)e^{-ikx}\, dx.$$

The sense of transition to the spectral representation lies in the fact that the evolution of the impulse in k-space is trivial. That is each spectral component simply oscillates at the frequency $\omega(k)$

$$\psi_k(t) = \psi_k(0)e^{-i\omega(k)t}.$$

Now we only have to provide the inverse Fourier transformation

$$\psi(x, t) = \int \psi_k(0) \exp[ikx - i\omega(k)t]\, dk \tag{5.25}$$

in order to obtain the solution of the problem.

This is an exact solution and is quite similar to (5.1). However, we now do not consider the spectrum of the initial impulse to be sufficiently narrow and therefore cannot avail ourselves of the advantages provided by the expansions (5.4) or (5.12). The solution (5.25) remains formal, i.e. it does not present an obvious picture of the initial impulse evolution until a corresponding integral is calculated.

We avail ourselves of the method of stationary phase in order to find the asymptote of the integral on the right-hand side of (5.25). Let us introduce the designation $\varphi(k) = kx/t - \omega(k)$. At large t the subintegral function oscillates rapidly everywhere, except close to the points $k = k_s$, where the phase is stationary. The position of these points is determined by the condition $d\varphi/dk = 0$, or

$$v_g(k_s) = x/t \tag{5.26}$$

where $v_g = d\omega/dk$, as before. It is clear that at large t the neighbourhoods of the stationary points make the dominating contribution to the integral.

Suppose, there is only one stationary point. (If such points are more than one their contributions are summed up.) Let us expand the functions $\psi_k(0)$ and $\varphi(k)$ in a Taylor series in the vicinity of the stationary point to within the first two nonvanishing terms. If $v_g'(k_s) \neq 0$, the integral is calculated and the solution (5.25) takes the form[3]

$$\psi(x, t) = \psi_{k_s}(0) \left[\frac{2\pi}{t\, |v_g'(k_s)|}\right]^{1/2} \exp\left\{i\left[k_s x - \omega(k_s)t - \frac{\pi}{4}\operatorname{sign} v_g'(k_s)\right]\right\}. \tag{5.27}$$

---

[3] If $v_g'(k_s) = 0$, but $v_g''(k_s) \neq 0$, then at large t the solution of (5.25) may be expressed by the Airy function. We recall that $v_g' = d^2\omega/dk^2$, etc.

So at large distances and large periods of time the arbitrary initial impulse transforms into a modulated quasi-sinusoidal wave. Its amplitude decreases with time as $1/\sqrt{t}$. In accordance with (5.26) the spectral components $\psi_k(0)$ of the initial impulse propagate along the world lines $x = v_g(k)t$ in two-dimensional space–time x–t. In other words, in the asymptotic limit there is an unfolded fan of spatio-temporal rays along which the spectrum of the initial impulse 'spreads out'.

5.2.2   Green's function of the propagation trace

We have considered the Cauchy problem when at $t = 0$ an impulse signal $\psi(x, 0)$ is given in the entire space $-\infty < x < \infty$. From the geophysical point of view the boundary problem is also of interest, i.e. at the point $x = 0$ the signal $\psi(0, t)$ is given, where $-\infty < t < \infty$; temporal evolution $\psi(x, t)$ at the observation point $x > 0$ is to be determined. The segment $[0, x]$ will be termed the propagation trace.

The boundary problem arises, for example, when investigating the signals excited by a lightning discharge. In this case the propagation trace is a section of the waveguide Earth–ionosphere and (or) a section of the geomagnetic field line connecting the source (lightning discharge) with the receiver. The receiver may be situated in the ionosphere on board the satellite, on the Earth's surface in the opposite hemisphere or even in the same hemisphere. In the latter case the signal passes twice the magnetospheric section of the propagation trace, i.e. back and forth.

One more example of setting the boundary problem is associated with the problem of extramagnetospheric origin of some types of geomagnetic pulsations. In this case it is natural to set the signal at the boundary of the magnetosphere and to study how it penetrates the Earth's surface.

The solution of the one-dimensional problem for the homogeneous half-space $x \geq 0$ is given by the Fourier integral

$$\psi(x, t) = \int_{-\infty}^{\infty} \psi_\omega(0) \exp[ik(\omega)x - i\omega t] \, d\omega \qquad (5.28)$$

where $k(\omega)$ is the branch of the dispersion equation, corresponding to the type of normal waves which propagate from the source to the observation point (compare (5.28) and (5.25)). If the perturbation is transferred along the trace by normal waves of several different types, the solution (5.28) should be written for each type with the corresponding dispersion law $k(\omega)$ and then all such solutions should be summed up. (A similar comment applies to (5.25).)

The frequency spectrum of the source

$$\psi_\omega(0) = \frac{1}{2\pi} \int_{-\infty}^{\infty} \psi(0, t) e^{i\omega t} \, dt$$

is determined by the form of the signal $\psi(0, t)$, set at the boundary $x = 0$. From the condition that $\psi(0, t)$ and $\psi(x, t)$ are real it follows that $\psi_{-\omega}(0) = \psi_\omega^*(0)$ and $k(-\omega) = -k^*(\omega)$. Then using the the Parseval theorem we obtain

$$\int_{-\infty}^{\infty} |\psi(x, t)|^2 \, dt = 2\pi \int_{-\infty}^{\infty} |\psi_\omega(0)|^2 e^{-2\mathrm{Im}k(\omega)x} \, d\omega.$$

The case when the δ-impulse is fed in at the input of the trace is of particular interest. Then the solution (5.28) is called the Green's function or the impulse response of the propagation trace. It is designated by

$$G(x, t) = \frac{1}{2\pi} \int_{-\infty}^{\infty} e^{ik(\omega)x - i\omega t} \, d\omega. \tag{5.29}$$

The Green's function defines the trace completely. If we succeeded in finding $G(x, t)$ then the response to the arbitrary input signal is calculated according to the formula

$$\psi(x, t) = \int_{-\infty}^{\infty} G(x, t - \tau)\psi(0, \tau) \, d\tau. \tag{5.30}$$

The Green's function is not easy to find even in the case of a homogeneous trace. Asymptotic evaluation can be made using the method of stationary phase. Paraphrasing all the above said in connection with the Cauchy problem, we find the stationary points $\omega = \omega_s$ by solving the equation

$$v_g(\omega_s) = x/t \tag{5.31}$$

where $v_g = c/n_g$, $n_g = \partial(\omega n)/\partial\omega$, $n = ck/\omega$. Then we calculate the first term of the asymptotic expansion of the integral (5.29). It has a form analogous to (5.27).

In contrast to (5.27), the trace length x is fixed, which corresponds to the ordinary conditions of geophysical observations. Therefore, we cannot speak of the spatio-temporal rays $x/t = $ constant, $t \to \infty$, $x \to \infty$. Instead we may speak of the asymptotics of spectral-temporal representation of the Green's function. The idea is as follows.

At the trace output of large, but fixed length x, the impulse signal represents a quasi-monochromatic wave with a slowly changing amplitude and frequency. The 'instantaneous' carrier frequency yields the relation (5.31) or, which is the same,

$$t = \left(\frac{x}{c}\right) n_g(\omega). \tag{5.32}$$

Let us construct the plane ω, t and plot all the points that yield the equation (5.32). They form a system of lines along which the 'ridges' of the Green's function run. Such a picture is called the dynamic spectrum. It may be improved if not only 'ridges' are plotted over the ω, t plane, but also other elements of the spectral-temporal 'relief'.

An evident generalization of the method lies in taking into account the longitudinal inhomogeneity of the propagation trace. Instead of (5.28) we should write

$$\psi(x, t) = \int_{-\infty}^{\infty} R(\omega, x)\psi_\omega(0)\exp[i\varphi(\omega, x) - i\omega t]\,\mathrm{d}\omega.$$

Here R is the amplitude factor of the signal attenuation and φ is the phase shift at the trace. Both values are sought by solving the problem of the propagation of a monochromatic wave. If the trace is smoothly inhomogeneous, then the WKB approximation may be used, i.e.

$$\varphi = \int_0^x k(\omega, x)\,\mathrm{d}x$$

and instead of (5.32) we have

$$t(\omega) = c^{-1}\int_0^x n_g(\omega, x)\,\mathrm{d}x. \tag{5.33}$$

This is the most important characteristic of the propagation trace. It is called the group delay of signals. Instead of $t(\omega)$ other designations: $t_g(\omega)$ or $\tau(\omega)$ are often used in the geophysical literature. We shall consider typical examples of frequency dependence of the group delay.

5.2.3 Tweaks, whistlers and pearls

We have already mentioned tweaks in Chapter 3. They appear after lightning discharges as a result of signal spreading at the horizontal traces in the Earth–ionosphere waveguide. Let us construct a simplified model of the tweak based on the dispersion law (5.16).

We rewrite (5.16) in the form of the formula for the effective refractive index

$$n(\omega) = \left(1 - \omega_s^2/\omega^2\right)^{1/2}. \tag{5.34}$$

Here $n = ck/\omega$, and according to the meaning of the problem k is a two-dimensional wave vector of horizontal propagation, so that $k = \sqrt{k_x^2 + k_y^2}$. The values $\omega_s = \pi cs/h$, $s = 1, 2, \ldots$ are the critical frequencies of the waveguide[4]. It should be noted that $nn_g = 1$. This useful ratio is equivalent to (5.17).

---

[4] If we replace ω_s by the Langmuir frequency ω_{0e} and the two-dimensional vector k by the three-dimensional one, then (5.34) may be understood as a refractive index of high-frequency waves in a plasma (see (5.18)). Therefore, the solution given below may be adapted in application to this case.

Omitting inessential multipliers we rewrite the Green's function in the form

$$G(t) \propto \int_{-\infty}^{\infty} e^{-i\nu t + i(r/c)\nu n(\nu)} \, d\nu. \tag{5.35}$$

Here r is the length of the trace, i.e. the distance along the waveguide from the lightning discharge to the point of signal reception. The waveguide is considered homogeneous. The function $n(\nu)$ is determined from the formula (5.34). We shall assume $s = 1$ in this formula, i.e. only one waveguide mode with the least critical frequency $\omega_1 = \pi c/h$ will be considered. The analysis of the other modes is carried out similarly.

Over the long trace the signal damps owing to the dissipation in the lower ionosphere. Over the short trace the dispersion spreading will manifest itself only at frequencies close to the critical frequency ω_1. Therefore, we fail to make an asymptotic estimate of the Green's function by means of the stationary phase method.

Let us find the dynamic spectrum $\Gamma(\omega, t)$ of the Green's function $G(t)$ using the following construction. We calculate the Fourier spectrum of the function $G(t)$. The result $\exp(i\nu n r/c)$ is multiplied by $\exp[-\alpha(\nu - \omega)^2]$, i.e. a window of $\sim \alpha^{-1/2}$ width centred at the frequency $\nu = \omega$ is cut in the spectrum. After that we make the inverse Fourier transformation

$$\Gamma(\omega, t) \propto \int_{-\infty}^{\infty} d\nu \exp\left\{-i\nu t + i\frac{r}{c}\nu n(\nu) - \alpha(\nu - \omega^2)\right\}. \tag{5.36}$$

At sufficiently large α the main contribution to the integral is made by the frequencies ν, close to ω. We designate $\xi = \nu - \omega$ and expand $n(\nu)$ into a series in powers of ξ. As a result we obtain

$$\Gamma(\omega, t) \propto \int_{-\infty}^{\infty} d\xi \exp\left\{i\xi\left[\frac{r}{c}n_g(\omega) - t\right] + \frac{\xi^2}{2}\left[i\frac{r}{c}n'_g(\omega) - \alpha\right]\right\}$$

where $n_g = \partial(\omega n)/\partial\omega$ and $n'_g = \partial n_g/\partial\omega$. Using the known value of the integral

$$\int_{-\infty}^{\infty} dx \, e^{-iAx - Bx^2} = \left(\frac{\pi}{B}\right)^{1/2} e^{-A^2/4B} \qquad \mathrm{Re}\, B > 0$$

we obtain

$$|\Gamma| \propto (\alpha^2 + \tau'^2)^{-1/4} \exp\left[\frac{-\alpha(t - \tau)^2}{2(\alpha^2 + \tau'^2)}\right] \tag{5.37}$$

where $\tau(\omega) = (r/c)n_g(\omega)$ and $\tau' = d\tau/d\omega$.

The maximum of $|\Gamma(\omega, t)|$ is reached at $t = \tau(\omega)$. The 'ridge' of the dynamic spectrum of the tweak is shown in figure 5.1. We see that the signal emitted by the lightning discharge is subject to frequency modulation. The instantaneous frequency asymptotically approaches the critical waveguide frequency.

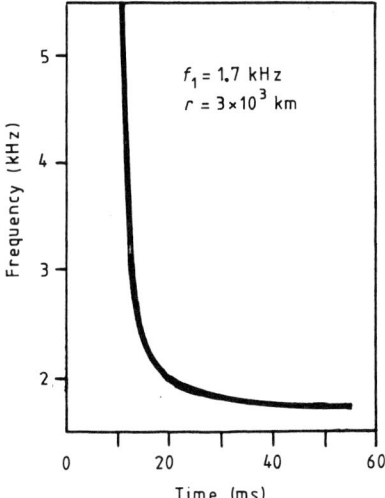

Figure 5.1. The dynamic spectrum of a tweak. The time is counted from the moment of the lightning discharge.

Whistlers may be analysed similarly. These signals are also excited by lightning, but in contrast to the tweaks they propagate not along the Earth's surface, but upwards into the ionosphere and the magnetosphere. The propagation trace is inhomogeneous, so (5.33) should be employed instead of (5.32).

We know two species and several varieties of whistler. Electron whistlers, or simply whistlers, refer to the first kind. The refractive index for them equals

$$n = \omega_{0e}/\sqrt{\Omega_e \omega}. \tag{5.38}$$

We recall that this formula is applicable to the longitudinal propagation of $\mathcal{R}$-waves within the frequency range $\Omega_i \ll \omega \ll \Omega_e$. The group refractive index n_g is connected with n in the following way

$$n_g = n/2.$$

This means that the group velocity is twice the phase velocity. Both phase and group velocities increase with increasing frequency. Therefore, a whistler is a signal of falling tone[5].

---

[5] Here we do not discuss the so-called nose whistlers that have more complex frequency modulation.

Let us examine figure 5.2. Here a spheric, practically coinciding in time with a lightning discharge, with a whistler following it, is plotted schematically. This whistler is a so-called long whistler. Its trajectory coincides with one of the field lines of the geomagnetic field and it passes along this trajectory twice, back and forth. The source of the signal and the receiver are located in the same hemisphere. In contrast to this, a short whistler appears if the source (lightning discharge) and the receiver are in opposite hemispheres. Then the whistler passes along the geomagnetic field line in one direction only, for example, from the southern hemisphere into the northern hemisphere. The frequency dependence of the time of group delay of long and short whistlers is well approximated by the formula $\tau = Df^{-1/2}$, and the coefficient D is usually called the dispersion. Typical values are $D \simeq 10\text{--}100 \ \text{s}^{1/2}$ within the range 1–6 kHz. The theory predicts that the dispersion of long whistlers is twice as large as the dispersion of short whistlers. The observations prove it.

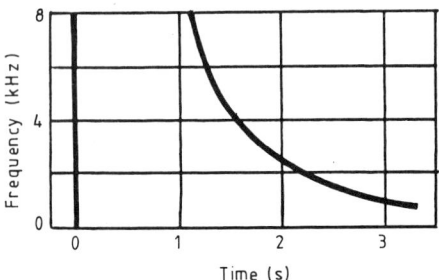

Figure 5.2. The dynamic spectrum of the spheric and long whistler.

The second species is formed by ion whistlers. Unlike the electron whistler, the energy of which is transferred by the $\mathcal{R}$-waves, the energy of the ion whistler is transferred by the $\mathcal{L}$-waves. At purely longitudinal propagation the group refractive index equals

$$n_g = \frac{\omega_{0p}(\Omega_p - \omega/2)}{\Omega_p^{1/2}(\Omega_p - \omega)^{3/2}}. \tag{5.39}$$

This formula is applicable to a plasma composed of electrons and ions of one sort (protons). It may also be applied in the multicomponent plasma provided ω is close to Ω_p (see section 4.1). It is obvious from (5.33) and noting (5.39) that the frequency of the proton whistler grows in time, approaching asymptotically the gyrofrequency of protons, i.e. the formula (5.39) is applicable for the analysis of asymptotic behaviour of the signal in the multicomponent plasma.

Let us study the wave train composed of $\mathcal{L}$-waves and propagating upwards along the geomagnetic field line. If the wave-train spectrum is situated above the minimal gyrofrequency of protons, then the wave train will not reach the other hemisphere, since there is a non-transparent band on the propagation trace. Such a wave train may be observed only on board a satellite[6].

Protons whistlers were discovered at frequencies of $\simeq 0.5$ kHz during the flight of the satellites *Injune-3* and *Alouette-1* at an altitude of about 1000 km. Figure 5.3 shows schematically how the proton whistler arises. First a short whistler appears. Its dispersion is of the order of $D \simeq 5$ s^{-1}. Low value of dispersion corresponds to a short propagation trace, running along the geomagnetic field line from the ionosphere to the point of observation. The proton whistler appears at a certain frequency. (This frequency, called the crossover frequency, is determined by the intersection of $\mathcal{R}$ and $\mathcal{L}$ branches of the dispersion equation, see exercise 5.2.1.) The proton whistler lasts for about 2 s.

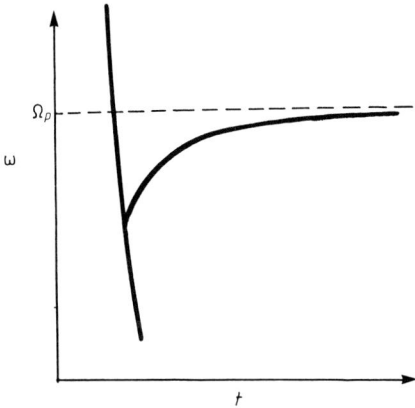

Figure 5.3. The proton whistler (signal of rising tone) and short electron whistler (signal of falling tone).

At night proton whistlers are an ordinary phenomenon. They appear with a frequency of up to 30 signals per minute within the altitude interval of 500–3000 km. In the daytime the frequency of appearance of proton whistlers approximately equals one signal per minute. Helium whistlers appear much more rarely than the proton whistlers, and oxygen whistlers appear more rarely still.

All the whistlers enumerated are excited by electric discharges in the atmosphere. At the same time self-excitation of ion whistlers in the magnetosphere caused by cyclotron instability of the plasma is possible. It is supposed that one of the species of geomagnetic pulsations within the range of Pc1, namely pearl pulsations, represents a periodical consequence of such whistlers. Sometimes they are called hydromagnetic whistlers.

---

[6] If special hypotheses are not introduced, then it follows from the theory of wave propagation that the ion whistler cannot return to the hemisphere from which it has been emitted, since the signal approaches the opaqueness band from the pole n^2 side and is completely absorbed there.

When thus interpreted, the frequency dependence of the repetition period of pearls is determined by the integral (5.33)

$$\tau(\omega) = 8\sqrt{\pi m_p}(R_E/B_E)L^4\sqrt{N_0}I(\omega)$$

$$I(\omega) = \int\limits_0^{z_0} \frac{[1 - \omega/2\Omega_p(z)]}{[1 - \omega/\Omega_p(z)]^{3/2}}(1 - z^2)^{3-s/2}\,dz$$

$$\Omega_p(z) = \Omega_0\frac{(1 + 3z^2)^{1/2}}{(1 - z^2)^3}$$

$$z_0 = (1 - 1/L)^{1/2}.$$

(5.40)

Here $R_E = 6.37 \times 10^8$ cm is the Earth's radius, $B_E \simeq 0.315$ G is the geomagnetic field at the equator, L is the McIllwain parameter of the magnetic field line which approximates the trajectory of the pearls, and Ω_0 and N_0 are the gyrofrequency and concentration of protons at the top of the trajectory. The geomagnetic field is supposed to be a dipole. The distribution of a plasma along the field line is taken in the form $N(z) = N_0(1 - z^2)^{-s}$, where $z = \sin\phi$ and ϕ is the geomagnetic latitude. Assuming that the plasma is formed of electrons and protons, we used the formula (5.39) when writing the integral (5.40). The integral is taken twice along the field line between the magnetic conjugate points in order to imitate bounce motion of the signal along the trajectory.

Let us introduce the parameter

$$D = \partial \ln \tau/\partial \ln \omega \tag{5.41}$$

which defines the dispersion of signals. It does not depend on N_0. The results of numerical calculations of the integral (5.40) show that D depends rather weakly on the parameter s. Therefore, it is usually assumed that $s = 4$, which supposedly corresponds to the character of plasma density decrease when moving away from the Earth along the geomagnetic field line. The dependence of D on x is also insignificant if $L > 2$. The dependence of D on the parameter ω/Ω_0 is essential and this is quite comprehensible, since the major contribution to dispersion is made by the equatorial section of the trajectory, where the ratio $\omega/\Omega_p(x)$ reaches its maximum value. The result of the numerical calculation is satisfactorily approximated by the formula

$$D \simeq 0.6\omega/(\Omega_0 - \omega) \tag{5.42}$$

where

$$\Omega_0 \simeq 3 \times 10^3 \, L^{-3} \, \text{s}^{-1}.$$

For the ion cyclotron waves $\omega < \Omega_0$ and, consequently, $D > 0$ always. At the end of the series of pearls, when the signals start damping rapidly, positive dispersion is observed fairly often. Processing of the dynamic spectra of such signals according to (5.41) gives the values $D \simeq 0.3$–1. Hence from (5.42) we obtain the estimate $\omega/\Omega_0 \simeq 0.33$–$0.6$. Now we may estimate the parameter L by the value of the carrier frequency. For example, at $\omega \simeq 6 \text{ s}^{-1}$ roughly we have $L \simeq 5$.

However, positive dispersion is not always observed. In the midst of the series of pearls the value D is usually indistinguishable from zero and at the beginning of the series often $D < 0$. These anomalies are obviously associated with the nonlinearity of pearls.

There also exist other types of frequency-modulated signals known for their anomalous behaviour. A wide variety of these signals are detected in polar caps. Signals of both rising and falling tones are observed within the range of Pc1. In contrast to middle and auroral latitudes, where rising tones predominate, the most widespread form of discrete signals in the polar caps are the bursts of falling tone.

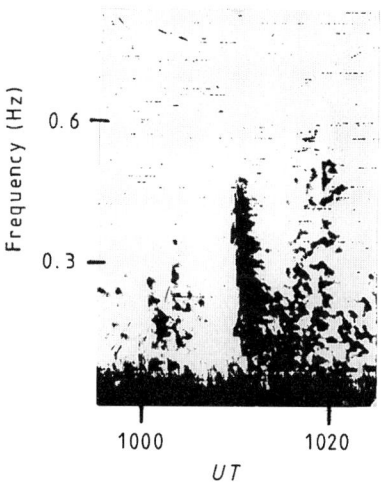

Figure 5.4. The discrete frequency-modulated signal registered at the South Pole (Vostok Observatory) 4 April 1973.

Falling bursts resemble whistlers, and rising ones resemble structural elements of the series of pearls. On these grounds they could be identified with $\mathcal{R}$- and $\mathcal{L}$- waves but such a simple interpretation encounters a number of difficulties. For example, in the case of falling tone bursts the sign of the second derivative frequency over time $(\mathrm{d}^2 f/\mathrm{d}t^2 < 0)$ usually does not agree with the dispersion law of the $\mathcal{R}$-waves in the MHD range $(\mathrm{d}^2 f/\mathrm{d}t^2 > 0)$. The dynamic spectra of signals sometimes have a distinctively anomalous character. At Vostok observatory (Antarctica) a great number of such signals were detected under conditions that do not differ much from the conditions of excitation of 'normal signals'. Take, for example, the strong discrete signal of the falling tone and the weak burst, resembling trigger radiation, that follows it (figure 5.4).

5.2.4 Propagation of waves in a moving plasma

Until now we have been speaking of the propagation of impulses and wave packets in the motionless plasma. However, in space the moving plasma represents an ordinary phenomenon. The solar wind blows in the interplanetary medium. A complicated picture of convective, drift and thermodiffusive motions of the plasma is observed in the magnetosphere. Finally, the plasma is in motion relative to the receiver of the electromagnetic signals on board the satellite or rocket.

In the moving plasma the dispersion law changes, additional group delay appears, modulation is distorted and, if the motion is inhomogeneous, the signal trace is deformed.

Let the plasma move as a whole with velocity u relative to the observer. We shall find the refractive index of the moving plasma assuming the refractive index of the motionless plasma to be known.

Obviously the relations $ck = \omega n(\omega, \theta)$ and $ck' = \omega' n'(\omega', \theta')$ hold. The prime refers to the reference system connected with the observer (laboratory system). The plasma is considered to be cold. Therefore, n, besides the frequency, depends only on the angle θ between k and B. The connection between ω, k, θ on the one hand and ω', k', θ' on the other is sought by the formulae of transformation of frequency and wave vector

$$\omega = \omega' - k \cdot u \qquad k = k' - (\omega/c^2)u.$$

Here u is the velocity of the concomitant reference system relative to the laboratory one, $u \ll c$. Expanding $n(\omega, \theta)$ into a series by powers $(\omega - \omega')$, $(\theta - \theta')$, we obtain with accuracy to terms of the first order in β

$$n' = n + \beta \left[\left(1 - n \frac{\partial \omega n}{\partial \omega} \right) \cos \varphi + \frac{1}{n} \frac{\partial n}{\partial \gamma} (\gamma \cos \varphi - \cos \psi) \right]. \tag{5.43}$$

Here $\beta = u/c$, $\gamma = \cos \theta$ and φ is the angle between u and k and ψ is the angle between u and B. In order to simplify the formula we substituted ω', θ' for ω, θ. In other words the frequency ω and the direction of propagation θ in the right-hand and left-hand sides of (5.43) refer to the laboratory reference system.

Equation (5.43) differs from the known Lorentz formula for the moving isotropic dielectrics by the second term in square brackets. In the course of slow wave propagation ($n \gg 1$) this term may be neglected. Then

$$n' = n[1 - (u/v_g) \cos \varphi] \tag{5.44}$$

where $v_g = c/n_g$ and $n_g = \partial \omega n/\partial \omega$.

We shall apply (5.44) to whistlers in order to demonstrate a curious paradox. Substituting (5.38) into (5.44) we find that $n'_g = n_g$, i.e.

$$v'_g(\omega) = v_g(\omega). \tag{5.45}$$

The paradox of equation (5.45) lies in the fact that owing to the definition of v_g as the velocity of energy transfer we expect that v_g and u should be simply added when passing into a moving (laboratory) reference system. In the nondispersive medium

$$v'_g = v_g + u. \tag{5.46}$$

This follows from (5.43) and agrees with intuitive knowledge.

On the face of it, the rule of velocity addition (5.46) should also always be found in the dispersive medium. But it is obviously necessary to take into account the character of the wave process and conditions of observations. If a quasi-monochromatic wave packet propagates, then the moving observer may not see the signal at all if he or she carries out the registration by means of a narrow-band receiver. If a wideband receiver is employed the observer will see a wave packet travelling at the velocity $v'_g = v_g + u$, as it should be. A different story arises if the wave represents a wideband impulse. Original relativity, expressed by (5.45), may appear in the dispersive medium.

From (5.45) it follows that the shape of the frequency modulation of a whistler $f = D^2 t^{-2}$ is invariant (i.e. it does not change its form) relative to the Galilean transformations. On the contrary, the frequency modulation of the ion whistlers is strongly distorted even at a relatively low travelling speed of the receiver relative to a plasma. Let us examine figure 5.5. The frequency dependence of the group velocity of $\mathcal{L}$-waves in the hydrogen–helium plasma is shown above. The group velocity is normalized by the Alfvén velocity, and the frequency is normalized by the gyrofrequency of ions He^+. Relative concentration of He^+ is denoted ξ. The frequency dependence of the correction δv_g, appearing when the plasma travels at the velocity u, is shown below. We see that δv_g increases essentially when ω approaches the gyrofrequencies Ω_{He^+} and Ω_p.

The asymptotics of the group delay of the proton whistlers

$$\tau(\Delta\omega) = \left(\frac{c_A}{\Omega_p} \left| \frac{\partial \Omega_p}{\partial l} \right| \right)^{-1} \left(\frac{\Omega_p}{\Delta\omega} \right)^{1/2} \left\{ 1 + \frac{u_\parallel}{2c_A} \left(\frac{\Omega_p}{\Delta\omega} \right)^{3/2} \right\} \tag{5.47}$$

is derived from equations (5.33), (5.39) and (5.44) by the expansion $\Omega_p(l)$ in a series in the vicinity of the point $l = 0$. Here $\Delta\omega = \Omega_p - \omega$, $u_\parallel$ is the projection of the satellite velocity over the signal trace and l is the distance along the trace (field line). The values c_A and Ω_p are taken at the point of observation ($l = 0$).

We should recall that according to our agreement $\Delta\omega$ on the right-hand and left-hand sides of (5.47) is a spectral gap, measured in the laboratory reference system. The same gap in the concomitant system must be positive according to (5.39). It is clear that according to the value and sign of $u_\parallel$ the gap $\Delta\omega$ may have any sign in reality. However, the conditions of derivation of (5.43) require that $\Delta\omega > 0$. This imposes a restriction over the dimensions of the gap in the concomitant system, i.e. it must exceed the value of the order of $\Omega_p(u_\parallel/c_A)^{2/3}$.

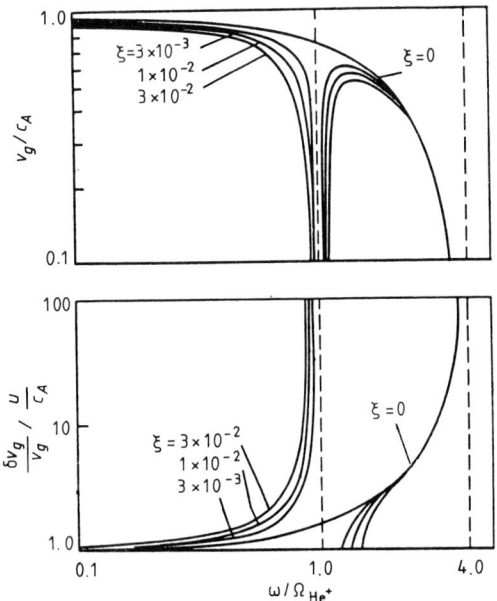

Figure 5.5. The group velocity of $\mathcal{L}$-waves in the hydrogen–helium plasma (upper panel) and corrections to the group velocity which arise in the moving plasma (lower panel).

If this condition is not fulfilled, then simple analytical estimates are impossible and we have to turn to numerical methods of analysis.

The values $u_\parallel$ and $\partial \ln \Omega_p / \partial l$ are calculated according to the data on the satellite trajectory within the framework of an appropriate model of the geomagnetic field. Taking this into account the observed dependence $\tau(\omega)$ contains, in essence, enough information to define both $\Omega_p(0)$ and $N_p(0)$ with the help of (5.47). It should be noted that simultaneous on-board measurements of the magnetic field allow us to remove completely the ambiguity in $\Omega_p(0)$ and consequently increase the accuracy of diagnostics of proton density.

Let us estimate the second term in the braces in (5.47). Cyclotron damping of waves limits the dimensions of the gap $\Delta\omega$. In the concomitant reference system the gap does not exceed the value $\Omega_p(v_T/c_A)^{2/3}$, where v_T is the thermal proton velocity. At sufficiently small observer velocity the value $\Delta\omega_{min}$ has the same order. Consequently, the maximum correction to the time of group delay $\delta\tau/\tau \sim u_\parallel/2v_T$. Under typical values of $u_\parallel \sim 3 \times 10^5$ cm s^{-1}, $v_T \sim 5 \times 10^5$ cm s^{-1} we have $\delta\tau/\tau \sim 0.3$, i.e. a quite noticeable value.

The result of the analysis testifies to the fact that the effect of the satellite movement should be taken into account in the course of diagnostics of proton concentration by the dispersion of the proton whistlers. When diagnosing the temperature of protons by the upper cut-off of the proton whistler spectrum the satellite movement should also be taken into account.

Exercise

Exercise 5.2.1.

Find the nose and crossover frequencies of whistlers.

Solution 5.2.1.

The nose frequency of a whistler is the frequency ω_n at which the time of the group delay (5.33) reaches its minimum value. The propagation trace is supposed to be fixed. If the trace is homogeneous then ω_n is determined by the conditions $\partial n_g/\partial\omega = 0$, $\partial^2 n_g/\partial\omega^2 > 0$, where $n_g = \partial\omega n/\partial\omega$, $n(\omega)$ is the refractive index of the $\mathcal{R}$-waves.

It is clear that within the applicability of (5.38) these conditions are not satisfied. If we omit the restriction $\omega \ll \Omega_e$, then instead of (5.38) we have

$$n = \omega_{0e}/[\omega(\Omega_e - \omega)]^{1/2}.$$

Hence it follows that the nose frequency of the electron whistlers is

$$\omega_n = \Omega_e/4.$$

Ion whistlers also have a nose frequency, but only when they propagate in a plasma which contains at least two kinds of ions with a different charge to mass ratio.

The frequency at which the branches of dispersive curves for the $\mathcal{R}$- and $\mathcal{L}$-waves intersect, i.e. $n_1(\omega_c) = n_2(\omega_c)$, is called the crossover frequency ω_c. According to (4.4) the intersection appears at $g(\omega_c) = 0$ or

$$\sum \frac{\omega_0^2 \Omega}{\Omega^2 - \omega_c^2} = 0$$

where the summation is carried out over particle species. One can easily make sure that the branches intersect only in the multicomponent plasma. Thus, in the hydrogen–helium plasma the crossover frequency approximately equals

$$\omega_c = \Omega_{\mathrm{He}^+}(1 + 7.5\xi)$$

where ξ is the relative concentration of He^+ ions and $\xi \ll 1$.

Bibliography

For section 5.1

The term 'dispersion' was introduced by Newton in connection with his experiments on the splitting of white light. The fundamental concept of group velocity v_g was introduced by Hamilton in 1839. The interpretation of v_g as the velocity of modulation propagation was given by Stokes and independently by Rayleigh back in the 1870s. At the same time Rayleigh suggested the term 'group velocity'. In 1874 Umov defined the group velocity as the velocity of energy transfer $v_g = S/W$, but his research was severely criticized and soon forgotten. Independent of this and independently of each other, Poynting and Heaviside introduced the idea of density of the electromagnetic energy flux in the mid-1880s.

Concerning the spreading of wave packets, besides the information presented above, one may find discussions in the books by Ginzburg (1971) and Whitham (1974). Exercise 5.1.1 was based on the book by Mikhailovskii (1974) where one may find reference to the first publications on absolute and convective instabilities of the plasma.

For section 5.2

The method of stationary phase put forward by Kelvin in 1887, is presented in many guides to the theory of waves, for example in the book by Bhatnagar (1979).

The frequency modulation of the whistlers was interpreted by Storey (1953). His work, which has become classical, stimulated the research that brought in many outstanding discoveries. Helliwell *et al* (1956) discovered nose whistlers. Carpenter (1963) discovered the plasmapause when investigating nose whistlers.

Proton whistlers were discovered by Smith *et al* (1964). The corresponding theory was suggested by Gurnett *et al* (1965). Gurnett and Shawhan (1966) put forward the method of plasma diagnostics using the data on the frequency modulation of the proton whistlers. Helium whistlers were discovered by Barrington *et al* (1966). Gurnett and Rodrigues (1970) reported on the observation of oxygen O^{++} whistlers (or, possibly, He_2^+ whistlers). Deuteron whistlers were discovered by Watanabe and Ondoh (1975). They also reported on the transequatorial propagation of ion whistlers.

In accordance with the idea suggested by Obayashi, the frequency modulation of pearls was interpreted within the framework of the theory of ion cyclotron waves (Jacobs and Watanabe 1964, Obayashi 1965). The model of the dynamic spectrum of pearls was constructed by analogy with the model of the whistling atmospherics. The difference between the prediction of the theory and the behaviour of the real signals served as a stimulus to search for a nonlinear mechanism of the frequency modulation of pearls. Obviously, this problem cannot be considered settled, however, the idea put forward by Feygin and Yakimenko (1969, 1970, 1971) is worth mentioning. In their opinion the key is the fact that Pc1 signals propagating through the active medium are subject to amplification and simultaneously modify the amplification coefficient (for example, due to the quasilinear relaxation of the distribution function of energetic ions). In other words, the signal propagates through the nonstationary medium. This nonstationarity may compensate for the broadening of signals so that the parameter of dispersion (5.41) will approach zero. At the end of a series of pearls the amplitude starts weakening because of the exhaustion of sources, and the signals spread like ordinary ion whistlers.

According to the literature, there exists the opinion that motion of the medium may allegedly be ignored when analysing frequency modulation of whistlers. As regards the whistling atmospherics, this opinion is not to be disputed. It is curious that motion should be disregarded not because the velocity is small but because the form of the dynamic spectrum of a whistler does not change at all with the transition into a moving reference system. However, if we speak of ion whistlers the movement of the observer should be taken into account (Guglielmi 1963, 1967a,b, 1968).

Exercise 5.2.1 is based on the works by Stix (1962), Helliwell (1965) and Gurnett and Shawhan (1966).

Chapter 6

Instability

Oscillations of the geomagnetic field appear either under the influence of external sources, or as an effect of self-excitation. We call the sources external to imply that their spatio-temporal structure does not depend with sufficient accuracy on the wave field that they excite. A lightning discharge may serve as a typical example; the current in it is almost independent of the field of Schumann oscillations. In the near-Earth plasma we may also single out the system of variable currents and consider it as given. For example, we may thus regard the pulsating electrojet, flowing along the auroral zone, the modulated beams of charged particles, which are sporadically injected during substorms, etc. At the same time self-excitation of waves takes place in the plasma due to instability[1].

The plasma is considered unstable if the amplitude of perturbations increases with time. The loss of stability may take place in an inhomogeneous plasma if the gradients of macroscopic parameters exceed some critical value. The spatially homogeneous plasma may be unstable at a nonequilibrium distribution of charged particles over the velocities. The character of growing perturbations is determined by specific conditions and is fairly diverse, which corresponds to a certain diversity of eigen oscillations and the types of deformations of plasma.

The action of the instability in the near-Earth plasma will be illustrated by some typical examples. But first let us present the necessary information from physical kinetics.

6.1 The method of kinetic equations

The diversity of excitation mechanisms of electromagnetic waves calls for the application of different methods to describe the dynamics of geophysical media. Above we have stated a number of methods for macroscopic description, when

---

[1] We shall discuss instability of a plasma, although any dynamic system may become unstable. For example, when the wind is strong enough, the plane surface of water becomes unstable and waves appear over the surface.

phenomenological parameters (conductivity, dielectric permeability, viscosity, elasticity modules, etc) are introduced into the theory more or less arbitrarily. The equations of macroscopic theory are used quite often as convenient rough models regardless of strict conditions of applicability. A more general microscopic approach, based on the application of kinetic equations allows one, first, to calculate phenomenological parameters proceeding from first principles, secondly, to prove the applicability of macroscopic equations and, finally, to describe in detail the dynamics and kinetics of nonequilibrious media.

The application of the methods of physical kinetics is difficult mathematically and is in reality not always justified. However, when the magnetospheric plasma is under consideration, the method of kinetic equations gives not only a general but in a number of cases the only possible description.

The kinetic equations in the collisionless approximation have the form

$$\frac{\partial f_a}{\partial t} + v \cdot \nabla f_a + \frac{e_a}{m_a}\left(E + \frac{1}{c}v \times B\right)\frac{\partial f_a}{\partial v} = 0. \tag{6.1}$$

Here $f_a(v, x, t)$ is the particle distribution function, i.e. an average number density of particles at the moment of time t in the point (x, v) of the phase space. The index a denotes the particle species (electrons, ions), e_a and m_a are the particle charge and mass. The fields $E(x, t)$ and $B(x, t)$ yield the Maxwell equations (3.32), where the charge density q and the current density j are expressed via the distribution functions

$$q = \sum e_a \int f_a \, dv$$

$$j = \sum e_a \int v f_a \, dv. \tag{6.2}$$

Summation is carried out over the particle species.

In such a description we suppose that the electric and magnetic fields as well as the charge and current densities are averaged over small volumes, containing quite a large number of particles. Equations (3.32), (6.1) and (6.2) form a self-consistent system that implies that the motion of plasma particles excites the electromagnetic field, which, in its turn, influences the motion of particles. The motion of a separate particle with mass m and charge e in a self-consistent field yields the equation

$$\frac{d^2 x}{dt^2} = \frac{e}{m}\left[E(x, t) + \frac{1}{c}\frac{dx}{dt} \times B(x, t)\right] \tag{6.3}$$

where $x(t)$ is the particle trajectory.

During slow large-scale motions when the characteristic time is much longer than the gyroperiod, and the characteristic dimension is much larger than the gyroradius of particles, we carry out averaging of particle trajectories over the fast Larmor rotation and proceed to the so-called drift kinetic equations. In the drift approximation the distribution function $f(v_\parallel, \mu, x, t)$ depends on six variables only, in contrast to the general case, where the distribution function depends on seven variables. Here $\mu = mv_\perp^2/2B$ is the magnetic moment of the particles, and $v_\parallel$ and $v_\perp$ are the longitudinal and transverse components of the particle velocity. From the drift kinetic equation we may proceed to a still simpler set of equations of magnetohydrodynamics with anisotropic pressure. It has the form (2.4), (2.6)–(2.8) with the following modifications. It should be supposed that $\sigma \to \infty$ in (2.7). In (2.8) the scalar p should be replaced by the symmetric tensor

$$p_{ij} = p_\perp(\delta_{ij} - B_i B_j/B^2) + p_\parallel B_i B_j/B^2$$

where $p_\perp$ is the plasma pressure in the transverse direction relative to the external magnetic field and $p_\parallel$ is the pressure in the longitudinal direction. The adiabatic equation (2.6) should be replaced by

$$\frac{d}{dt}\left(\frac{p_\parallel B^2}{\rho^3}\right) = 0 \qquad \frac{d}{dt}\left(\frac{p_\perp}{\rho B}\right) = 0. \tag{6.4}$$

The physical meaning of (6.4) lies in the fact that in the absence of collisions there is no energy exchange between the transverse and longitudinal degrees of freedom. When compressed in the longitudinal direction the plasma behaves as a one-dimensional gas, and when compressed in the transverse direction it behaves as a two–dimensional gas.

However, let us take up the initial set of kinetic equations. Using them, we may generalize the expression for the dielectric permeability tensor, found above in the cold plasma approximation (see (3.46))

$$\varepsilon_{\alpha\beta} = \delta_{\alpha\beta} + \sum \frac{\omega_0^2}{\omega^2} \sum_{s=-\infty}^{\infty} \langle \zeta G_{\alpha\beta} \rangle. \tag{6.5}$$

Here

$$G_{\alpha\beta} = F_\alpha^* F_\beta L - e_\alpha e_\beta F_z^2 M$$

$$F = \left\{\frac{s}{a} J_s, i J_s', \frac{v_z}{v_\perp} J_s\right\}$$

$$L = v_\perp \left\{\omega \frac{\partial}{\partial v_\perp} - k_z v_z \left[\frac{\partial}{\partial v_\perp} - \frac{v_\perp}{v_z} \frac{\partial}{\partial v_z}\right]\right\}$$

$$M = v_\perp(\omega - k_z v_z - s\Omega)\left[\frac{\partial}{\partial v_\perp} - \frac{v_\perp}{v_z} \frac{\partial}{\partial v_z}\right].$$

The function $\zeta = \zeta(\omega - k_z v_z - s\Omega)$ in accordance with the Landau rule has the form

$$\zeta(x) = \frac{P}{x} - i\pi \delta(x)$$

where P stands for the principal value. The brackets in (6.5) signify the averaging over the unperturbed particle distribution function

$$\langle g \rangle = \int g f v_\perp \, dv_\perp \, dv_z.$$

It is supposed that the distribution function $f(v_\perp, v_z)$ depends only on the longitudinal v_z and the transverse $v_\perp$ components of particle velocity and is normalized so that $\langle 1 \rangle = 1$. The coordinate systems in the ordinary space and in the velocity space are oriented as shown in figure 6.1. The other notations are as follows: $J_s = J_s(a)$ are the Bessel functions, $a = k_\perp v_\perp / \Omega$, $e = (0, 0, 1)$ and $s = 0, \pm 1, \pm 2, \ldots$. Summation in (6.5) is carried out over s and over the particle species.

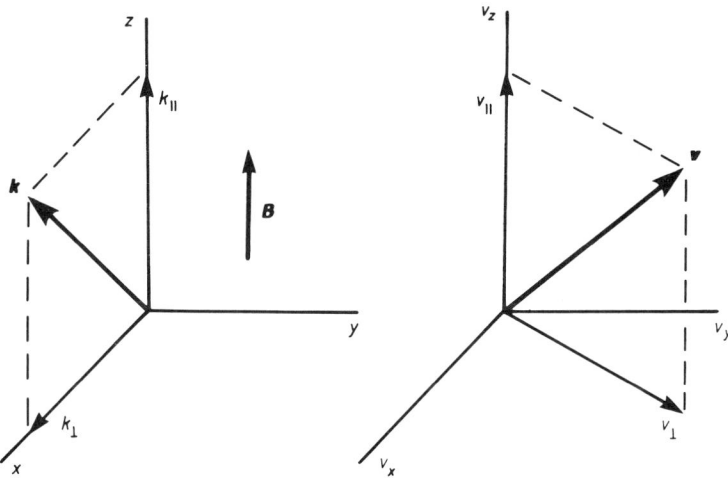

Figure 6.1. Orientation of coordinate systems in the ordinary space (left) and in the velocity space (right).

The dielectric permeability tensor (6.5) describes all the electrodynamic properties of homogeneous, stationary, nonrelativistic, collisionless plasma in the linear approximation. It is deduced by means of linearization and integration of a self-consistent set of equations (3.32), (6.1) and (6.2).

With the help of (6.5) we may investigate the stability of the homogeneous unbounded plasma with respect to infinitely small perturbations. The problem of stability will be considered as a particular case of the problem of the spectrum of eigen oscillations. In the homogeneous infinite plasma the spectrum of oscillations is found by investigating the roots of the dispersion equation

$$\text{Det}\left\{ k^2 \delta_{\alpha\beta} - k_\alpha k_\beta - \frac{\omega^2}{c^2} \varepsilon_{\alpha\beta}(\omega, k) \right\} = 0. \tag{6.6}$$

At the given real k the frequencies $\omega(k)$ derived from (6.6), in general, will be complex. The plasma is unstable in the case where at least one of the eigenfrequencies has a positive imaginary part $\gamma \equiv \mathrm{Im}\,\omega > 0$. In this case the wave amplitude increases exponentially in time $|b| \propto \exp(\gamma t)$. The value γ is termed the growth rate. The wave energy is derived from the kinetic energy of plasma particles.

The exchange of energy between the waves and particles is most effective under the resonance condition

$$\omega = s\Omega + k_z v_z \tag{6.7}$$

where v_z is the projection of the particle velocity on the external magnetic field and $s = 0, \pm 1, \pm 2, \ldots$. Equation (6.7) means that the wave frequency $\omega' = \omega - k_z v_z$ either equals zero ($s = 0$), or is divisible by the gyrofrequency ($s \neq 0$) in the reference system where the Larmor centre of the particle is stationary. If $s = 0$, we speak of the Cherenkov interaction of a particle with a wave. If $s \neq 0$, we speak of the cyclotron interaction under the conditions of normal ($s > 0$) or anomalous ($s < 0$) Doppler effect.

Whether the waves increase or damp is determined completely by the tensor $\varepsilon_{\alpha\beta}(\omega, k)$, which in its turn depends on the initial distribution of charged particles over velocities. On choosing the specific expressions for $f(v_\perp, v_z)$ of electrons and ions we substitute them into (6.5). Then we substitute (6.5) into (6.6) and get a transcendental equation to seek ω at the given k. On the whole its solution is associated with calculation difficulties. Let us indicate a useful method of approximate calculation.

The matrix in braces in (6.6) will be denoted $M_{\alpha\beta}$. Let us separate Hermitian and anti-Hermitian parts in $\varepsilon_{\alpha\beta}$

$$\varepsilon_{\alpha\beta} = \varepsilon'_{\alpha\beta} + i\varepsilon''_{\alpha\beta}$$

where

$$\varepsilon'_{\alpha\beta} = (\varepsilon_{\alpha\beta} + \varepsilon^*_{\beta\alpha})/2 \qquad \varepsilon''_{\alpha\beta} = (\varepsilon_{\alpha\beta} - \varepsilon^*_{\beta\alpha})/2i.$$

We shall consider the wave vector k to be real and the frequency ω complex ($\omega \to \omega + i\gamma$). Assuming that $\gamma \ll \omega$ and $\varepsilon''_{\alpha\beta} \ll \varepsilon'_{\alpha\beta}$ we make an expansion

$$\varepsilon_{\alpha\beta}(\omega + i\gamma) \approx \varepsilon'_{\alpha\beta}(\omega) + i\varepsilon''_{\alpha\beta}(\omega) + i\gamma \frac{\partial \varepsilon'_{\alpha\beta}}{\partial \omega}\bigg|_{\gamma=0}.$$

Substituting this into $M_{\alpha\beta}$, we obtain

$$M_{\alpha\beta} = M'_{\alpha\beta} + iM''_{\alpha\beta}$$

where

$$M'_{\alpha\beta} \approx k^2 \delta_{\alpha\beta} - k_\alpha k_\beta - (\omega/c)^2 \varepsilon'_{\alpha\beta}$$

$$M''_{\alpha\beta} \approx -(\omega/c)^2 \varepsilon''_{\alpha\beta} - (\gamma/c^2)\partial \omega^2 \varepsilon'_{\alpha\beta}/\partial \omega. \tag{6.8}$$

At first we completely neglect the anti-Hermitian part of $\varepsilon_{\alpha\beta}$. Then the Maxwell equation takes the form

$$M'_{\alpha\beta}(\omega, k)E_\beta = 0.$$

From the condition of the existence of a nontrivial solution of these equations, we find $\omega_\sigma(k)$, where σ designates the type of a normal wave. At this stage we may introduce the polarization vectors a^σ that yield the condition of orthonormalization

$$a_\alpha^{\sigma^*}(\delta_{\alpha\beta} - k_\alpha k_\beta/k^2)a_\beta^\nu = \delta_{\sigma\nu} \tag{6.9}$$

and the relation

$$k^2\delta_{\sigma\nu} = (\omega/c)^2 a_\alpha^{\nu^*}\varepsilon'_{\alpha\beta}a_\beta^\sigma \tag{6.10}$$

where $\omega = \omega_\sigma(k)$ (see exercise 3.2.1).

In order to find $\gamma_\sigma(k)$ let us use the following approximation for $\varepsilon''_{\alpha\beta}$. We multiply the vector equation $M_{\alpha\beta}a_\beta^\sigma = 0$ by $a_\alpha^{\sigma^*}$. Since the matrices $M'_{\alpha\beta}$ and $M''_{\alpha\beta}$ are Hermitian, we have $a_\alpha^{\sigma^*}M_{\alpha\beta}a_\beta^\sigma = 0$, whence using (6.8) and (6.10) we get

$$\gamma = -\frac{\omega}{2}\left\{n\frac{\partial \omega n}{\partial \omega}\right\}^{-1}\varepsilon''_{\alpha\beta}a_\alpha^*a_\beta. \tag{6.11}$$

Here $n = ck/\omega$ and σ is omitted.

The applicability of (6.11) is restricted by the condition that the anti-Hermitian part of the dielectric permeability tensor has to be small enough. This condition is yielded, for example, when there is a background plasma that may be considered cold and a small admixture of energetic particles. If the concentration of energetic particles is small, and their velocity scatter is sufficiently large, then $\omega(k)$ is determined by the parameters of the background plasma, and only $\gamma(k)$ depends on the distribution of energetic particles in accordance with (6.11).

If the plasma is inhomogeneous, the analysis of instability becomes much more complicated. Furthermore, we shall confine ourselves to considering the situations where a geometro-optical concept may be used more or less reliably. Here we just note that in general the inhomogeneity of the plasma causes, first, the appearance of new types of wave, and, secondly, the modification of the dispersion characteristics of those types of waves that may propagate in the homogeneous plasma. From the physical point of view these changes are connected with the existence of various drifts in inhomogeneous plasma. A so-called drift instability may appear as a result. As in the case of the homogeneous plasma, there appears a sum over cyclotron harmonics in the dispersion equation. However, if $\omega \ll \Omega_i$, then the main contribution is made by the Cherenkov term ($s = 0$). The instability develops most rapidly when a resonant condition of the type

$$\omega \simeq k_\perp \cdot v_D \tag{6.12}$$

is yielded, where v_D is the velocity of the transverse drift. For example, in the Larmor drift $v_D \sim v_T(r_B/a)$, where v_T is the thermal velocity of particles, $r_B = v_T/\Omega$ is the gyroradius, and a is the characteristic scale of the plasma inhomogeneity across the magnetic field. In the case of Alfvén waves we have $\omega = k_\parallel c_A$. Comparing this with (6.12) we find that $k_\perp \gg k_\parallel$, since $v_D \ll c_A$ in the magnetosphere.

6.2   Self-excitation of waves in a transverse waveguide

In section 4.3 we described the toroidal wave channel under the arch of the plasmasphere, surrounding the Earth. There is no ordinary mechanism of dissipation, i.e. absorption of waves at the ends of geomagnetic field tubes. Therefore, even a fairly small enhancement above marginal stability of radiation belt ions may lead to the exponential growth of the wave intensity.

   We shall examine self-excitation of waves under purely transverse propagation. This simplification is justified by the fact that the waveguide axis is perpendicular to the field lines of the geomagnetic field. It turns out that at $k_z = 0$ it is necessary to take into account the contribution of all cyclotron harmonics in expression (6.5) for $\varepsilon_{\alpha\beta}$, and under certain conditions the growth rate has a maximum at large values of s. The necessary condition of instability at high ion harmonics is the nonmonotonic dependence of the distribution function over the transverse velocity, i.e. $\partial f/\partial v_\perp > 0$.

   The analysis of the angular dependence of the growth rate shows that γ is an even function of α, and

$$\partial^2 \gamma/\partial \alpha^2|_{\alpha=0} < 0$$

where $\alpha = \tan^{-1}(k_z/k_\perp)$. In other words, the waves grow most rapidly under transverse propagation. This fact serves as an additional basis for the model of strictly transverse propagation.

6.2.1   The dispersion equation

If $k_z = 0$, then the general dispersion equation (6.6) separates into two: for the ordinary wave

$$n^2 - \varepsilon_{zz} = 0 \tag{6.13}$$

and for the extraordinary wave[2]

$$\varepsilon_{xx}(\varepsilon_{yy} - n^2) + \varepsilon_{xy}^2 = 0. \tag{6.14}$$

An ordinary wave is purely transverse, it is polarized linearly ($E \parallel B$) and is of no interest to us since it propagates only at high frequencies. An extraordinary

---

[2] The ordinary wave was termed so because the magnetic field does not influence the propagation of this wave in a cold plasma, i.e. it propagates as an 'ordinary' wave in a plasma devoid of the external magnetic field.

wave is polarized elliptically in the plane perpendicular to the external magnetic field (figure 6.2). Note that it refers to the vector $\boldsymbol{E}$. The vectors $\boldsymbol{b}$ and $\boldsymbol{D}$ are polarized linearly. This wave is termed a fast magnetosonic wave in the low-frequency limits in the cold plasma.

At $k_z = 0$ the components of the dielectric permeability tensor we are interested in are

$$\varepsilon_{xx} = 1 + \sum \frac{\omega_0^2}{k_\perp^2 \omega} \sum_{s=-\infty}^{\infty} \frac{s^2 \Omega^2}{\omega - s\Omega} \int_0^\infty dv_\perp \frac{\partial f_\perp}{\partial v_\perp} J_s^2$$

$$\varepsilon_{xy} = -\varepsilon_{yx} = i \sum \frac{\omega_0^2}{k_\perp \omega} \sum_{s=-\infty}^{\infty} \frac{s\Omega}{\omega - s\Omega} \int_0^\infty dv_\perp \frac{\partial f_\perp}{\partial v_\perp} v_\perp J_s J_s' \qquad (6.15)$$

$$\varepsilon_{yy} = 1 + \sum \frac{\omega_0^2}{\omega} \sum_{s=-\infty}^{\infty} \frac{1}{\omega - s\Omega} \int_0^\infty dv_\perp \frac{\partial f_\perp}{\partial v_\perp} v_\perp^2 J_s'^2.$$

Here $J_s = J_s(a)$, $a = k_\perp v_\perp / \Omega$ and $f_\perp = \int_{-\infty}^{\infty} f \, dv_z$.

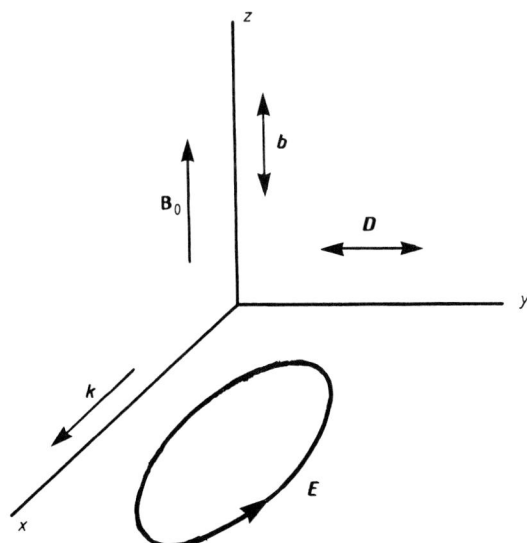

Figure 6.2. Polarization of the magnetosonic wave for transverse propagation.

Bearing in mind the supplements to the toroidal waveguide under the arch of the plasmasphere, we shall consider that the plasma consists of the cold two-component background plasma and a small admixture of energetic protons of the radiation belt. In the range of frequencies $\Omega_i \ll \omega \ll \Omega_e$ from (6.15) there follow the expressions

$$\varepsilon \equiv \varepsilon_{xx}^0 \simeq 1 - \frac{\omega_{0i}^2}{\omega^2} + \frac{\omega_{0e}^2}{\Omega_e^2} \qquad g \equiv -i\varepsilon_{xy}^0 \simeq -\frac{\omega_{0i}^2}{\Omega_i \omega}. \qquad (6.16)$$

Suppose $\varepsilon_{xx} = \varepsilon + (4\pi i/\omega)\sigma_{xx}'$, where σ_{xx}' describes the contribution of energetic particles. Other components of the tensor $\varepsilon_{\alpha\beta}$ may be calculated in the cold plasma approximation. Thus, (6.14) takes the form

$$\frac{g^2}{\varepsilon - n^2} - \varepsilon = \frac{4\pi i}{\omega} \sigma_{xx}'. \qquad (6.17)$$

Substituting expressions (6.15) and (6.16) into (6.17) we obtain the dispersion equation

$$D(\omega, k) = -\frac{\Omega_i^2 \omega_{0i}'^2}{\omega \omega_{0i}^2} \sum_{s=-\infty}^{\infty} \frac{s^2}{\omega - s\Omega_i} \int_0^\infty dv_\perp \frac{\partial f_\perp}{\partial v_\perp} J_s^2(a) \qquad (6.18)$$

where

$$D(\omega, k) = \frac{\omega_{0i}^2}{c^2 \Omega_i^2} - \frac{k^2}{\omega^2} \left(1 - \frac{\omega^2}{\Omega_e \Omega_i} \right).$$

When deriving (6.18) we used the inequalities

$$\Omega_i^2 \ll \omega^2 < \Omega_e \Omega_i \qquad \Omega_e^2 \ll \omega_{0e}^2 \qquad N' \ll N$$

and introduced the designation $\omega_{0i}' = (4\pi e^2 N'/m_i)^{1/2}$, where N' is the energetic proton number density.

Let us express the infinite sum by harmonics through the product of the Bessel functions

$$\sum_{s=-\infty}^{\infty} \frac{J_s^2}{\omega - s\Omega_i} = \frac{\pi \Omega_i}{\sin(\pi \omega/\Omega_i)} J_{\omega/\Omega_i} J_{-\omega/\Omega_i}$$

and bring the dispersion equation to the form

$$D(\omega, k) = -\pi \frac{\omega/\Omega_i}{\sin(\pi \omega/\Omega_i)} \left(\frac{N'}{N} \right) \int_0^\infty dv_\perp \frac{\partial f_\perp}{\partial v_\perp} J_{\omega/\Omega_i}(a) J_{-\omega/\Omega_i}(a). \qquad (6.19)$$

6.2.2 Instability growth rate

We shall consider the wavenumber k to be given and real, and the frequency complex, i.e. $\omega \to \omega + i\gamma$, and $\gamma \ll \omega$. The solution of the dispersion equation, which gives the dependencies $\omega(k)$ and $\gamma(k)$, may be found in the general case only numerically. We shall avail ourselves of the small parameter (N'/N) and obtain the analytical expressions for $\omega(k)$ and $\gamma(k)$ by the method of successive iterations.

In the zero approximation we have

$$\omega^2 = \frac{\Omega_e \Omega_i}{1 + \omega_{0e}^2/c^2 k^2}. \tag{6.20}$$

If $k \ll \omega_{0e}/c$, then $\omega \simeq c_A k$; if $k \to \infty$, then $\omega \to \Omega_{LH}$, where $\Omega_{LH} = (\Omega_e \Omega_i)^{1/2}$ is the lower hybrid resonance frequency.

The expression for the growth rate may be found in the first approximation by (N'/N). The presence of $\sin(\pi\omega/\Omega_i)$ in the denominator in the right-hand side of (6.19) indicates that the growth rate has its maximum near the harmonics of the proton cyclotron frequency. If $\omega \simeq s\Omega_i$, where s is integer, and $\gamma \ll \Omega$, then from (6.19) we can easily obtain

$$\left(\frac{\gamma}{\omega}\right)^2 \simeq \frac{N'}{N} \frac{\omega^2}{k^2} \int_0^\infty J_s^2 \left(\frac{kv_\perp}{\Omega_i}\right) \frac{\partial f_\perp}{\partial v_\perp} dv_\perp. \tag{6.21}$$

The relationship between k and ω in (6.21) is determined by (6.20) at $\omega \simeq s\Omega_i$. We see from (6.21) that the instability appears at $\partial f_\perp/\partial v_\perp > 0$, i.e. for a nonmonotonic distribution function of energetic protons. Under these conditions magnetosonic waves with a discrete spectrum are excited, for which $\omega \simeq s\Omega_i$. We may show that the width of a separate spectral band is of the order of $\Delta\omega \sim \gamma$.

A nonmonotonic distribution of energetic particles in the magnetosphere of the Earth is not exotic. On the contrary, it is well known that there exists a so-called loss-cone effect. It may be easily shown that a distribution function of the loss-cone type is nonmonotonic over transverse velocities and consequently it is unstable relative to the excitation of magnetosonic waves at the harmonics of the proton gyrofrequency.

An appropriate distribution function should be selected for the numerical estimate of the growth rate. A function of the form

$$f_\perp(v_\perp) = \frac{2}{w^2 - \bar{w}^2} \left[e^{-v_\perp^2/w^2} - e^{-v_\perp^2/\bar{w}^2} \right] \tag{6.22}$$

is fairly simple to calculate the integral on the right-hand side of (6.21) on the one hand, and on the other, this function models a nonmonotonic dependence of distribution of energetic protons in the magnetosphere over $v_\perp$.

Substituting (6.22) into (6.21) we find at $w^2 \gg \bar{w}^2$, $(kw)^2 \gg \Omega_i^2$ and

$$\left(\frac{\gamma}{\omega}\right)^2 \simeq 2\frac{N'}{N} \left(\frac{\omega}{kw}\right)^2 \left[\exp\left(-\frac{k^2\bar{w}^2}{2\Omega_i^2}\right) I_s \left(\frac{k^2\bar{w}^2}{2\Omega_i^2}\right) - \frac{\Omega_i}{\sqrt{\pi}kw} \right]. \tag{6.23}$$

Here I_s is the modified Bessel function. If $(k\bar{w})^2 \gg \Omega_i^2$, then

$$\left(\frac{\gamma}{\omega}\right)^2 \simeq \frac{2s}{\sqrt{\pi}} \left(1 - s^2\frac{m_e}{m_i}\right)^{3/2} \frac{N'c_A^3}{N\bar{w}w^2}. \tag{6.24}$$

The growth rate as a function of s has a wide maximum at the harmonics $s \simeq 21, 22$ i.e. in between the proton gyrofrequency Ω_i and the lower hybrid resonance frequency Ω_{LH}.

If the nonmonotonic behaviour of $f_\perp(v_\perp)$ has a loss-cone nature, then $\bar{w} \simeq 2wL^{3/2}$. At $L \sim 4$, $c_A/w \sim 0.2$–0.3, $N'/N \sim 10^{-2}$–5×10^{-3} we get the estimate $\gamma/\Omega_i \sim 10^{-1}$–$5 \times 10^{-2}$ near the maximum of the spectrum. The dependence of γ on ω, calculated with the help of (6.24), is shown in figure 6.3. Here the width of each resonance is only several times larger than the width of the line depicting the spectrum ($\Delta\omega \sim 0.1\Omega_i$).

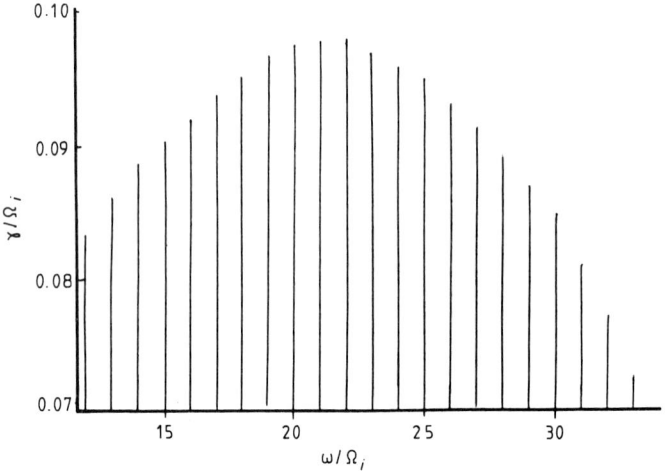

Figure 6.3. Dependence of the growth rate on frequency.

6.2.3   Formation of nonmonotonic proton distribution during a magnetic storm

The leakage of energetic particles into the loss cone is a permanent mechanism for the formation of nonmonotonic distributions. Other mechanisms begin acting sporadically in the magnetosphere. For example, at a 'sudden' injection of particle clouds, separation by energies, stipulated by the dependence of the gradient drift velocity on the energy, takes place. As a result, the cloud is depleted by the small-energy particles at the forefront, i.e. a nonmonotonic distribution is formed. Naturally, such a distribution will be nonstationary. We shall dwell on this in section 6.4, but here we shall discuss an interesting mechanism for the appearance of stationary nonmonotonic distribution[3]. While

---

[3] In the magnetosphere the corresponding conditions are realized during magnetic storms. Consequently, it would be more correct to speak of a quasi-stationary distribution.

the 'loss-cone' and the 'injection' mechanisms are equally effective for protons and electrons, the mechanism that we shall consider now works only for protons.

The action of this mechanism is based on the separation of particles by energy in the course of the electric drift in an inhomogeneous field. Since the velocity of the gradient drift depends on the energy whereas that of the electric drift does not, the area of accessibility for particles of different energies is different, and a nonmonotonic distribution may appear in some points of space. It is interesting to point out that if the co-rotation and convection electric field are taken into account, nonmonotonic distribution may be obtained without any special assumptions about the source of particles, but just by supposing there is such a source with the energy spectrum wide enough at the periphery of the magnetosphere.

The physical essence of the process may be uncovered by analysing drift trajectories of protons. For the transverse propagation of magnetosonic waves the details of the distribution function over longitudinal velocities are of no importance. We may choose the distribution $f(v_\parallel, v_\perp) = \delta(v_\parallel) f(v_\perp)$, to imitate distribution of the general form when calculating the growth rate by (6.21). Then it is sufficient to analyse the trajectories of particles, lying in the equatorial plane. They are determined by the integral of energy

$$\mathcal{E} = \mu B + e\phi = \text{constant} \tag{6.25}$$

where $\mu = mv_\perp^2/2B$ is the transverse invariant, B is the geomagnetic field at the equator and ϕ is the potential of electric field. The geomagnetic field is approximated by the dipole field, i.e. $B = M/r^3$, where $M = 8.06\times10^{25}$ G cm^{-3} is the magnetic moment of the Earth. The convection electric field E is considered stationary, homogeneous and directed from the morning side of the magnetosphere to the evening side. In addition, we shall take into account the electric field, which depends on the rotation of the Earth. It is directed radially (towards the Earth) and is equal to the modulus $\Omega_E M/cr^2$, where $\Omega_E = 7.3 \times 10^{-5}$ s^{-1} is the angular velocity of the Earth's rotation. The electric potential is

$$\phi = -\frac{\Omega_E M}{cr} - Er \sin \varphi \tag{6.26}$$

where φ is the angle measured to the west from the midnight meridian.

Substituting (6.26) into (6.25) and introducing dimensionless variables we obtain

$$\mathcal{E} = \frac{\mu}{L^3} - \frac{1}{L} - EL \sin \varphi. \tag{6.27}$$

Here the energy is expressed in units $e\phi_E = 92$ keV, where $\phi_E = M\Omega_E/cR_E$ is the electric field potential of rotation over the Earth's surface; the electric field is expressed in the units ϕ_E/R_E, and the magnetic moment μ in the units $e\phi_E R_E^3/M$.

The form of trajectories determined by (6.27) is investigated numerically (see bibliography for this section.) For our purpose it will be sufficient to analyse the behaviour of singular points and separatrices at different values of μ.

The singular points are determined by equations $\partial\mathcal{E}/\partial\varphi = 0$, $\partial\mathcal{E}/\partial L = 0$. We may easily prove that there exist three singular points a, b, c lying in the plane of the dawn–dusk meridian ($\varphi = \pm\pi/2$). The coordinates of the singular points in the dusk sector

$$L_{a,b} = \frac{1}{\sqrt{2E}} \left(1 \pm \sqrt{1 - 12\mu E}\right)^{1/2} \tag{6.28}$$

and in the dawn sector

$$L_c = \frac{1}{\sqrt{2E}} \left(-1 + \sqrt{1 + 12\mu E}\right)^{1/2}. \tag{6.29}$$

The points a and c are the node points and the point b is of the centre type[4]. Singular self-intersecting curves (separatrices), which divide the area into sectors with different trajectory topologies, pass through a and c.

Figure 6.4 shows separatrices and singular points in the plane of the geomagnetic equator at different μ. The azimuthal component of the drift velocity is equal to zero along the dotted line connecting b and c. Specific situations are arranged according to the increase of μ_n, $n = 1, 2, \ldots 6$. In the first case the magnetic momentum is zero: $\mu_1 = 0$. There is one separatrix, passing through the point a with coordinate

$$L_a = 1/\sqrt{E}. \tag{6.30}$$

The co-rotation area ('plasmasphere') lies inside the separatrix. Here the particles move along closed orbits counter-clockwise. The convection area is situated outside the separatrix. Here the particles drift along the infinite trajectories from the night side to the day side.

At $\mu > 0 \, (n = 2)$ there appear two new separatrices, passing through the point c, and a singular point b. Gradient drift prevails inside the dotted line, linking b and c. The co-rotation zone is located between the dotted line and the separatrix passing through point a. Two separatrices, passing through c, bound the area with the closed trajectories that do not envelope the Earth (the particles drift around the point b). It is clear that the particles cannot penetrate from infinity into the magnetosphere deep enough: their trajectories are limited by the convection zone.

With the increase of μ the point c moves away from the Earth. The external separatrix, passing through c, and the separatrix, passing through a, merge at a

[4] There exists only one singular point of the node type for electrons in the dusk sector. Electrons have simpler trajectories than protons and for them the mechanism of formation of nonmonotonic distribution under consideration does not work. This is due to the fact that the azimuthal components of the gradient drift velocity and electric drift velocity in the co-rotation field have the same signs for electrons and opposite signs for protons.

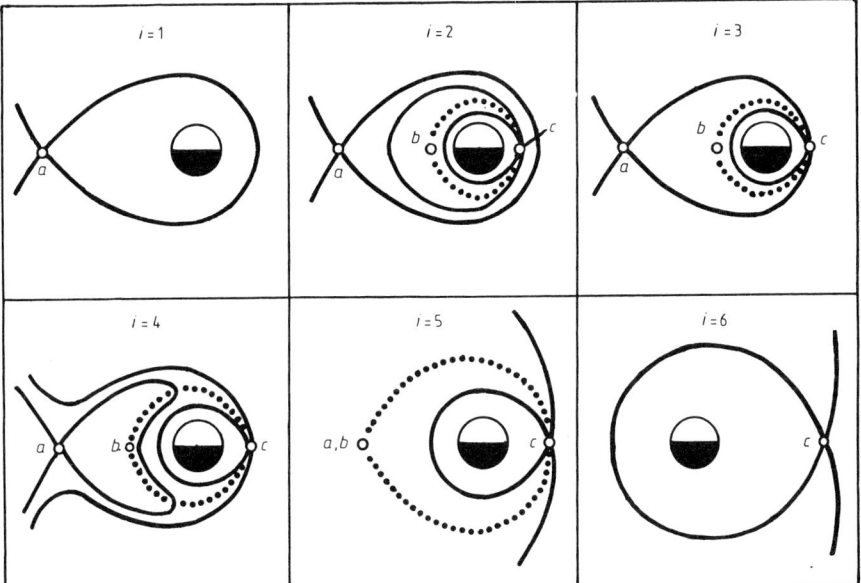

Figure 6.4. Topology of drift proton trajectories in the plane of the geomagnetic equator. The plot is based on the results of the paper by Chen (1970) with the kind permission of the author; a, b, c are the singular points; solid lines are the separatrices. The azimuthal component of the drift velocity is equal to zero along the dotted line.

certain critical value of μ_3 ($n = 3$). When $\mu > \mu_3$ a qualitative reconstruction of the picture of motion takes place. The external separatrix opens and the particles from infinity may penetrate deep into the magnetosphere ($n = 4$). Further increase of μ causes merging at first ($n = 5$), and then leads to the disappearance of points a and b. The evening sector of the magnetosphere becomes accessible for the particles from infinity. However, the forbidden area, limited by the internal separatrix ($n = 6$) is expanded in this case.

The picture under consideration allows qualitative comprehension of the formation mechanism of the nonmonotonic distribution of protons in the evening sector of the magnetosphere ($\varphi = \pi/2$). Let the source of particles be located at the periphery of the magnetosphere. The particles may penetrate inside the plasmasphere ($L < L_a$) only at $\mu > \mu_*$. The exact value of μ_* cannot be obtained analytically. However, it is clear that in any case $\mu_3 < \mu_* < \mu_5$. Taking into account that the increase of μ corresponds to the growth of transverse energy at the fixed magnetic shell, it is clear that the energy spectrum of protons, penetrating into the evening sector of the magnetosphere, will be limited from below.

Limitation of the spectrum from above is also obvious: expansion of the internal separatrice takes place with an increase of energy, and the given fixed point $(L, \pi/2)$ is found in the forbidden zone.

Let us make an approximate estimate of the parameters of proton distribution in the evening sector of the magnetosphere. When estimating the maximum energy we ignore co-rotation and employ the known Alfvén formula for the dimensions of the forbidden zone

$$\varepsilon_{\perp \max} \simeq 3.3 E L. \qquad (6.31)$$

Here $\varepsilon_\perp = m v_\perp^2 / 2$. The minimum energy $\varepsilon_{\perp \min} = \mu_* / L^3$ will be evaluated in the following way. Since $\mu_* < \mu_5 = 1/12E$, then $\varepsilon_{\perp \min} < 1/12EL^3$ (see (6.28)). On the other hand, $\mu_* > \mu_3$. The value of μ_3 is found from the condition of merging of separatrices. In accordance with (6.27) this condition has the form

$$\mathcal{E}(L_a, \pi/2) = \mathcal{E}(L_c, -\pi/2).$$

This equation may be solved with respect to L_c, and then μ may be found using the formula $\mu \simeq L_c^2/3$, which follows from (6.29) at $L_c^2 E \ll 1$. The connection between L_a and L_c has the form

$$L_c^2 \simeq L_a^2 (1 - E L_a^2).$$

For the condition $(L_c/L_a)^2 \ll 1$, which is satisfied at the moment of merging of separatrices, the problem has an analytical solution

$$\mu_3 \simeq 1/27E.$$

Thus we obtain the following estimates of minimum energy from above and from below

$$(27L^3 E)^{-1} < \varepsilon_{\perp \min} < (12L^3 E)^{-1}. \qquad (6.32)$$

Since $\varepsilon_{\perp \max}$ decreases, and $\varepsilon_{\perp \min}$ increases when approaching the Earth, there exists a minimum parameter L_0, determined by the condition $\varepsilon_{\perp \min} = \varepsilon_{\perp \max}$. Using (6.31) and (6.32) we obtain

$$L_0 \simeq (0.35 - 0.4)/\sqrt{E}. \qquad (6.33)$$

The particles do not penetrate over the distances $L < L_0$. At the point L_0 the energetic spectrum has the form $\sim \delta(\varepsilon_\perp - \varepsilon_{\perp 0})$, where

$$\varepsilon_{\perp 0} \simeq 1.2\sqrt{E}. \qquad (6.34)$$

With the growth of L the spectrum of energies widens upwards and downwards (see (6.31) and (6.32) respectively).

So a 'wedge-shaped' structure of the energetic spectrum of protons is formed in the evening sector. Two conditions are necessary for that: a sufficiently strong electric field E and a source of particles with a sufficiently rich spectrum of energies at the magnetosphere periphery. It is essential that the 'edge' of the wedge penetrates inside the plasmasphere ($L_0 < L_a$) and the distribution function near the 'edge' has a δ-shaped form. Both circumstances favour magnetosonic wave excitation at high harmonics of proton gyrofrequency.

The 'wedge-shaped' structure of the proton distribution in the evening sector of the equatorial plane was observed during a magnetic storm on 24 February 1972. According to the *Explorer-45* data, at $L_0 \simeq 4.6$ the energy of particles was approximately 15–20 keV. These values differ by a factor of 1.5–2 from the theoretical estimate with the help of the formula

$$\varepsilon_{\perp 0} \simeq 0.45/L_0 \tag{6.35}$$

which follows from (6.33) and (6.34). We note that during the storm on 24 February 1972 the plasmapause was at a distance $L_a \simeq 4.9$, i.e. protons with nonmonotonic distribution over transverse velocities actually penetrated the plasmasphere.

Let us estimate the instability growth rate near L_0, where the distribution function has the form $f \simeq \text{constant} \times v_\perp^{-1} \delta(v_\perp - v_{\perp 0})$

$$\left(\frac{\gamma}{\omega}\right)^2 = -\frac{N'}{N} \frac{\omega^2}{k_\perp^2 v_{\perp 0}} \frac{\mathrm{d} J_s^2}{\mathrm{d} v_\perp}\bigg|_{v_\perp = v_{\perp 0}}. \tag{6.36}$$

Here, as well as in equation (6.21), $\omega \simeq s\Omega_i$ and $k_\perp \simeq s(\omega_{0i}/c)(1 - s^2 m_e/m_i)^{-1/2}$. Instability appears when $J_s(a)J_s'(a) < 0$. We shall take the values $L_0 \simeq 4.6$, $\varepsilon_{\perp 0} \sim 20$ keV, $N' \sim 0.1$–0.3 cm to be typical for the storm on 24 February 1972. If we also assume $N \sim 10^2$ cm^{-3} and $s \sim 5$–10, then $a \gg 1$ and the asymptotic expansion of the Bessel functions may be applied. Then from (6.36) it follows that $\gamma \sim 3$ s^{-1} in the frequency range $\omega \sim 150$–300 s^{-1}. The instability rapidly develops, within a fraction of a second.

6.2.4 Comparison with the observational data

The emission in the range $10 < f < 100$ Hz was observed on board the *OGO-3* satellite from the internal side of the plasmapause near the equatorial plane within $\pm 2°$ along the latitude. The waves propagated almost perpendicular to the geomagnetic field. The deviation from the perpendicular propagation does not exceed $\sim 1°$. The polarization of the magnetic field oscillations is almost linear, and the vector b is almost parallel to B.

The localization of emission, and the direction of propagation and polarization are in agreement with the results of qualitative theory of magnetosonic wave propagation in the toroidal waveguide under the arch of plasmasphere.

Self-excitation of waves at high harmonics of the proton gyrofrequency is possible in this waveguide. The theory predicts, firstly, the increase of waves in the range $\Omega_i < \omega < (\Omega_e \Omega_i)^{1/2}$ and, secondly, the discrete structure of the spectrum (see figure 6.3).

Let us examine figure 6.5 with the emission spectrum observed at *OGO-3*. We see that there exists emission between the proton gyrofrequency and the lower hybrid resonance frequency. Narrow spikes in the emission spectrum are of great interest. We assume that this peculiarity of the spectrum is caused by the specific instability mechanism, leading to the growth of waves in the narrow bands in the vicinity of the harmonics of the proton gyrofrequency. The analysis of the spectra, similar to the one shown in figure 6.5, testifies to the fact that within the accuracy of measurements the distance between the adjacent bursts actually coincides with the local proton gyrofrequency.

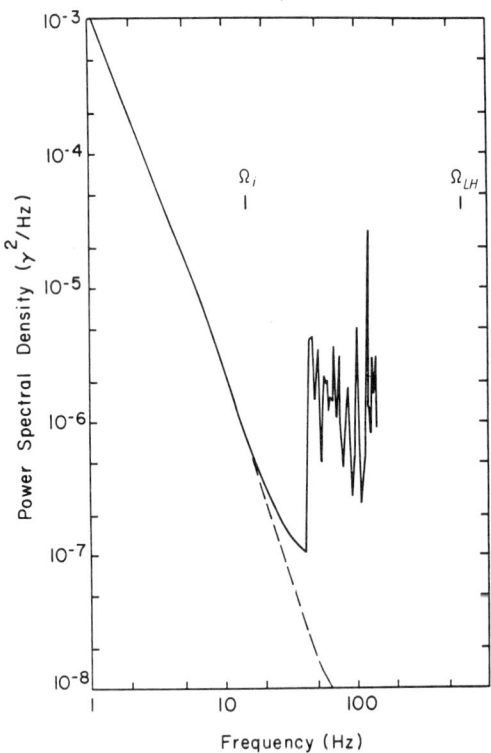

Figure 6.5. Spectrum of electromagnetic waves observed on board the *OGO–3* satellite (Russell *et al* 1970).

The visual proof of this interpretation was provided by the measurements of geoelectromagnetic waves at the satellites *IMP-6* and *Hawkeye-1*. The fragments of trajectories, where the emission with a discrete spectrum was observed, are indicated by the dashed lines in figure 6.6. The dynamic spectrum of oscillations is shown in figure 6.7. We see a clear lined structure of the spectrum in the range from about 100 to 250 Hz. There is no doubt that we are dealing with the radiation at the harmonics of the proton gyrofrequency. As a matter of fact, the local proton gyrofrequency at the moment of observation ($f_{Bi} \simeq 16$ Hz) coincides with the distance between adjacent lines in the spectrum ($\Delta f \simeq 14$ Hz) within the measurement errors ($\sim 10\%$).

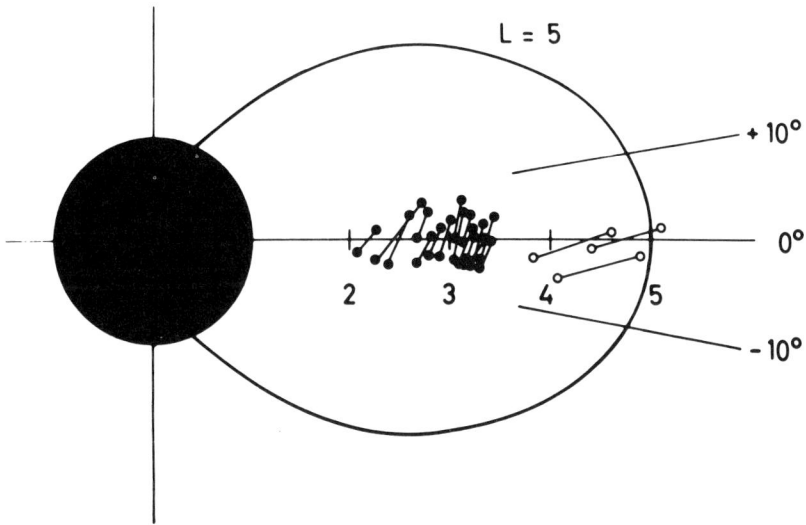

Figure 6.6. Fragments of the trajectories of *IMP-6* and *Hawkeye-1* (dark circles) and *OGO-3* (light circles) at which specific emission in the range $\sim 10^2$ Hz was observed (Gurnett 1976).

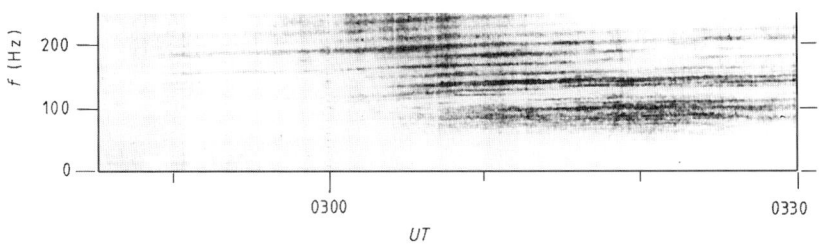

Figure 6.7. Dynamic spectrum of the emission at the harmonics of proton gyrofrequency observed on board the *IMP-6* satellite (Gurnett 1976).

6.3 Self-excitation of longitudinal waveguides

At a critical distribution of radiation belt particles an increase of the Alfvén waves takes place in longitudinal waveguides just as in the case of magnetosonic waves in the transverse waveguide.   Two specific features of longitudinal waveguides hamper theoretical research. First of all, in contrast to the transverse waveguide, the longitudinal one cannot be considered regular even in the rough approximation.   Irregularity is related to longitudinal inhomogeneity of the plasma distribution and magnetic field, to the variability of the waveguide axis curvature, to the presence of nontransparency bands on the axis, etc. Moreover, in contrast to the transverse waveguide, the longitudinal waveguide rapidly loses energy due to the leakage of waves from the waveguide ends into the ionosphere. Some energy is supposed to return into the waveguide after the reflection from the ionosphere, but the probability of this hypothesis is based on the observational data of Pc1 pulsations rather than on accurate calculation.

Both these specific features make the problem of self-excitation of longitudinal waveguides global. We cannot restrict ourselves by the analysis of the roots of the local dispersion equation here as has been done in section 6.2 when considering a transverse waveguide, and yet this is how we start 'to improve' the local solution with waveguide irregularity and losses at the ends taken into account.

6.3.1   The local dispersion equation

At $k_\perp = 0$ the general dispersion equation (6.6) splits into two independent equations

$$\varepsilon_{zz} = 0 \tag{6.37}$$

$$c^2 k_z^2 = \omega^2(\varepsilon_{xx} \mp i\varepsilon_{xy}). \tag{6.38}$$

Under longitudinal propagation the components of the dielectric permeability tensor (6.5) have the form

$$
\varepsilon_{xx} = \varepsilon_{yy} = 1 + \frac{1}{4}\sum \frac{\omega_0^2}{\omega^2} \sum_{s=\pm 1} \int \frac{v_\perp^2 \, dv_\perp dv_z}{\omega - k_z v_z - s\Omega}
$$
$$
\times \left\{ \frac{\partial f}{\partial v_\perp} - \frac{k_z v_z}{\omega}\left[\frac{\partial f}{\partial v_\perp} - \frac{v_\perp}{v_z}\frac{\partial f}{\partial v_z} \right]\right\} \tag{6.39}
$$

$$
\varepsilon_{xy} = -i\varepsilon_{yx} = \frac{i}{4}\sum \frac{\omega_0^2}{\omega^2} \sum_{s=\pm 1} s \int \frac{v_\perp^2 \, dv_\perp dv_z}{\omega - k_z v_z - s\Omega}
$$
$$
\times \left\{ \frac{\partial f}{\partial v_\perp} - \frac{k_z v_z}{\omega}\left[\frac{\partial f}{\partial v_\perp} - \frac{v_\perp}{v_z}\frac{\partial f}{\partial v_z} \right]\right\} \tag{6.40}
$$

$$
\varepsilon_{zz} = 1 + \sum \frac{\omega_0^2}{\omega} \int_{-\infty}^{\infty} \frac{v_z dv_z}{\omega - k_z v_z}\frac{\partial F}{\partial v_z} \tag{6.41}
$$

where

$$
F(v_z) = \int_0^\infty f(v_z, v_\perp) v_\perp \, dv_\perp \qquad \int_{-\infty}^{\infty} F \, dv_z = 1.
$$

A contribution to ε_{xx} and ε_{xy} is made only by cyclotron terms, $s = \pm 1$, and to ε_{zz} only by the Cherenkov term ($s = 0$). The poles in the subintegral expressions (6.39)–(6.41) should be passed from below (from above) at $k_z > 0$ ($k_z < 0$).

Equation (6.37) describes potential waves of the Langmuir and ion-acoustic type. Polarization of waves is linear, the vector E is parallel to the wave vector (and the external magnetic field). The magnetic field does not influence the propagation of these waves, as seen from (6.41). They are of no interest to us and will not be considered hereafter.

Equation (6.38) describes the propagation of two circularly polarized transverse waves with rotation of the electric and magnetic vectors (if we look along B) to the left (upper sign) and to the right (lower sign). It is quite clear that in the low-frequency limit these are the Alfvén and magnetosonic waves.

Substituting (6.39) and (6.40) into (6.38) we bring the dispersion equation to the form

$$c^2 k_z^2 = \omega^2 + \frac{1}{2} \sum \omega_0^2 \int \frac{v_\perp^2 \, dv_\perp \, dv_z}{\omega - k_z v_z \mp \Omega} \left\{ \omega \frac{\partial f}{\partial v_\perp} - k_z v_z \left[\frac{\partial f}{\partial v_\perp} - \frac{v_\perp}{v_z} \frac{\partial f}{\partial v_z} \right] \right\}. \tag{6.42}$$

Note that the terms with $s = +1$ make a contribution to the equation for the $\mathcal{L}$-waves, and the terms with $s = -1$ contribute to the equation for the $\mathcal{R}$-waves.

The rule of rounding the pole in (6.42) when integrating over v_z may be written as

$$\frac{1}{z} = \frac{P}{z} - i\pi \delta(z). \tag{6.43}$$

Here the symbol P means the principal part of the integral and $z = \omega - k_z v_z \mp \Omega$. Hence it is evident that the imaginary part of the frequency (growth rate γ) is determined by the particles which are in cyclotron resonance with the wave

$$v_z = (\omega \mp \Omega)/k_z. \tag{6.44}$$

The sign and value of γ depend on the expression

$$\left[\frac{v_\perp}{v_z} \frac{\partial f}{\partial v_z} - \frac{\partial f}{\partial v_\perp} \right]_{v_z = (\omega \mp \Omega)/k_z}. \tag{6.45}$$

In the case of an isotropic distribution of particles, when (6.45) is equal to zero, we always have $\gamma < 0$, i.e. the waves are attenuating. On the other hand, if the distribution is fairly anisotropic, then growth of $\mathcal{R}$- or $\mathcal{L}$-waves may occur, depending on the sign of (6.45).

Let us consider electron cyclotron resonance for the whistlers. It should be kept in mind that whistlers propagate in the range $\Omega_i \ll \omega < \Omega_e$ and refer to the $\mathcal{R}$-wave type (magnetosonic branch). Substituting $\Omega = -\Omega_e$ into (6.44) and employing the dispersion equation (4.15) we prove that the longitudinal velocity of resonance particles is directed in the opposite direction to the wave propagation: $k_z v_z < 0$. The resonant velocity is

$$v_z \simeq c \left(\frac{\Omega_e}{\omega} \right)^{1/2} \frac{\Omega_e}{\omega_{0e}}. \tag{6.46}$$

Analysis of (6.42) shows that the instability occurs in the case of negative anisotropy, i.e. when (6.45) is negative.

In the MHD frequency range, which is limited by the ion gyrofrequency at $k_\perp = 0$, the electron cyclotron resonance is inefficient for the following reason. Combining (6.44) with the dispersion laws (4.12) and (4.13) we get the estimate of the resonance velocity

$$v_z \simeq c \left(\frac{m_i}{m_e} \right)^{1/2} \frac{\Omega_e \Omega_i}{\omega_{0e} \omega}.$$

At $\omega < \Omega_i$ and typical values of B and N, it corresponds to the ultra-relativistic energies of electrons. The density flux of ultra-relativistic particles in the magnetosphere of the Earth is negligibly small, and instability does not occur[5].

The situation is different in the case of cyclotron interaction of protons with the MHD waves. The resonant velocity is approximately

$$v_z \simeq c \Omega_i^2 / \omega_{0i} \omega \tag{6.47}$$

and for the waves in the Pc1 range corresponds to typical velocities of the radiation belt protons. This leads to an increase of magnetosonic waves at positive anisotropy, when $v_z k_z > 0$. If the anisotropy of protons is negative, then the Alfvén waves increase. For them $v_z k_z < 0$, as seen from (6.44).

Before we proceed with the analysis, let us briefly discuss two problems. The first one concerns the applicability of the approximation of longitudinal propagation. If we abstract from methodical considerations as well as for reasons of simplicity then the fact that the growth rate of cyclotron instability, caused by the anisotropy of particle distribution, has a maximum at $k_\perp = 0$ may serve as a reason to single out strictly parallel propagation. It may seem that one more argument lies in the fact that we have considered that the longitudinal waveguides duct the waves along the geomagnetic field lines. However, the modes with $k_\perp = 0$ are absent among the waveguide modes. The condition $k_\perp = 0$ should be replaced by a weaker condition $k_\perp \ll k_z$, but we shall not do so in order not to complicate the theory.

The second problem that will be considered in what follows concerns the distribution of particles in the radiation belt.

---

[5] Although we have taken advantage of the formulae of nonrelativistic theory for the approximate estimate of the velocity of resonant particles, the conclusion will not be changed if we take account of the relativistic effects in the explicit form. The conclusion will not be changed either if we use more general dispersion laws instead of (4.12) and (4.13).

6.3.2 Mechanism of anisotropy formation

The velocity distribution of energetic particles is anisotropic in the radiation belts. On the one hand, there is a loss cone in the geomagnetic trap, as in any other trap with magnetic mirrors. The particles with small pitch-angles perish in dense layers of the atmosphere. On the other hand, the very character of the formation of the external radiation belt leads to anisotropy. In fact, the magnetic moment

$$\mu = mv_{\perp}^2/2B$$

and the longitudinal invariant

$$J = \int v_{\parallel}\, dl$$

are conserved during the transport of particles from the periphery into the deep magnetosphere. The integration is carried out along the geomagnetic field line between the mirror points. At large pitch-angles ($\alpha = \tan^{-1}(v_{\perp}/v_{\parallel}) \simeq \pi/2$) it follows from the conservation of magnetic moment that $v \simeq v_{\perp} \propto L^{-3/2}$, i.e. the energy increases as $\varepsilon_{\alpha=\pi/2} \propto L^{-3}$ with the decrease of the parameter L. At small pitch-angles ($\alpha \ll 1$) the longitudinal invariant is approximately $J \simeq vl$, where l is the field-line length and, consequently, $\varepsilon_{\alpha=0} \propto L^{-2}$.

Therefore, when transported into the magnetosphere the particles accelerate, and their distribution becomes anisotropic, and on average transverse energy is larger than longitudinal energy. Hence the value of (6.45) is negative.

6.3.3 Instability growth rate

Energetic particles of the radiation belt form only a small admixture to the background plasma. In accordance with this fact, let us divide the distribution function in (6.42) into two parts: $f \to f_0 + f$. Far from the poles of the index of refraction the background plasma may be considered cold, i.e. we may suppose that $f_0 = 2v_{\perp}^{-1}\delta(v_z)\delta(v_{\perp})$. We shall not specify distribution function of energetic particles $f(v_z, v_{\perp})$ at present, but we shall take into account the fact that the concentration of these particles N' is much smaller than that of the cold plasma N. Then the dispersion equation (6.42) in the first approximation by the small parameter (N'/N) will take the form

$$k^2 - \frac{\omega^2}{c^2}n_{\pm}^2(\omega) = -\mathrm{i}\frac{\pi\omega_0'^2\Omega}{c^2k^2}\int_0^\infty v_{\perp}^2\, dv_{\perp}\left[\frac{kv_{\perp}}{\Omega}\frac{\partial f}{\partial v_z} \pm \frac{\partial f}{\partial v_{\perp}}\right]_{v_z=u} \tag{6.48}$$

where

$$n_{\pm}^2 = \varepsilon_{\perp} \pm g \qquad u = (\omega \mp \Omega)/k \qquad \omega_0' = (4\pi e^2 N'/m)^{1/2}.$$

Here we have taken advantage of (6.43), and we have neglected the contribution of energetic particles to the expression for the refractive index $n_{\pm}(\omega)$. Therefore, the refractive index, the real part of the frequency, polarization and the group velocity of the waves are determined by the cold plasma. Only the growth rate depends on the distribution of energetic particles in the given approximation.

To find the growth rate we consider the wavenumber k to be real, and the frequency complex, i.e. $\omega \to \omega + i\gamma$. In the approximation of weak instability ($\gamma \ll \omega$) we may solve (6.48) using the method of successive iterations by $\gamma \propto N'$. In the zero approximation we get

$$\omega n_\pm(\omega) = ck. \tag{6.49}$$

In the first approximation[6]

$$\gamma = \frac{\pi \omega_0'^2 \Omega v_g}{4c^2 k^2} \int\limits_0^\infty v_\perp^2 \, dv_\perp \left[\frac{kv_\perp}{\Omega} \frac{\partial f}{\partial v_z} \pm \frac{\partial f}{\partial v_\perp} \right]_{v_z = u} \tag{6.50}$$

where v_g is defined by the formulae (5.19) and (5.20) if $n_\pm(\omega)$ is defined by (4.5).

The problem may be posed in a different way when $\omega = \mathrm{Re}\,\omega$ is fixed and a complex wavenumber $k = k' + ik''$ is sought. The value k'' describes spatial wave amplification ($k'' < 0$) or attenuation ($k'' > 0$). At $k'' \ll k'$ the real part of the wavenumber k' is determined by (6.49), and the coefficient of amplification k'' is connected with the growth rate (6.50) by the formula

$$k'' = -\gamma/v_g. \tag{6.51}$$

Let us specify the form of the distribution function. Knowing that the distribution of energetic particles in the radiation belt is anisotropic we shall take as a model the Maxwell distribution with different temperatures along and across the magnetic field

$$f = \frac{1}{\sqrt{2\pi}} \left(\frac{m}{T_\parallel} \right)^{1/2} \left(\frac{m}{T_\perp} \right) \exp\left\{ -\frac{m}{2} \left[\frac{v_z^2}{T_\parallel} + \frac{v_\perp^2}{T_\perp} \right] \right\}. \tag{6.52}$$

Substituting this into (6.50) we get

$$\frac{\gamma}{\Omega} = -\frac{\sqrt{\pi}}{2} \frac{\omega_0'^2 v_g}{c^2 k^2 w_\parallel} \left[\frac{\omega}{\Omega} \frac{w_\perp^2}{w_\parallel^2} \mp \left(\frac{w_\perp^2}{w_\parallel^2} - 1 \right) \right] \exp\left[-\left(\frac{\Omega \mp \omega}{k w_\parallel} \right)^2 \right] \tag{6.53}$$

where $w_{\parallel,\perp} = (2T_{\parallel,\perp}/m)^{1/2}$. The instability obviously arises under the condition

$$\frac{\omega}{\Omega} < \mp \frac{\Delta}{1 - \Delta}. \tag{6.54}$$

Here $\Delta \equiv 1 - T_\perp/T_\parallel$. In the case of ions ($\Omega = \Omega_i$), for $\Delta < 0$ $\mathcal{L}$-waves increase, and for $\Delta > 0$ $\mathcal{R}$-waves increase. In the case of electrons ($\Omega = -\Omega_e$)

---

[6] Expression (6.50) may also be obtained using (6.11).

for $\Delta < 0$ $\mathcal{R}$-waves increase in the range of whistlers. For $\Delta > 0$ theoretically $\mathcal{L}$-waves may increase. However, electrons with ultra-relativistic energies will be necessary for that. Otherwise the growth rate is exponentially small.

Instability may be absolute or convective. At absolute instability the amplitude exponentially increases in time at every point of space. And if the instability is of convective character, then at the propagation of the wave packet the amplitude remains limited at any fixed point. This difference is essential for us, since the wave packet, propagating in the longitudinal waveguide, fairly rapidly escapes the region of the radiation belt.

Let us write down the criterion of absolute instability in the form

$$\gamma_m > \left(\tfrac{1}{4}\right) v_g^2 \delta (\alpha^2 + \delta^2)^{-1} \tag{6.55}$$

where we used the same symbols as in exercise 5.1.1. If $k \ll \omega_{0i}/c$ and $c_A/w_\parallel \ll |1 - T_\parallel/T_\perp|$, then the growth rate is approximately

$$\gamma(k) \simeq \mp \frac{\sqrt{\pi}}{2} \frac{\omega_{0i}'^2 \Omega_i^2 \Delta}{\omega_{0i} w_\parallel c k^2} \exp\left(-\frac{\Omega_i^2}{w_\parallel^2 k^2}\right). \tag{6.56}$$

The maximum γ corresponds to

$$k_m \simeq \Omega_i / w_\parallel \tag{6.57}$$

and approximately equals

$$\gamma_m \simeq \mp \frac{\sqrt{\pi}}{2e} \frac{\omega_{0i}'^2 w_\parallel \Delta}{\omega_{0i} c}. \tag{6.58}$$

Here e is the base of natural logarithms. The group velocity v_g, as well as the values of α and δ are

$$v_g \simeq c(\Omega_i/\omega_{0i}) \qquad \alpha \simeq \mp(\Omega_i/2)(c/\omega_{0i})^2 \qquad \delta \simeq 2\gamma_m/k_m^2. \tag{6.59}$$

Substituting (6.58) and (6.59) into (6.55) we obtain the criterion of absolute instability

$$\beta_\parallel' |\Delta| > 1 \tag{6.60}$$

where $\beta_\parallel' = 4\pi m_i w_\parallel^2 N'/B^2$.

In the radiation belt $\beta_\parallel' |\Delta| \ll 1$, i.e. the absolute instability is not realized. If we eliminate extreme situations, then locally the radiation belt is the amplifier and not the generator of MHD signals. However, since the amplified signal is partly reflected from the ionosphere and comes back into the amplifier, there arises positive feedback, and the radiation belt may switch over to the generation regime. To analyse such a possibility we have to take account of the inhomogeneity of the magnetosphere and the effect of reflection from the ionosphere and find the integral coefficient of the signal amplification.

6.3.4 Longitudinal inhomogeneity

If $k'' < 0$, then the wave amplitude increases $\exp(-k''\Delta l)$ times over the section Δl of the regular waveguide. Replacing this value by

$$\exp\left(-\int k''\,dl\right) \tag{6.61}$$

we obtain an estimate of amplification in an irregular waveguide provided the waveguide parameters change little over the wavelength. In other words, we use the approximation of geometric optics in order to take into account the longitudinal inhomogeneity of the waveguide.

When reflected from the ionosphere the wave amplitude decreases $1/R$ times, where R is the coefficient of reflection. Consequently, the damping decrement of the wave packet, oscillating along the longitudinal waveguide between the conjugate points in the northern and southern hemispheres, is

$$\frac{1}{\tau}\ln\frac{1}{R^2} \tag{6.62}$$

where

$$\tau = \int\frac{dl}{v_g} \tag{6.63}$$

is the period of the wave packet repetition. Note that for simplicity we have ignored the difference between the coefficients of reflection from the northern and southern ends of the waveguide.

Combining (6.51), (6.61) and (6.62) we get the integral growth rate

$$\bar{\gamma} = \frac{1}{\tau}\left[\int\frac{\gamma}{v_g}\,dl - \ln\frac{1}{R^2}\right]. \tag{6.64}$$

Here the integration is carried out twice along the field line between the conjugate points. The local growth rate is determined by the formula (6.53).

We have obtained the most important formula of the theory of longitudinal waveguides. It allows us to investigate the conditions of self-excitation of the waveguide and find the spectrum of growing waves. Self-excitation arises at $\bar{\gamma}(\omega) > 0$. Transition into an overcritical state is possible both as the local growth rate increases, and as losses decrease when waves are reflected from the ionosphere, i.e. with the growth of R.

Let us estimate the resonance frequency and maximum growth rate for the Alfvén waves. We have proved that these waves are subject to local enhancement due to the cyclotron interaction with the protons of the external radiation belt. If the longitudinal waveguide crosses a fairly active part of the radiation belt, and if the ionosphere at the ends of the waveguide is a fairly good reflector, then the Alfvén waves are subject to global enhancement, i.e. self-excitation of the waveguide takes place.

Let the waveguide be located inside the plasmasphere. Then we may take advantage of the strong inequality $c_A/w_\parallel \ll 1$. Outside the plasmasphere the strong inequality is substituted for a weak one. An analytical estimate becomes impossible. However, the estimates, given below, are also valid here by their order of magnitude.

A simple analysis of equations (6.51) and (6.56)–(6.58) testifies to the fact that the local coefficient of enhancement, as a function of l, is maximum at the waveguide equator ($l = 0$) and is a function of frequency so that at the resonance frequency

$$\omega \simeq \Omega_i(c_A/w_\parallel) \tag{6.65}$$

where Ω_i and c_A are taken at the point $l = 0$.

In order to estimate the integral on the right-hand side of (6.64) let us assume $N(l) \propto B(l)$. Then the dependence of the subintegral expression on l at the frequency (6.65) will be mainly determined by the factor

$$\exp\{-[B(l)/B(0)]^3\}.$$

Expanding $B(l)$ into a series in the vicinity of the point $l = 0$

$$B(l) \simeq B(0)[1 + (9/2)(l/R_E L)^2]$$

we find the length of a field-line segment, where the local coefficient of enhancement decreases by e^{-1}

$$l_{\text{ef}} \simeq 0.24 R_E L.$$

Now we choose the upper sign in (6.58) in accordance with the type of the wave and obtain an estimate of the integral in (6.64)

$$\frac{4l_{\text{ef}}\gamma_m}{c_A} \simeq 1.3 \times 10^{-10} L^4 J_\parallel \left(\frac{T_\perp}{T_\parallel} - 1\right) \tag{6.66}$$

where $J_\parallel = N'w_\parallel$. The dimension coefficient in (6.66) corresponds to the estimate of the flux of energetic protons $J_\parallel$ in the units $\text{cm}^{-2}\,\text{s}^{-1}$.

Let us use (6.66) for the numerical estimation of the critical value R_c of the coefficient of reflection from the ionosphere. At $L \simeq 4$ the ratio of temperatures of protons in the radiation belt is $T_\perp/T_\parallel \simeq 2$, the average longitudinal velocity is $w_\parallel \simeq 3 \times 10^8$ cm s^{-1}, concentration is $N' \simeq 0.1$ cm^{-3} and the flux $J_\parallel \simeq 3 \times 10^7$ cm^{-2} s^{-1}. Hence the magnitude of (6.66) approaches unity. The value R_c is defined from the condition $\bar{\gamma} = 0$: $R_c \simeq 0.6$. If $R > 0.6$, then at the presented values of parameters the regime of self-excitation of the Alfvén waves ensues.

Let us supplement the set of parameters with the following values: $\Omega_i \simeq 50$ s^{-1}, $c_A \simeq 5 \times 10^7$ cm s^{-1}, $R \simeq 0.8$. Then, using (5.40), (5.42) and (6.64)–(6.66) we obtain

$$\omega \simeq 8.4 \text{ s}^{-1} \qquad \tau\omega \simeq 10^3 \qquad D \simeq 0.12 \qquad \bar{\gamma}/\omega \simeq 5 \times 10^{-4}. \tag{6.67}$$

6.3.5   The mystery of 'pearls'

Ion cyclotron instability of the radiation belt obviously leads to the excitation of geomagnetic pulsations of the 'pearls' type. In sections 4.4 and 5.2 we described briefly the properties of pearls, referring to the frequency ω, repetition period τ and the dispersion D. Let us supplement this list with information on the growth rate.

The integral growth rate may be estimated by the rising stage of a series of pearls (figure 6.8). On average over many events of this kind $\gamma_1 \approx 5 \times 10^{-3}$ s^{-1}. One may find the damping decrement using the rate of decrease of the amplitude at the end of the series of pearls. On average $\gamma_2 \simeq 3 \times 10^{-3}$ s^{-1}. Evidently the growth rate ensured by the signal travelling through the active part of the radiation belt regardless of the losses in the ionosphere $\gamma = \gamma_1 + \gamma_2 \simeq 8 \times 10^{-3}$ s^{-1}. These numbers were obtained by processing several dozens of series of pearls, the frequency of which is $f \simeq 0.75$ Hz on average. Therefore, $\gamma/f \approx 10^{-2}$ and $\gamma_1/f \approx 6.7 \times 10^{-3}$. Both values reasonably agree with the estimates (6.66) and (6.67).

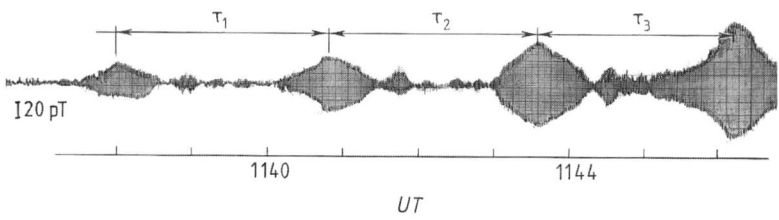

Figure 6.8. Rising stage of a series of pearls (Sogra Observatory 13 November 1968). One can see the progressive decrease of the signal repetition period with increasing amplitude.

So, the experimental data testify to the fact that the problem of excitation of pearls on the whole is solved satisfactorily by means of the theory of ion cyclotron instability of the radiation belt. The frequency, repetition period and growth rate, measured experimentally, agree with the theoretical evaluations. However, the two most important properties of pearls, anomalous dispersion and discrete excitation, are still something of a mystery.

The fact that the dispersion D is not constant in the course of a series of pearls and may even assume negative values was considered in section 5.2. We know of clever attempts to solve this paradox, but the question remains unclear, since the problem of anomalous dispersion cannot be separated from another question: why does a series of pearls consist of separate periodically repeating wave packets?

So we try to understand how discrete signals arise in a continuously acting source.

Attempts were made to explain the discrete excitation by the properties of the initial conditions. It was supposed, for example, that at a certain moment there appears a priming wave packet which grows in the radiation belt and, reflecting periodically from the ends of the longitudinal waveguide, is observed in the form of a series of pearls. This hypothesis seems doubtful. It is not clear why other priming impulses do not appear during dozens of minutes, and moreover, why thermal fluctuations do not enhance and do not fill in the gaps between separate structural elements of oscillations.

Another approach to the problem of discreteness is as follows. First of all the decisive effect of such factors as the presence of a priming impulse or specific initial distribution of particles is excluded. To the best of our judgment, discreteness is an asymptotic property of the solution of the nonlinear equation of transport and not the consequence of special initial conditions. In other words, the stability of a discrete structure during hundreds of passes of a wave packet through the active part of the radiation belt should be sought in the properties of the equations describing the process themselves rather than in the properties of the initial conditions.

A major part in the origin of periodic succession of discrete signals may be played by a simple effect, i.e. the speed of travel of resonant particles across the longitudinal waveguide due to the azimuthal drift of particles in the geomagnetic field. If the length of time that the particles are present within the limits of the wave field is shorter than some critical time, then there appears to be self-modulation of the growth rate, and the wave field breaks up into separate packets.

The stress in this scenario is made on self-modulation of the growth rate. An alternative scenario contains an assumption on the self-modulation of the decrement, which appears as a result of modification of the ionosphere under the influence of pearls.

Thus, the pearl pulsations discovered by Harang and Sucksdorf back in the 1930s is still an excellent means for the study of different manifestations of ion cyclotron instability of the radiation belt. It is not surprising that hundreds of experimental and theoretical papers are dedicated to the investigation of pearls. However, the dominating property of pearls—their discreteness—is still a mystery.

6.4   Origin of pulsations of rising frequency

In the two preceding sections we have discussed the cyclotron instability of a plasma relative to electromagnetic waves, propagating strictly along and strictly across the external magnetic field. When propagating at an arbitrary angle to the external magnetic field the instability analysis is limited by calculation difficulties. Altogether, it is advisable to apply numerical methods. A certain simplification arises in the case of the Cherenkov instability, i.e. at $s = 0$ in the resonance ratio (6.7). We shall investigate this case as applied to the Earth's magnetosphere and present arguments in favour of the hypothesis that, owing to the Cherenkov instability of protons injected into the magnetosphere during the magnetic storms, MHD oscillations of rising frequency are excited in the evening sector. Sometimes we term these oscillations 'hydromagnetic howling' in our attempt to draw an analogy with whistle and howling wind during storms in the troposphere. Another term in common use is IPDP which is the abbreviation for 'interval of pulsations of diminishing period'.

We shall begin with a short digression in order to outline the position of the IPDP in the general picture of appearance and evolution of geomagnetic pulsations.

6.4.1   General picture of waves

Absolutely quiet conditions in the magnetosphere correspond to the streamlining of the geomagnetic dipole by the strictly stationary solar wind. Even in this ideal case one should expect the electromagnetic waves to appear. According to estimates, laminar streamlining is unstable, and waves appear over the surface of the magnetosphere, just as waves appear over a water surface when the wind is strong.

In reality the quiet solar wind contains different inhomogeneities that run over the magnetosphere and are an additional source of oscillations. Permanent oscillations in the daylight hours and sporadic impulses at night are characteristic for quiet conditions. This difference in oscillation regimes undoubtedly reflects the asymmetry of the magnetosphere relative to the plane of the morning–evening meridian. Under quiet conditions 'hydromagnetic hiss' of low intensity is not infrequent in the auroral zone.

Strong perturbations of magnetosphere are called geomagnetic storms. The wave picture during different phases of the storm is roughly as follows. 8 min 20 s after the chromosphere flash, which engenders the processes causing the storm, the ionosphere over the day side of the Earth is perturbed by the UV radiation of the Sun. As a result, there follows the reconstruction of the system of ionospheric currents and a weak magnetic signal arises. The storm begins 30–60 hours after the front of the interplanetary shock wave has arrived at the surface of the magnetosphere. At this moment the frequencies of oscillations, which existed till then, increase abruptly. Moreover, the collision of the interplanetary shock wave with the magnetosphere engenders multiple oscillations of different form and duration. Then the initial phase of the storm follows. It is characterized by a progressive increase of frequency and chaotic modulation of the amplitude of oscillations.

During the main phase of the storm the field is sharply perturbed at all scales. The waves typical for the auroral zone under quiet conditions start penetrating into middle latitudes. The most active period of the main phase— a so-called magnetospheric substorm—is accompanied by 'hydromagnetic howlings' before the midnight hours. These signals are of rising tone with a frequency difference over two–three octaves. Howling is usually preceded by a short powerful impulse.

In the post-storm period, or, as it is put, during the recovery phase, separate impulses keep on appearing at the night side of the Earth, but their intensity weakens. Gradually quiet conditions return over the day side. Permanent oscillations last from morning till night. Their period gradually increases. It is at this period that so-called 'giant pulsations', possessing regular and rather attractive form, may be observed. About a week after the storm a great number of hydromagnetic whistlers are excited. Evidently, the excitation of these whistlers is the most long-lasting after-effect of the storm.

An example of an IPDP or hydromagnetic howling is presented in figure 6.9. Here we see a dynamic spectrum of oscillations. Dark vertical stripes are the pulsations of Pi1B, usually testifying to the impulsive injection of energetic particles from the tail plasma layer into the magnetosphere in the vicinity of the midnight meridian. Obviously, a short powerful impulse at 16.45 UT corresponds to the moment of injection that has generated an IPDP. After this impulse there emerges wideband radiation with slowly increasing average frequency. Within ~ 20 min the frequency increased from fractions of hertz to ~ 1 Hz. The velocity of frequency drift $df/dt \sim 5 \times 10^{-2}$ Hz min^{-1}. The amplitude of oscillations is 0.1–1 nT. Against the background of the noise band we see discrete elements, brighter than the background, that resemble pearls. In contrast to the pearls, these signals arrive at the magneto-conjugate points simultaneously.

The effect of the 'western frequency drift', discovered during synoptical observations at observatories distributed in longitude, is noteworthy. The effect consists of a longitudinal dependence of the spectral evolution, i.e. the given frequency $f = $ constant is observed at the eastern observatories earlier than at the western observatories. We may say that the frequency 'travels' along the azimuth from the east to the west. The velocity of the 'western frequency drift' is of the order of $\dot{\varphi}_\omega \sim 2$ deg min^{-1} on average.

It is natural to explain the effect by the travel of the generation region along the longitude. The western drift direction provides a definite indication of the type of radiating particles (ions), and the size of the drift velocity enables us to estimate their energy approximately.

The key problem in the IPDP theory is the problem of the mechanism for frequency growth. Let us demonstrate that the nonstationarity of the spectrum and other properties of IPDPs find their natural explanation within the framework of the Cherenkov instability. The nonmonotonic character of the distribution function necessary for the instability to appear is framed in the process of azimuthal drift and spreading of the cloud of injected protons.

6.4.2 The mechanism for the formation of unstable distributions

Let us examine the evolution of the distribution of particles in the dipole magnetosphere at a fixed magnetic shell. Suppose the injection occurs at $t = 0$,

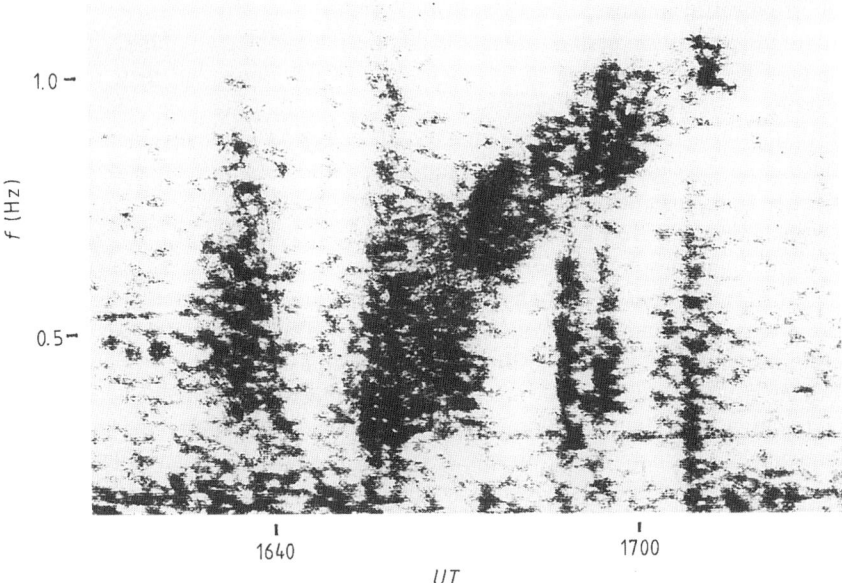

Figure 6.9. Geomagnetic pulsations of rising frequency in the evening sector (LT = UT + 3h).

and the distribution function has the form

$$f(\mu, v, \varphi) = F(\mu, v)\Phi(\varphi).$$

Here $\mu = v_\perp^2/B$, v is the velocity, B is the geomagnetic field and φ is the azimuth counted westwards from the midnight meridian. The distribution will be normalized as follows

$$\int\limits_{-\infty}^{\infty} dv_\parallel \int\limits_{0}^{\infty} F v_\perp dv_\perp = N'\Delta\varphi \qquad \int\limits_{-\pi}^{\pi} \Phi d\varphi = 1$$

where N' is the concentration of the injected particles at the equator and $\Delta\varphi$ is the azimuthal extent of the injection region.

Within a short period of time (of the order of the longitudinal oscillations of particles) intermixing will occur and a quasi-stationary distribution along the field lines will appear. In general this distribution will have the form of two oppositely moving beams. We shall simplify the problem by assuming that there are no beams. Moreover, the initial distribution will be considered to be Maxwellian

$$F = \left(2N'\Delta\varphi/\sqrt{\pi}w^3\right)\exp\left(-v^2/w^2\right). \tag{6.68}$$

The azimuthal dependence of injection will be chosen in the form

$$\Phi = \sum_{n=-\infty}^{\infty} \delta(\varphi + 2\pi n) \tag{6.69}$$

where the condition of periodicity $\Phi(\varphi + 2\pi n) = \Phi(\varphi)$ is taken into account.

The kinetic equation, averaged by the gyroperiod and the period of longitudinal oscillations, has the form

$$\partial f/\partial t + \langle \dot\varphi \rangle \partial f/\partial \varphi = 0. \tag{6.70}$$

The averaged drift velocity is approximately

$$\langle \dot\varphi \rangle = (3m_i c R_E L/2eM)(v_{\perp 0}^2 + p v_{\parallel 0}^2) \tag{6.71}$$

where M is the magnetic moment of the Earth, $v_{\parallel 0}$, $v_{\perp 0}$ are the longitudinal and the transverse velocities at the equator and p is a parameter that characterizes the pitch-angle dependence of the drift velocity. In the dipole magnetic field $p \simeq 0.85$. The general solution (6.70) has the form

$$f = f(\mu, v, \varphi - \langle \dot\varphi \rangle t). \tag{6.72}$$

The initial distribution (6.68) is stable. However, at $t > 0$ during the drift the particles are separated by energies and pitch-angles. As a result a nonmonotonic ($\partial f/\partial v > 0$) and anisotropic ($\partial f/\partial v_\perp \ne \partial f/\partial v_\parallel$) distribution is formed, and its parameters are dependent on time. Instability of this distribution will cause excitation of low-frequency electromagnetic waves with a nonstationary spectrum.

6.4.3 Analysis of the growth rate

Let us calculate the local instability growth rate according to (6.11) with regard to (6.5) and (6.68)–(6.72). The Cherenkov term ($s = 0$) will predominate if the perturbation of the field is the magnetosonic waves. At $\omega \ll \Omega_i$ the refraction index and the polarization vector have the form

$$n = \omega_{0i}/\Omega_i \qquad a = (0, 1, 0) \tag{6.73}$$

where $\omega_{0i} = (4\pi e^2 N/m_i)^{1/2}$, $\Omega_i = eB/m_i c$, N is the background plasma number density and $N \gg N'$. Here the Cartesian system of coordinates is used and orientated so that $B = (0, 0, B)$ and $k = (k_\perp, 0, k_z)$. For Cherenkov resonance we have

$$\varepsilon''_{yy} = -\frac{4\pi e^2}{m_i \omega} \int_{-\infty}^{\infty} \frac{dv_z}{v_z} \frac{\partial g(v_z)}{\partial v_z} \delta(\omega - k_z v_z) \tag{6.74}$$

$$g(v_z) = \pi \int_0^\infty f(v_z, v_\perp) J_1^2 \left(\frac{k_\perp v_\perp}{\Omega}\right) v_\perp^3 dv_\perp.$$

Combining (6.11) and (6.68)–(6.74) we obtain the spatio-temporal dependence of the growth rate

$$\gamma \simeq -\sqrt{\pi} \frac{N'}{N} \frac{\Omega_i^2 \Delta\varphi}{|k_z| w} \frac{1}{\tau} \sum_{n=0}^\infty \exp(-\psi_n/\tau) z_n J_0(z_n) J_1(z_n) \qquad (6.75)$$

where

$$\tau = \dot\varphi_\perp t \qquad \psi_n = \varphi + 2\pi n \qquad \dot\varphi_\perp = 3 m_i c R_E L w^2 / 2 e M$$

$$z_n = (k_\perp w / \Omega_i)(\psi_n/\tau - \omega^2 / k_z^2 w^2)^{1/2}.$$

For simplicity the terms of order $(1 - p)/p \ll 1$ in (6.75) were neglected.

While deriving (6.75) it was assumed that the parameters of the particle distribution vary sufficiently slowly with the time. The corresponding criterion has the form $(1/\gamma)(\partial\gamma/\partial t) \ll t_{NL}^{-1}$, where t_{NL} is the characteristic time of the nonlinear processes which saturate the oscillation amplitude. Assuming $t_{NL} \sim 1/\gamma$ and using (6.75) we rewrite the criterion of adiabaticity in the form $\gamma \gg \dot\varphi_\perp/\varphi$.

This criterion is satisfied with ample margins at $\varphi \sim 1$ if the condition cn the local growth rate is obeyed, i.e if γ^{-1} is much less than the time l/c_A during which the waves escape the instability region. Here l is the characteristic dimension of this region. After some manipulations with the formula (6.75) the criterion of locality will take the form

$$N'w \gg cB/4\pi e l \Delta\varphi.$$

If the magnetic field in the instability region is of the order of $B \sim 300$ nT and $l \sim R_E$, $\Delta\varphi \sim 0.3$ then the flux should be $N'w \gg 5 \times 10^7$ cm^{-2} s^{-1}.

Two more conditions of the applicability $\gamma \ll \omega \ll \Omega_i$ lead to the inequalities $c_A \ll w$ and

$$N'\varepsilon \ll B^2/8\pi$$

where $\varepsilon = m_i w^2/2$. It is assumed that all three above-mentioned conditions are obeyed.

At $\tau < 2\pi$, i.e. within a time interval less than the period of rotation of the bulk of particles around the Earth, the principal term in (6.75) is the first term of the series ($n = 0$). The instability arises in the frequency bands where $J_0 J_1 < 0$. The zeros of the Bessel functions serve as the boundaries of the bands. Within the first band we have $2.4 < z < 4.0$, and in the second band $5.6 < z < 7$, etc. The growth rate decreases with increasing the number of the band ν as $1/\nu$. Within the first band the growth rate has a maximum at

$$\omega/\Omega_i \simeq (c_A/w)\sqrt{\tau/\varphi} \qquad (6.76)$$

$$k_\perp/k_z \simeq (w/c_A)\sqrt{\varphi/\tau} \tag{6.77}$$

and

$$\gamma/\Omega_i \simeq (N'w\Delta\varphi/Nc_A)(\varphi/\tau^2)\exp(-\varphi/\tau). \tag{6.78}$$

Here the condition $w \gg c_A$ was taken into account and multipliers of the order of unity were omitted.

From (6.76) it follows that the emission frequency increases with time as $\omega \propto \sqrt{t}$. Then concerning (6.78) at the fixed longitude φ, the growth rate as a function of time has a maximum at $\tau_m = \varphi/2$. The dependence of the steepness of the frequency growth on φ has the form

$$(\partial\omega/\partial t)_m \propto \varphi^{-1}. \tag{6.79}$$

The growth rate as a function of φ has a maximum at $\varphi_m = \tau$. From this it follows that the region of maximum amplification moves along the azimuth at the velocity

$$(\partial\varphi/\partial t)_m \simeq \dot\varphi_\perp. \tag{6.80}$$

Let us make some estimates of the emission parameters in the evening sector ($\varphi \simeq \pi/2$). We assume $L \simeq 6$, $c_A \simeq 10^8$ cm s^{-1}, $w \simeq 3 \times 10^8$ cm s^{-1}, $N' \simeq 0.3$ cm^{-3} and $\Delta\varphi \simeq 20°$. The given value of c_A corresponds to $N \simeq 10$ cm^{-3} and the value w to protons with energy ~ 50 keV. Regarding the value $\Delta\varphi$, see exercise 9.3.4. Using (6.76)–(6.80) we find the maximum growth rate $\gamma \sim 0.6$ s^{-1}, the time of reaching the maximum growth rate $\Delta t \sim 20$ min, the frequency $f \sim 0.5$ Hz, the steepness of the frequency increasing $\partial f/\partial t \sim 0.1$ Hz min^{-1} and finally the 'westward drift' velocity $\partial\varphi/\partial t \sim 2.4$ deg min^{-1}. It interesting to note the set of parameters produced by the few initial assumptions and the surprisingly close correspondence they have to the observational data of IPDPs, especially if one takes into account the simplicity of the model.

6.4.4 Discussion

The relatively simple model considered describes IPDP pulsations with a high degree of accuracy. Let us enumerate the main consequences of the theory which find experimental confirmation or may serve as the subject of the following searches.

The theory explains the nonstationarity of the oscillation spectrum or, at least in general features, describes the dependence of the frequency on time. At $w \gg c_A$ we have $d\omega/dt > 0$, $d^2\omega/dt < 0$, i.e. the dynamic spectrum has a bell-like structure. In many cases it corresponds to observations. However, sometimes we have $d\omega/dt > 0$, $d^2\omega/dt^2 > 0$. The theory gives indirect indications of such a possibility; concerning (6.75) at $w \sim c_A$ we have $\omega \propto (\tau_0 - \tau)^{-1}$, where $\tau_0 = (w/c_A)^2\varphi$, i.e. $d^2\omega/dt^2 > 0$. However, at $w \sim c_A$ the condition for the applicability of the model is violated and for the study of such a situation one has to use numerical methods.

The experimentally determined 'westward drift frequency' velocity corresponds to the theoretical estimation at reasonable values of the injection parameters. In the framework of the theory one may explain the observable dependence of the steepness of the frequency increasing with local time: the further to the west from the midnight meridian the observation point is located the less steeply the frequency increases with time.

The theory takes into account the possibility of multiple generation of IPDPs which is a result of the rotation of the injected particles' cloud around the Earth. It is expected that after each rotation the dynamic spectrum of IPDPs becomes smoother. Really, from (6.76) it follows that at fixed frequency $df/dt \propto t^{-1}$.

What is principally new is the conclusion of the possibility of IPDP generation in two or more frequency bands. The growth rate decreases with the increase of the number of the band. This explains the absence of multiband emissions under ordinary conditions. Nevertheless it is reasonable to try to find the multiband IPDPs since they may supply information on extreme conditions in the magnetosphere, on the specific features of the generation mechanism, etc.

The energetic particle flux during a magnetospheric storm is so strong that the excitation of IPDPs does not need positive feedback. In the case of Pc1 this is realized by the wave reflection from the waveguide ends and the return of the waves to the amplification region. During IPDP excitation the magnetosphere is possibly in a state of absolute instability or (in the case of convective instability) the amplification is so strong that the wave amplitude increases up to observable values within one pass through the amplification region.

Let us note that the structural elements of Pc1 undergo frequency modulation similar to the frequency modulation of IPDPs. According to the point of view of many researchers, the Pc1 frequency modulation is caused by the wave packet spreading of ion cyclotron waves. However, it is useful to take into account other possible explanations.

Let us imagine the wave packet in one of the longitudinal waveguides. It undergoes amplification in the near equatorial segment of the trajectory, causes strong pitch-angle particle diffusion and moves to one of the ends of the field line. New energetic particles come into the waveguide due to the azimuthal drift of protons so that the appearance of protons in the waveguide goes through the side (east) boundary. When the wave packet, reflected from the ionosphere, comes back to the near equatorial segment it falls into the amplifier which has modulated over the frequency amplification coefficient. Really, due to the separation of particles by energy during the azimuthal drift, the waveguide is filled by more energetic particles which amplify relatively low frequencies and then by less energetic particles which amplify relatively high frequencies. In the framework of this scenario Pc1 pulsations may be termed 'micro-IPDP'.

Exercise

Exercise 6.4.1

Calculate and analyse the Alfvén wave growth rate in a cold plasma with an admixture of energetic protons whose distribution function is described by (6.68)–(6.72).

Solution 6.4.1

At $\omega \ll \Omega_i$ the refractive index and polarization vector of the Alfvén wave are

$$n \simeq \frac{\omega_{0i}/\Omega_i}{|\cos\theta|} \qquad a = \left\{ \frac{1}{|\cos\theta|}, 0, 0 \right\}$$

where θ is the angle between $\boldsymbol{k}$ and $\boldsymbol{B}$. Cherenkov interaction in this approximation is absent. Cyclotron interaction leads to instability with a growth rate

$$\gamma \simeq -\frac{\sqrt{\pi}}{2} \frac{N'}{N} \frac{\Omega_i^2 \Delta\varphi}{|k_z|w} \frac{1}{\tau} \sum_{n=0}^{\infty} \exp\left(-\frac{\psi_n}{\tau}\right) \sum_{s=-\infty}^{\infty} \frac{s^2}{z_n^2} \frac{\mathrm{d}J_s^2}{\mathrm{d}z_n}.$$

Here

$$z_n = (k_\perp w/\Omega_i)(\psi/\tau - s^2\Omega_i^2/k_z^2 w^2)^{1/2}.$$

The other notations are as used in (6.5) and (6.11).

During cyclotron interaction not only nonmonotonicity but also anisotropy of distribution plays a part. Here the terms associated with the anisotropy are neglected for simplicity. This is valid if the following inequalities are obeyed

$$1 - p \ll 1/s \qquad (1-p)/\sqrt{p} \ll c_A/w \ll 1.$$

The second inequality is fairly strong since in the dipole magnetosphere $p \simeq 0.85$. Thus the given formula for γ only approximately describes the spatio-temporal dependence of the growth rate.

Let us suppose $\tau < 2\pi$. Then we may limit the analysis to the first term in the expansion over n. The instability takes place at $J_s' J_s < 0$, $\mathrm{Im}\, z = 0$. The second condition leads to the inequality $|k_z| > |s|\,(\Omega_i/w)\sqrt{\tau/\varphi}$. The emission is multiband as in the case of magnetosonic waves. If $(2\Omega_i/w)\sqrt{\tau/\varphi} > |k_z| > (\Omega_i/w)\sqrt{\tau/\varphi}$, we have only terms with $|s| = 1$. The frequency in this case is of the order (6.76) and the growth rate

$$\gamma/\Omega_i \simeq (N'\Delta\varphi/N)(\varphi/\tau^3)^{1/2} \exp(-\varphi/\tau).$$

The growth rate is approximately (w/c_A) times less than (6.78).

At fixed longitude the growth rate has a maximum at $\tau_m \simeq (2/3)\varphi$. The region of maximum amplification moves along the longitude at a velocity of $(\mathrm{d}\varphi/\mathrm{d}t)_m \simeq (\frac{1}{2})\dot{\varphi}_\perp$.

6.5 Pulsations of extramagnetospheric origin

For a long time the predominant opinion was that geomagnetic pulsations arose in the magnetosphere and at its boundary, i.e. at the magnetopause. This standpoint was presented in all the reviews and monographs published before 1970. However, it is not as obvious as it may at first seem. For example, types of pulsations could be MHD waves penetrating into the magnetosphere from the solar wind.

The following objection was put forward against this idea. On its way from the solar wind into the magnetosphere the wave must overcome a complicated system of obstacles: the near-Earth shock wavefront and the turbulent plasma layer, separating this front from the magnetopause, and, finally, the magnetopause itself, across which there are sharp jumps of pressure, density and velocity of plasma as well as the jump of the magnetic field intensity. It was considered that even if the MHD wave did not attenuate in the turbulent layer, it would completely reflect from the magnetopause in any case and would not penetrate into the magnetosphere.

However, these considerations were never supported by rigorous treatment. In fact, a rigorous calculation of such a problem can hardly be carried out if we want to take adequate account of the fairly high complexity of the medium in which the waves propagate. So we should turn to the observations themselves for the answer.

Below we consider an experimental proof of the extramagnetospheric origin of the main type of geomagnetic pulsations, observed everywhere on the day side of the Earth, and present arguments in favour of the analogous origin of some other types of pulsations, observed over the polar caps.

6.5.1 The main type of geomagnetic pulsation

Every day numerous observatories distributed over the globe register nearly sinusoidal oscillations of the magnetic field with periods from 10 to 45 s. According to the nomenclature, approved by the XIII General Assembly of the International Union of Geodesy and Geophysics, they are indicated by the symbol Pc3.

Here we present some brief information on the morphology of Pc3. The typical value of the carrier frequency is $f \simeq 0.04$ Hz. The relative band width is $\Delta f / f \simeq 0.1$. The amplitude is highly variable in time and space ($b = 0.1$–5γ). Oscillations, registered at a given observatory, last for several hours: oscillations start early in the morning and are over in the afternoon. This modulation is evidently associated with the 24 hour rotation of the Earth around its axis.

An instantaneous picture of the Pc3 distribution over the Earth's surface is as follows. The longitudinal extent of the region occupied by pulsations is $\sim 120°$ at mid and $\sim 60°$ at high latitudes. The distribution is characterized by the general growth of the amplitude from low to high latitudes, against the background of which two maxima stand out—high-latitude ($\Phi \simeq 68°$) and sub-auroral ($\Phi \simeq 60°$). The former may be best traced in the morning and the latter in the afternoon sector. The period of the oscillations is approximately equal over the entire Earth's surface.

The polarization of Pc3 in the horizontal plane is elliptic as a rule. At mid-latitudes the rotation of the horizontal projection of the vector b is counter-clockwise in the morning hours and clockwise in the afternoon hours. At high latitudes counter-clockwise rotation predominates (i.e. it is counter-clockwise if we look downwards in the northern hemisphere and upwards in the southern hemisphere).

With the enhancement of the general level of geomagnetic activity the period of oscillations decreases from $T \simeq 30$ s at $K_p = 1$ to $T \simeq 10$ s at $K_p = 6$. In addition, a deformation of the latitude profile of the amplitude takes place. This deformation may be described by the displacement of the maximum on the profile from $L \simeq 5$ at $K_p = 1$ to $L \simeq 3$ at $K_p = 6$. (Here $L = \cos^{-2} \Phi$ is the parameter of the magnetic shell).

A special property of the morphology is the relation between the characteristics of geomagnetic pulsations and the parameters of the interplanetary medium at the orbit of the Earth. Because of this relation the extramagnetospheric origin of Pc3 was discovered. The essence of the discovery lies in the fact that electromagnetic waves in the Pc3 range observed on the ground are continuously arriving at the Earth from the interplanetary medium. After long discussions and numerous tests this concept is now generally accepted.

The radical change in the views on the origin of Pc3 is interesting to follow. After the fundamental studies by Dungey it was known that the spectrum of MHD oscillations of the magnetosphere contains continuous and discrete components. It seemed natural to identify the carrier frequency of quasi-sinusoidal oscillations of Pc3 with one of the discrete eigenfrequencies of the spectrum. On the basis of this proposition it was inferred that the variations in Pc3 frequency are caused by changes in the dimensions of the magnetospheric resonator, which, in their turn, are caused by the variability of the solar wind.

The connection between the Pc3 frequency and the dimension of the magnetosphere was actually detected in the early 1960s, but the coefficient of correlation turned out to be small. Moreover, correlations of the properties of Pc3 with many other parameters were soon found: velocity, density and temperature of the solar wind, the position of the outer boundary of the radiation belt, the critical frequency of the ionosphere and so on. Finally, in early 1970s the connection between the frequency of Pc3 and the interplanetary magnetic field was found. While other correlations may possibly be explained based on the theory of magnetospheric origin, this correlation would not fit the framework of such a theory. An alternative hypothesis was put forward: the carrier frequency is determined by the spectrum of an extramagnetospheric source of oscillations, rather than by the spectrum of eigenfrequencies of the magnetospheric resonator. This hypothesis includes the radical supposition of the penetration of MHD waves from the interplanetary medium into the magnetosphere. It admits an experimental test and explains many of the properties of Pc3 which seemed mysterious before.

Let us enumerate the simple, reliably established morphological characteristics of Pc3. They are (1) frequency range, (2) permanence, (3) amplitude maximum at midday hours, (4) relative stability of frequency and considerable variability of amplitude, (5) disappearance of oscillations at transverse orientation of the interplanetary magnetic field as regards the Sun–Earth line. These properties, known for a long time, did not serve to formulate the theory of the extramagnetospheric origin of Pc3 and cannot serve as its foundation as they are. Moreover, each of them taken separately could be explained by some other appropriate theory, and, in fact, this was done. But there is no other theory, besides the theory of extramagnetospheric generation of Pc3, which could easily explain all the enumerated properties.

The new approach to the origin of Pc3, although it seemed to appear suddenly, was set in motion by the article by Fairfield (1969) dedicated to MHD waves in front of the magnetosphere in the so-called upstream region. This is a special region of the solar wind 'connected' with the bow shock by the field lines of the interplanetary magnetic field. It is supposed that the waves are excited due to the cyclotron instability of the beam of protons reflected from the bow shock and spiralling back along the field lines of the interplanetary magnetic field in a general direction towards the Sun (figure 6.10).

In addition to the upstream region, two more sources of MHD waves are known outside the magnetosphere. These are the Sun and the free stream solar wind. However, it is the upstream region that is of particular interest to us, since it is a powerful source of MHD waves in the Pc3 frequency range. Sometimes we succeed in comparing the waves in the upstream region directly with the Pc3 pulsations on the ground (figure 6.11), but it does not happen often. Therefore to prove the extramagnetospheric origin of Pc3 an indirect criterion was applied.

The starting point for the formation of such a criterion lies in the fact that the medium on the propagation path is stationary or almost stationary. Then the ground observer may register pulsations with frequency

$$f \simeq V/\lambda \tag{6.81}$$

where V is the solar wind velocity, and λ is the wavelength in the upstream region. Here we have used the Doppler shift and have taken into account that $V \gg c_A$. In addition the condition of cyclotron resonance is also employed

$$\lambda \simeq u/f_B \tag{6.82}$$

where $f_B = eB/2\pi m_p c$ is the frequency of gyrorotation of protons in the interplanetary magnetic field, u is the average longitudinal velocity of reflected protons and $u \gg c_A$ (figure 6.12). Combining (6.81) and (6.82) we have

$$f \simeq gB \tag{6.83}$$

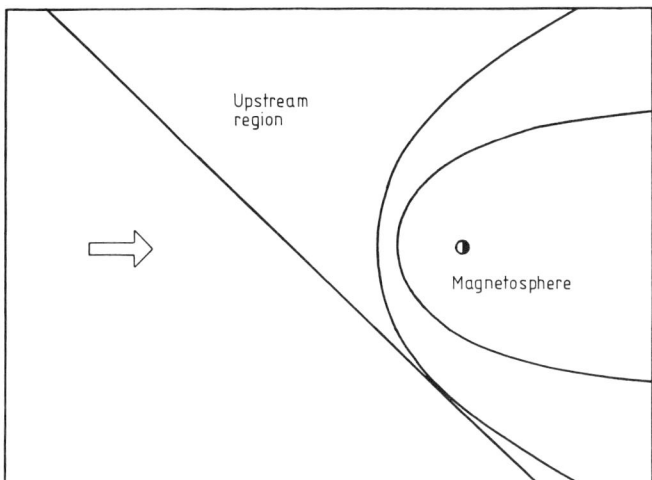

Figure 6.10. Schematic presentation of the near-Earth space in the equatorial plane. The arrow indicates the solar wind direction. The straight line stands for the interplanetary magnetic field line which is tangential to the shock front at the given moment. The other lines, which have already crossed the shock front (not shown on the figure), are draped around the magnetopause. Some fraction of the solar wind particles is reflected from the bow shock and moves along the interplanetary magnetic field lines from the right to the left in the upstream region. (After Russell and Hoppe (1983) with slight modification.)

where $g = eV/2\pi m_p u$. At typical values of $V = 400$ km s^{-1} and $u = 1000$ km s^{-1} we have $g = 6$ mHz nT^{-1}. On average $B \simeq 5$ nT, hence $f \simeq 30$ mHz, which corresponds to the Pc3 range. According to our hypothesis the carrier frequency of the geomagnetic Pc3 pulsations must increase with increasing interplanetary magnetic field.

The simple considerations presented above were confirmed by the analysis of the roots of the dispersion equation for the MHD waves in front of the magnetosphere (see exercise 6.5.4). The analysis proved the dependence of the resonance frequency on the interplanetary magnetic field and also gave information on the coefficient of wave amplification. It is noteworthy that the coefficient of amplification has a maximum at $k_\perp = 0$, i.e. for longitudinal propagation. This conserves the quasi-monochromaticity of oscillations and allows us to obtain the formula $f = 6B$ from the resonance condition with regard to the Doppler shift; this formula may be used to test the theory.

For the experimental test a so-called f–B diagram was constructed using the results of observations of pulsations on the ground and the data on the interplanetary magnetic field. Each point in figure 6.13 represents hourly average values of f and B. Small crosses indicate events during which oscillations with two different frequencies were observed simultaneously, i.e. a small circle on the oblique branch corresponds to each small cross on the horizontal branch.

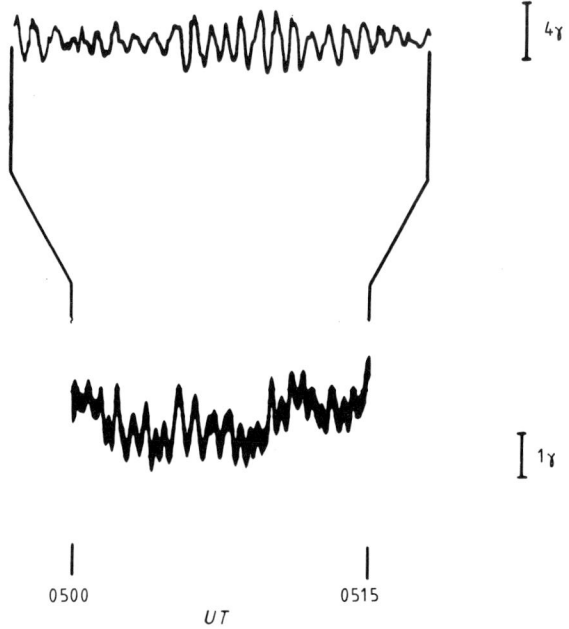

Figure 6.11. The comparison of magnetic pulsations in the solar wind and on the ground. Upper panel—data from *Explorer-34* (Fairfield 1969). Lower panel—the fragment of the magnetogram registered at a mid-latitude observatory (26 October 1967). The oscillation frequencies coincide with fairly high accuracy: $\Delta f \simeq 1$ mHz at $f \simeq 28$ mHz.

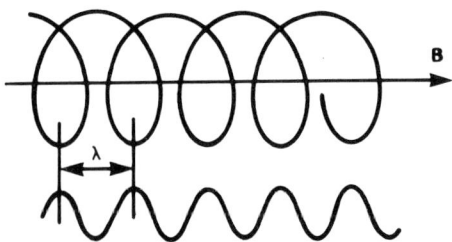

Figure 6.12. Illustration of the cyclotron resonance. The trajectory of a charged particle in the magnetic field B has the form of a spiral. The wavelength at resonance coincides with the spiral pitch if the particle velocity along the field considerably exceeds the phase velocity of the wave.

The points of the oblique branch are approximated by the formula $f = gB$, $g \simeq 5.7$ mHz nT^{-1}, and the correlation coefficient $r = 0.78 \pm 0.02$. This dependence has been confirmed by many research groups and convincingly

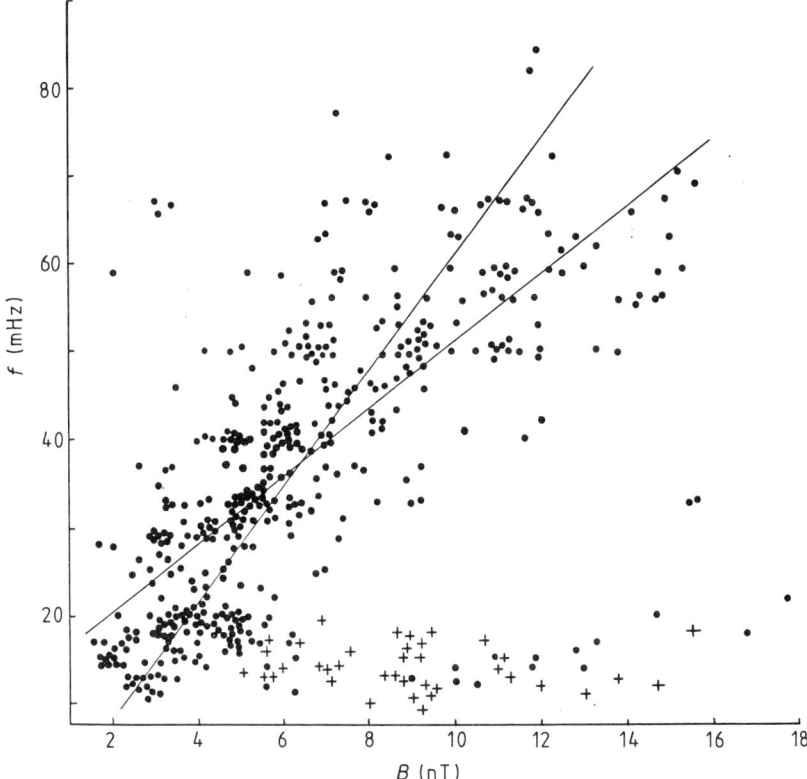

Figure 6.13. The dependence of the carrier frequency of the geomagnetic pulsations on the value of the interplanetary magnetic field. The upper group of dots corresponds to pulsations of extramagnetospheric origin ('oblique branch').

testifies to the extramagnetospheric origin of Pc3. The coefficient of the relation between f and B, according to all known measurements, is

$$g = 5.8 \pm 0.3 \text{ mHz nT}^{-1}. \tag{6.84}$$

Here an average value and the scatter (by sextiles) are indicated.

Let us turn back to the five morphological properties of Pc3, enumerated above, and comment on them from the position of the hypothesis on the extramagnetospheric generation. The first point does not need any explanation. The second is evidently connected with the permanence of the beam of reflected protons. The third property is explained by the preferential orientation of the field lines of the interplanetary magnetic field, creating asymmetry of the beam of reflected protons relative to the plane of the midday meridian. The preferential orientation stands for the orientation according to the Parker spiral, so that the pre-midday maximum of the Pc3 amplitude appears because of the axial rotation of the Sun and the Earth, and the Earth rotates in one particular direction[7]. The fourth property as well as the third are not accidental, but have an explicit physical reason. The predominant type of magneti; field variations in the solar wind are the long-period Alfvén waves, creating a variable picture of the orientation of the interplanetary field and leaving its magnitude almost invariable. Hence, there follows relative stability of Pc3 frequency, which depends on the magnitude, and the amplitude variability, which depends on the orientation of the interplanetary magnetic field. Finally, the fifth property is explained by the fact that there are no reflected particles in front of the magnetosphere at azimuthal orientation of the interplanetary magnetic field, and the MHD waves are not excited.

Why do we call these 'the main type' of Pc3 pulsations? The simplest explanation is as follows: pulsations of other types appear and Pc3 pulsations disappear only at some special and rather rare conditions in the near-Earth medium. If this explanation is insufficient we may draw a formal analogy with the main type of magnetospheric convection. Both Pc3 and the 'main type' of convection are particularly sensitive to the variability of the interplanetary magnetic field. Consequently, Pc3 may be called the main type in some narrow but interesting sense: this is the only type of geomagnetic pulsation, the extramagnetospheric origin of which has been proved definitely.

6.5.2   On the dual nature of Pc2 pulsations

Pc2 pulsations ($T = 5$–10 s) arise at mid-latitudes during magnetic storms. It was supposed that Pc2 are excited in the radiation belt by the oxygen ions during the main phase of the magnetic storm, in the same way as Pc1 pulsations are excited by the protons during the recovery phase. This point of view was indirectly confirmed by the comparison of ground magnetograms of Pc2 with the satellite measurements of the energetic O^+ fluxes. However, the correlation of Pc2 with the magnetic storms and increased O^+ fluxes does not eliminate another possibility for Pc2 appearing at mid-latitudes. It cannot be ignored that Pc2 pulsations may be MHD waves penetrating into the magnetosphere from interplanetary space, just like Pc3. In fact, for Pc3 pulsations, which are excited outside the magnetosphere, (6.83) applies. During a magnetic storm not only does the O^+ flux in the radiation belt increase, but there is also an increase of B in front of the magnetosphere relative to the background level. At $B > 17$ nT according to (6.83) we have $f > 0.1$ Hz, which corresponds to the range of Pc2 if we extrapolate (6.83) into this range. Logically, we cannot exclude the possibility of such extrapolation, and thus a question arises: where and how are the Pc2 excited: either in the radiation belt by the oxygen ions, just as Pc1 are

---

[7] At Venus, which rotates in the reverse direction, an afternoon maximum of the amplitude should have been observed.

excited by the trapped protons or in the upstream region by the reflected protons similar to the case of Pc3.

The problem may be solved using experimental data. For that purpose, it is natural to examine the different reaction of oscillations to the change of interplanetary magnetic field in these two cases. In the first one it is expected that oscillations will not react to the change of B, and in the second case the change of B must cause the proportional variation of the Pc2 frequency.

The selection of the data for the analysis was carried out in the following way. We started by choosing a pair of adjacent-hour intervals where $B \geq 20$ nT from the data book by Couzens and King (1986). Then, pairs where adjacent values of B differ by not less than 20% were chosen. Finally, the intervals during which Pc2 were not observed were eliminated. The majority of the events left took place during daylight hours at high geomagnetic activity.

On an $f-B$ diagram the events form a diffuse cloud of points. The coordinates of the cloud's centre are in the relation $f/B \simeq 6$ mHz nT^{-1}. If we compare this with (6.83), we may infer that Pc2 is the mere prolongation of the oblique branch for Pc3 into another range of the $f-B$ diagram. However, such an inference would be hasty. Morphologically Pc2 pulsations are distinctly separated from Pc3 pulsations. Therefore, we might as well see the internal structure of Pc2 without any reference to the connection of these pulsations with the pulsations from the adjacent range. The method of the $f-B$ diagram is too rough. In order to improve it we shall apply the following technique.

Let us connect pairs of points adjacent in time with arrows, indicating the succession of events. A fragment of the modified diagram is shown in figure 6.14 (use of all the material would have encumbered the picture so the material is presented in a simpler form). Inspection of figure 6.14 suggests that there are two peaks in the distribution of arrows determined by their orientation: one horizontal and the other oblique. This might signify that there are at least two independent sources of oscillations in the Pc2 range.

Let us introduce the angle

$$\alpha = \tan^{-1}[(B_2 + B_1)(f_2 - f_1)/(B_2 - B_1)(f_2 + f_1)]$$

describing the orientation of the arrow. The coordinates of the initial and final points are marked with indices 1 and 2. The distribution of all the events over α is presented in figure 6.15. We see that the distribution is anisotropic. Two directions are singled out: one (0–180°) corresponds to almost horizontal arrows, and the other (60–240°) to oblique arrows in figure 6.14. So, the assumption made when analysing the fragment of the $f-B$ diagram is confirmed.

Let us define the permeability $\chi = \partial f/\partial B$, taking the values $\chi = g$ under the dependence (6.83) and $\chi = 0$ when there is no reaction of oscillations to the change of B. Instead of χ we may also use the relative permeability

$$\xi = \partial \ln f/\partial \ln B$$

which takes the values $\xi = 1$ at (6.83) and $\xi = 0$ when there is no reaction.

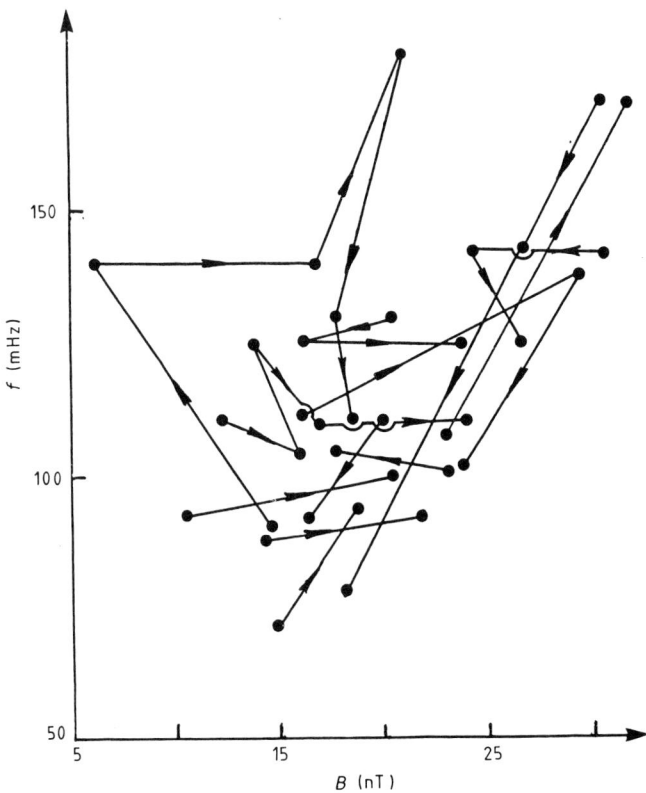

Figure 6.14. A fragment of the modified f–B diagram.

The distribution of the events over $\xi = \tan\alpha$ is presented in figure 6.16. We see two explicit maxima, notably exceeding the average level. Within the accuracy of statistical fluctuations, the position of one of them coincides with the permeability at the linear response (6.83), and the other appears when there is no reaction of oscillations to the change of B (arrows in figure 6.16). We may suppose that reflected protons in the upstream region are responsible for the first maximum and oxygen ions in the radiation belt for the second maximum. However, any such interpretation is much more hypothetical than the formal result and thus needs further confirmation.

The problem of dual origin arises not only relative to Pc2, but also relative to Pc4. On the f–B diagram, Pc4 pulsations form a horizontal branch which may seem to testify definitely to their magnetospheric origin. However, in the left lower corner, where the horizontal and oblique branches intersect, the f–B diagram gives no clue about the source of Pc4 (see figure 6.13). This part of the f–B diagram is characterized by small f and small B (approximately 10 mHz

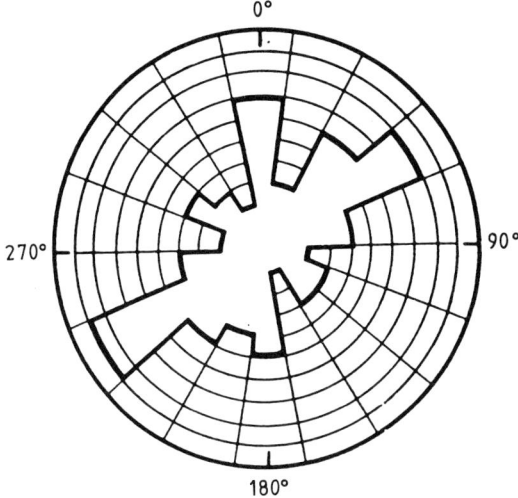

Figure 6.15. The distribution of events over different values of angle α.

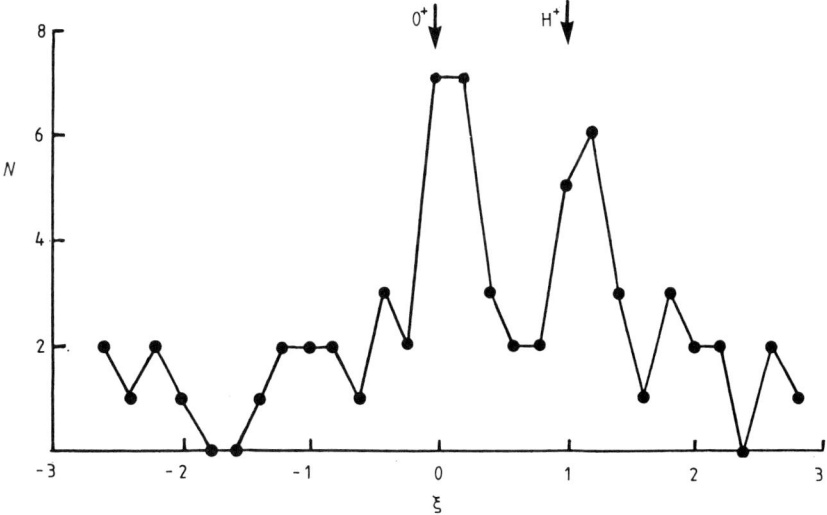

Figure 6.16. The distribution of events over the value ξ for Pc2 pulsations at large values of the modulus of interplanetary magnetic field ($B > 17$ nT).

and 2 nT respectively). There arises a suspicion that at small B at least some part of Pc4 penetrates into the magnetosphere from the upstream region.

Figure 6.17 shows the distribution of events according to the values $g = f/B$ and $\chi = \Delta f/\Delta B$. The term 'event' means the appearance of two adjacent-hour intervals, where $B \leq 3$ nT, $|\Delta B| \geq 0.2$ nT, and both intervals contain Pc4. Here ΔB is the jump of B during the transition from the first interval to the second, and Δf is the change in Pc4 carrier frequency. The distribution over g (left-hand side of figure 6.17) does not differ essentially from the analogous distribution for Pc3, but, as mentioned above, it does not give grounds to attribute Pc4 (at extremely small B) completely to extramagnetospheric oscillations.

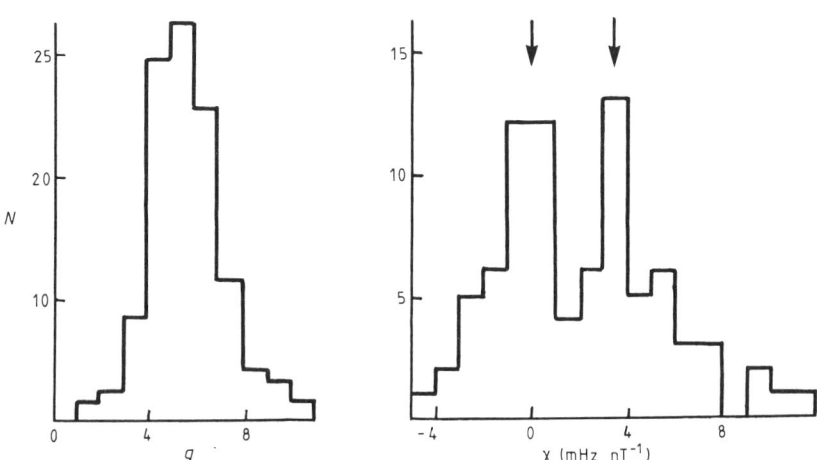

Figure 6.17. The distribution of events over the values g (left) and χ (right) for Pc4 pulsations at extremely small values of the interplanetary magnetic field ($B \leq 3$ nT).

The distribution of events over the magnitude of permeability χ will present itself in the form of a two-humped function with the maxima at $\chi = 0$ and $\chi = 3.5$ mHz nT^{-1} (indicated by arrows at figure 6.17). Hence it follows that at small B the Pc4 pulsations do not form a homogeneous complex. In this respect they are similar to Pc2 (large f and B, see figure 6.16). As in the case of Pc2, we understand that there are two varieties of Pc4 at small B, one of them being of magnetospheric and the other of extramagnetospheric origin, since from our elementary point of view the change of B only affects f if pulsations are excited in the upstream region by the beam of reflected protons.

6.5.3 Hydromagnetic aureola of planets

So, the magnetosphere of the Earth is fringed with a certain wave structure, or rather with a hydromagnetic halo, somehow similar to the light blue luminescence over the flame of a mine lamp. It arises when reflected protons appear in the solar plasma flow, flowing over the magnetosphere. This, in its turn, always happens when the field lines of the interplanetary magnetic field are directed at an oblique angle to the direction of flow. The main type of geomagnetic pulsation is one of the ground manifestations of the hydromagnetic aureola.

Not only the Earth, but also some other bodies of the solar system possess fairly powerful magnetic covers. It is not surprising that they also have similar wave structures. The hydromagnetic aureola have been detected with the help of satellites over Mercury, Venus and Jupiter. These observations are of great importance for the physics of planets. For geomagnetism the analysis of wave phenomena in front of the magnetospheres of different planets is of particular interest as it presents the possibility of testing the theory of Pc3 generation independently.

Hoppe and Russell (1982) constructed an $f-B$ diagram on which they plotted the measurement results of the MHD wave frequency f versus the modulus of the interplanetary magnetic field B in front of the magnetospheres of Mercury, Venus, the Earth and Jupiter (figure 6.18). It turned out that within the accuracy of measurement errors the experimental points are well approximated by the formula $f = gB$, with $g \simeq 5.8$ mHz nT^{-1}. In other words we have obtained the same dependence that was formerly established by the ground observations of Pc3.

Let us pay attention to the difference between the $f-B$ diagrams obtained for Pc3 (figure 6.13) and those obtained for MHD waves in front of the magnetospheres of planets (figure 6.18). In the first case the variation of f is caused by the temporal variation of B and in the second case it is caused by the spatial variation (from planet to planet).

Naturally, the conditions of streamlining of planets by the solar wind differ considerably. Thus, Mercury's radius of curvature bow shock is 0.5, and that of the Jupiter is 500 (measured in the Earth's radii). The angle between the direction of the solar plasma flow and the field lines of the interplanetary magnetic field increases on average from 20° at Mercury to 80° at Jupiter. Concentration of solar wind particles and the magnitude of the interplanetary magnetic field also changes considerably from planet to planet. Nevertheless, the coefficient of proportionality, g, between f and B, according to observations in the vicinity of planets, turned out to be the same as that found from observations of Pc3. The permanence of g signifies most probably that we are dealing with general physical properties. The mechanism of cyclotron excitation of MHD waves by the flux of reflected protons is invariant in a certain sense relative to the change of conditions of streamlining in wide limits.

6.5.4   Serpentine emission

We avoided difficult problems, arising within the framework of the theory of the extramagnetospheric origin of Pc3, and yet there is one problem that cannot be avoided. We proved that on the basis of all the measurements carried out up to the present the scatter of g does not exceed 10–20%. This is a surprisingly small value in the following sense.

In accordance with the elementary theory (6.81)–(6.83) the parallel scale

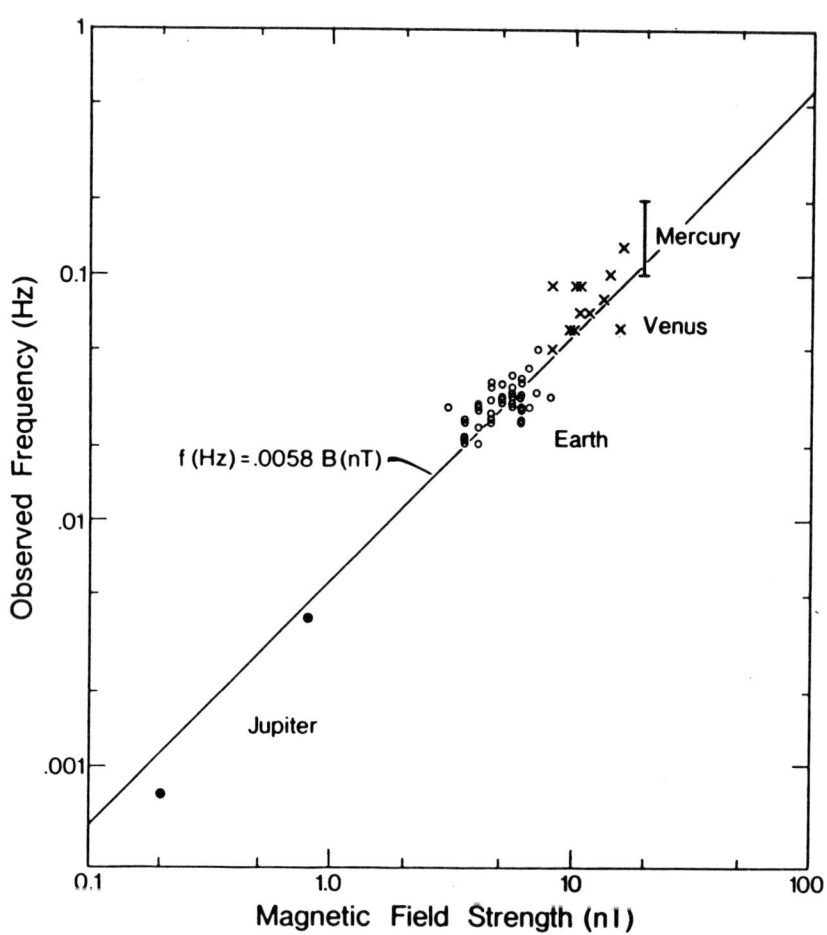

Figure 6.18. The dependence of the frequency of MHD waves on the value of the interplanetary magnetic field in the upstream regions of the different planets. The straight line indicates the dependence $f = 5.8\,B$, which approximates the experimental data (Russell and Hoppe 1983).

$\lambda_\parallel$ of the wave in front of the magnetosphere is given by the condition of cyclotron resonance, and the perpendicular scale $\lambda_\perp$ is formally infinite. If we take it literally, then, when waves are pushed down at the magnetosphere, a motionless observer should discover the Doppler dependence $g \propto \cos\psi$, where ψ is the angle between the field lines of the interplanetary magnetic field and the velocity of the solar wind. In fact $\lambda_\perp$ is limited. However, it is supposed explicitly or inexplicitly that $\lambda_\parallel \ll \lambda_\perp$ and consequently the dependence of g on ψ is considered sufficiently strong, but then the variations of ψ should have given scatter in g exceeding that observed in the experiment.

The attempt to find the dependence of g on $\cos\psi$ in the explicit form gives the following result. At $\cos\psi < 0.4$ and $\cos\psi > 0.8$ we have $g \simeq 4.6$ mHz nT^{-1} and $g \simeq 5.8$ mHz nT^{-1} respectively. By extrapolation we get $g \sim 6$ mHz nT^{-1} at $\psi \to 0$ and $g \sim 4$ mHz nT^{-1} at $\psi \to \pi/2$.

The result may be formulated in the following way: at any orientation of the interplanetary magnetic field the slope of the correlation changes not more than 1.5 times, which would not fit with the elementary theory. This interesting and not absolutely clear result is confirmed indirectly by observations in the vicinity of Mercury ($\cos\psi \sim 0.9$) and Jupiter ($\cos\psi \sim 0.17$), discussed above.

The elementary theory should be supplemented with new assumptions. Unfortunately though, the assumptions we have are not very reliable. The simplest one, for our purposes, is that there exists a perpendicular scale $\lambda_\perp$ comparable with a parallel scale $\lambda_\parallel$. Since the linear theory does not contain a perpendicular scale we may easily suppose that the localization of the wave field is the result of the nonlinear process.

However, it seems reasonable not to turn down the assumption of the elementary theory that $\lambda_\perp \gg \lambda_\parallel$. So a compensation mechanism improving $\lambda_\parallel$ for variations of ψ should be taken into consideration. It may be, for example, a corresponding dependence on ψ of the parallel velocity of reflected protons exciting MHD waves in front of the magnetosphere.

We are now reaching a general conclusion which, unfortunately, has an unwelcome aspect: until we have found the cause of stability of g under variations of ψ, the elementary theory (6.81)–(6.83) will as before play the role of nothing more than a guiding idea in favour of the concept of the extramagnetospheric origin of Pc3.

Although the theory of Pc3 has not produced any explanation of the stability of g for variations of ψ, it has opened the door to a possible solution, and, in general terms, it has considerably enriched the discussion of the problem. One of the questions is: if not Pc3, then is there possibly some other kind of geomagnetic pulsation that possesses a deep modulation of the carrier frequency, and is this modulation in causal relation with the variation of the orientation of field lines of the interplanetary magnetic field with time? Let us try to give grounds for a positive answer to this question. It is clear that we are discussing the waves, arising in the free solar wind, i.e. outside not only the magnetosphere, but the upstream region as well.

It is common knowledge, that the pressure of the interplanetary plasma is anisotropic, with $p_\parallel > p_\perp$ at the Earth's orbit (see exercise 6.5.1). At $p_\parallel > p_\perp$ the fire-hose instability may appear, leading to an increase of long-wave hydromagnetic perturbations ($\omega' \ll \Omega_p$), and cyclotron short-wave instability ($\omega' \sim \Omega_p$, where ω' is the frequency in the attendant reference system). The threshold of the cyclotron instability is lower, and the growth rate is higher than for the fire-hose instability. Evidently, the cyclotron instability leads to a restriction for the pressure anisotropy and to an equalling of the temperatures of electrons and protons in the solar wind. Consequently, the investigation of cyclotron instability is of vital importance for the physics of the interplanetary medium.

In the reference system connected with the Earth the wave frequency is

$$\omega = \boldsymbol{k} \cdot \boldsymbol{V} + \omega'$$

Waves with growth rate $\gamma \sim \omega' \sim \Omega_p$ increase under cyclotron instability at $k_\perp = 0$, $k_\parallel \simeq \omega_{0p}/c$, where $\omega_{0p} = (4\pi e^2 N/m_p)^{1/2}$ (see exercises 6.5.2 and 6.5.3). For typical parameter values $\Omega_p \ll (V/c)\omega_{0p}$, so that

$$\omega \sim \omega_{0p} \frac{V}{c} \cos \psi \qquad (6.85)$$

where ψ is the angle between $\boldsymbol{V}$ and $\boldsymbol{B}$.

The variations of the solar wind density and velocity must lead to modulation of the emission frequency. However, the greatest contribution to the modulation is expected from the variations in the orientation of the interplanetary magnetic field. These variations are caused by the transfer of the wave-like structure elements of the interplanetary field relative to the observer. It is essential that small-scale inhomogeneities of the interplanetary magnetic field are observed not only during perturbations but also under extremely quiet conditions. If $\cos \psi$ changes from 1 to 0, then the emission frequency should vary from $\omega_{\max} \sim (V/c)\omega_{0p}$ to $\omega_{\min} = 0$. At $V \sim 400$ km s^{-1}, $N \sim 5$ cm^{-3} and $B \sim 5\gamma$ we have $\omega_{\max} \sim 4$ s^{-1}.

Thus, we expect a deep modulation of the received emission. In addition, the emission should be quasi-continuous, since the pressure anisotropy is a constant feature of the solar wind.

Pulsations possessing these features were detected at Vostok Observatory in the Antarctic. Their dynamic spectrum has the form of a twisting dark stripe (figure 6.19). That is why the pulsations were called 'serpentine emission' (SE).

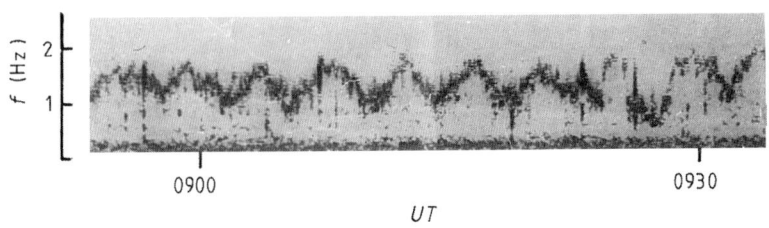

Figure 6.19. Dynamic spectrum of the serpentine emission (Vostok Observatory, Antarctic, 30 January 1968).

The most important distinguishing feature of SE is a deep modulation of the carrier frequency. In addition, the oscillations last continuously for many hours, sometimes for days. Permanence and deep frequency modulation correspond to the properties expected. Hence, it is reasonable to suppose that the emission arrives at the polar cap from the interplanetary medium.

The frequency range of SE agrees with the approximate estimate of the range of cyclotron instability of the interplanetary plasma. However, as far as we know, there is no information in the literature on the observation of SE in the interplanetary medium itself. Is it possible that SE may arise inside the magnetosphere? This is hard to believe, since there is no appropriate parameter, the variability of which could conserve such deep modulation of the oscillation frequency. One should think that these oscillations do exist in the interplanetary medium, but in order to detect them a more careful search is necessary.

6.5.5 Discrete signals

Besides serpentine emission, many other interesting and unusual kinds of pulsations were registered at the Vostok Observatory. For example, sometimes without apparent cause a wideband burst of noises arose, the dynamic spectrum of which resembles a haystack. One may encounter quasi-periodic successions of narrow-band signals with a ribbed structure of the dynamic spectrum. Figure 5.4 presents a powerful discrete signal of the falling tone and a weak burst, resembling a triggered emission, that follows.

Can these pulsations reach the Earth from the solar wind in the same way as Pc3 and SE? The analysis of observations allows us to regard a positive answer to this question as being probable enough, at least in respect of discrete signals of falling tone. But in order not to fall victims to stereotype thinking in the hope that having succeeded once we may succeed again, let us be careful and consider our inferences as only preliminary.

The discrete signals (DSs) in the frequency range 0.1–1 Hz appear sporadically in the polar caps in the form of a single signal or in small groups of wave trains appearing with an interval of 1–2 days. One may observe the signals of the rising ($\dot{f} > 0$) and falling ($\dot{f} < 0$) tones. In contrast to mid- and auroral latitudes, where the rising tones predominate, the most general form of DSs in the polar caps is a signal of falling tone. According to the observations at the Vostok Observatory the characteristic values of the maximum and minimum frequencies, frequency drift velocity, duration and amplitude are the following: $f_{max} = 0.3–0.6$ Hz, $f_{min} = 0.1–0.4$ Hz, $|\dot{f}| = 0.1–0.3$ Hz min^{-1}, $\Delta t = 2–3$ min and $b = 10–100$ mγ.

If judged by the steepness of the dynamic spectrum $\dot{f}$, then a DS travels a long path before it arrives at the Earth. But where does this path lie? Or where is the source of signals situated? Different opinions have been expressed in the literature: the source of DSs is situated (a) in the tail of the magnetosphere, (b) in the day cusps, (c) in front of the magnetosphere, (d) at the magnetopause. Here we shall describe a simple probability model for the critical test of the third of these possibilities.

Let us suppose that at least some part of a DS penetrates into the polar caps of the Earth from the upstream region, filled with reflected protons (similar to the discrete signals which have been registered in this region on board satellites.) Let us introduce the index $\sigma = \text{sign}(B_x B_z)$, taking two values $\sigma = \pm 1$ depending on the direction of field lines of the interplanetary magnetic field. Here B_x, B_z are the field components in the solar–ecliptic coordinate system. The field lines contacting the bow shock deviate northwards from the ecliptic plane if $\sigma = +1$ and southwards if $\sigma = -1$. Let us now make a probable assumption that on reaching one polar cap DSs do not propagate into the other polar cap. Then obvious geometric considerations suggest that DSs will appear more often at $\sigma = +1$ than at $\sigma = -1$ at the northern polar cap, and vice versa at the southern polar cap. This feature (we term it the σ-criterion) allows us to estimate the number of signals arriving at the Earth from the upstream region.

Figure 6.20 illustrates this idea. Fine lines inside the magnetosphere represent the geomagnetic field lines. Open lines appear from the northern (N) and southern (S) polar caps. The solid line represents the magnetopause, and the dashed line the bow shock. The field lines of the interplanetary magnetic field (straight parallel lines in the figure) change their orientation with time. Favourable ($\sigma = 1$) and unfavourable ($\sigma = -1$) orientations for the penetration of DSs from the upstream region into the northern polar cap are presented. In the southern polar cap the value $\sigma = -1$ corresponds to the favourable orientation and $\sigma = 1$ to the unfavourable one.

A fairly small collection of sonograms of discrete signals, registered at the Vostok (the Antarctic) and Tule (the Arctic) Observatories was used for the investigation. We also used the well-known catalogue by King to find the value of σ for each event[8].

Instead of σ we introduce σ^* in the following way:

$$\sigma^* = \begin{cases} \sigma & \text{if } N \\ -\sigma & \text{if } S. \end{cases}$$

Obviously, $\sigma^* = +1$ always when the criterion is yielded independently of whether the signal was registered in the northern or southern polar cap. Now we may present a test on the probability examination of the hypothesis within the framework of the Bernoulli classical scheme with two outcomes. But let us first put in the right order all the events in time. Note that other reasonable ways of regulation, for example shuffling events using a table of random numbers, will not change the result.

Let n_+ and n_- be the numbers of DSs appearing at $\sigma^* = +1$ and $\sigma^* = -1$ respectively in the succession n of events, where $n = n_+ + n_-$. Let us introduce the value $\Delta = (n_+ - n_-)/n$ and construct the plot of $\Delta(n)$. The irregular behaviour of $\Delta(n)$ at small n gives way to smooth fluctuations, which weaken with the growth of n (see figure 6.21). Finally a tendency to some limiting value

---

[8] An 'event' describes the appearance of one of the alternative outcomes ($\sigma = \pm 1$) in the hour interval of observations at the given observatory, containing at least one DS.

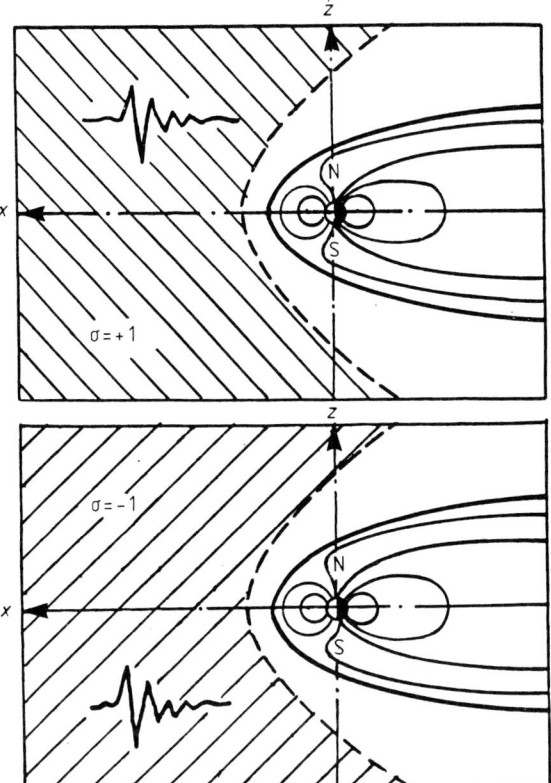

Figure 6.20. Cross section of the near-Earth space in the plane of the midday meridian. The Sun is located on the left, far beyond the plot. Two possible orientations of the interplanetary magnetic field lines are presented.

$\Delta \simeq +0.54$ is formed. With a probability of 0.95 the confidence interval for an average Δ is 0.47–0.62.

The principal result from this is that at least half of all the DSs reach the Earth evidently from the upstream region.

Let us draw preliminary conclusions. Until recently it was considered that all geomagnetic pulsations are of magnetospheric origin or arise at the boundary of the magnetosphere. The situation changed after the search for the ground manifestations of wave processes arising in the solar wind in front of the magnetosphere. Experimental proof of the extramagnetospheric origin of Pc3 is the principal result of the work carried out on this problem.

However, the possibilities for further research are far from being exhausted. In fact we are on the threshold of a rich and mysterious world. The next question concerns the origin of different kinds of pulsations in the polar caps. We have

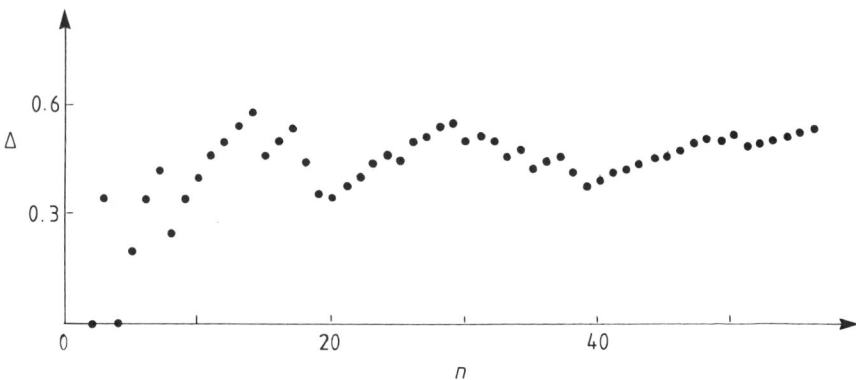

Figure 6.21. The plot of the function $\Delta(n)$.

yet to understand from what medium discrete and continuous signals with deep modulation of the carrier frequency arrive at the Earth.

Exercises

Exercise 6.5.1.

Show that adiabatic expansion of the solar corona, as it is, is a permanent mechanism for the formation of anisotropy of the interplanetary plasma.

Solution 6.5.1.

It is well known that the solar wind represents a radial plasma flow arising due to continuous expansion of the solar corona. The magnetic field is carried into the interplanetary medium together with the plasma, and the Sun rotation shapes the field lines into Archimedean spirals.

The problem may be solved at different levels of rigour with different degrees of elaboration. Here we shall restrict ourselves to a qualitative estimation of the effect, taking advantage of the adiabatic equations within the framework of the Chew–Goldberger–Low approximation (see (6.4)):

$$p_\parallel \propto \rho^3/B^2 \qquad p_\perp \propto \rho B.$$

At a fairly large distance from the Sun the radial dependence of the wind velocity V may be neglected. Then, in the stationary spherically symmetric model the dependence $\rho \propto r^{-2}$, where r is the distance from the Sun, follows from the continuity equation $\nabla \cdot \rho V = 0$. From the equation $\nabla \cdot B = 0$ we obtain $B_r \propto r^{-2}$. From the frozen-in condition $\nabla \times (V \times B) = 0$ it follows that $B_\varphi = -B_r \Omega_\odot r/V$ and consequently $B = |B_r|(1 + \Omega_\odot^2 r^2/V^2)^{1/2}$, where $\Omega_\odot = 2.9 \times 10^{-6}$ s^{-1} is the angular velocity of the Sun's rotation. Combining the given formulae we obtain the radial dependence of the pressure ratios. At $r \ll V/\Omega_\odot$ the ratio $p_\parallel/p_\perp$ increases as r^2, and at the inverse inequality it decreases as r^{-1}. The maximum is reached at $r_m = \sqrt{2}V/\Omega_\odot$. At the characteristic value of $V \simeq 4 \times 10^7$ cm s^{-1} we have $r_m \sim 1$ AU, i.e. $p_\parallel > p_\perp$ at the orbit of the Earth.

The information on the pressure anisotropy is used when analysing the hydrodynamic fire-hose instability. When analysing kinetic (cyclotron) instability more detailed information concerning anisotropy of the temperatures of ions and electrons is necessary. According to the observations in the unperturbed solar wind near the Earth's orbit $T_{\parallel p} \simeq 2T_{\perp p}$, $T_{\parallel e} \simeq 1.2T_{\perp e}$.

Exercises 6.5.2.

Find the growth rate of the cyclotron instability of the interplanetary plasma under strictly longitudinal propagation.

Solution 6.5.2.

Assume for simplicity that the plasma consists of electrons and one species of ions (protons). The particle distribution will be described by a function of the form (6.52). Substituting (6.52) into (6.39), (6.40) we find that at $k_\perp = 0$ the oscillation spectrum is determined by the dispersion equation

$$\omega^2 - c^2 k^2 = \sum \omega_0^2 \left\{ \frac{\omega}{\omega \mp \Omega} Z(z) + \Delta[1 - Z(z)] \right\}. \tag{1}$$

Here $\Delta = 1 - T_\perp/T_\parallel$; $z = (\omega \mp \Omega)/kw_\parallel$. The function

$$Z(z) = \frac{z}{\sqrt{\pi}} \int_{-\infty}^{\infty} \frac{e^{-t^2}}{z - t} \, dt$$

has the following asymptotes

$$Z(z) \simeq 1 + \frac{1}{2z^2} + \frac{3}{4z^2} + \ldots - i\sqrt{\pi}z e^{-z^2} \tag{2}$$

at $|z| \gg 1$, $|\mathrm{Re}z| \gg |\mathrm{Im}z|$, and

$$Z(z) \simeq 2z^2 - \frac{4z^4}{3} + \ldots - i\sqrt{\pi}z e^{-z^2} \tag{3}$$

at $|z| \ll 1$.

Let for electrons $\omega \ll \Omega_e$, $k_z w_{\parallel e} \ll \Omega_e$. Then $|z_e| \gg 1$ and the first two terms on the right-hand side of (2) may be used instead of $Z(z_e)$. This corresponds to the hydrodynamic description of electrons. Taking into account (1) we have

$$k_z^2 c_A^2 (1 - \beta_{\|e} \Delta_e) - \omega^2 \frac{\Omega_i}{\Omega_i \mp \omega} - \Omega_i^2 \left(\Delta_i - \frac{\omega}{\omega \mp \Omega_i} \right) \left[Z \left(\frac{\omega \mp \Omega_i}{k_z w_{\|i}} \right) - 1 \right] = 0$$

(4)

where $\beta_{\|e} = 4\pi N T_{\|e}/B^2$.

In the solar wind $\beta_{\|i} \lesssim 1$, and so the inequality $|z_i| \gg 1$ is yielded in the long-wave limits ($k \ll \omega_{0i}/c$) with ample margins. Therefore, protons may also be described in the hydrodynamic approximation. Then

$$\omega^2 = k^2 c_A^2 (1 - \beta_{\|e} \Delta_e - \beta_{\|i} \Delta_i).$$

(5)

At

$$\beta_{\|e} \Delta_e + \beta_{\|i} \Delta_i > 1$$

the roots of (5) are purely imaginary, and one of them corresponds to the aperiodic growth of perturbations. At $\beta_{\|e} = \beta_{\|i}$ and $\Delta_e = \Delta_i$ we obtain the known hydrodynamic criterion for the fire-hose instability.

In the absence of the conditions for the fire-hose instability the perturbation growth may arise only for the short wavelengths ($k \sim \omega_{0i}/c$). Let us first define the instability threshold. Considering ω and k to be real and equating the imaginary terms in (4) to zero we obtain

$$\omega_* = \mp \Omega_i \Delta_i / (1 - \Delta_i).$$

(6)

Similarly we have

$$k_* = \omega_{0i} \Delta_i / c \sqrt{(1 - \Delta_i)(1 - \beta_{\|e} \Delta_e)}$$

(7)

provided the real terms of the left-hand side of (4) are equal to zero and the relationship (6) is taken into account.

Now let us estimate the growth rate $\gamma(k)$ near the threshold of stability. It should be noted that at $T_{\|i} > T_{\perp i}$ and $\beta_{\|i} < 1$ the inequality $z_*^2 > 2$ takes place at least when the condition of fire-hose stability is yielded. Here $z_* = (\omega_* \mp \Omega_i)/k_* w_{\|i}$. At $z \simeq z_*$ this allows us to make an estimate with the following terms of the asymptotic series

$$Z(z) \simeq 1 + 1/2z^2 - i\sqrt{\pi} z \exp(-z^2).$$

Substituting this into (4) we obtain

$$\gamma(k) \simeq \alpha(k_* - k) \exp[-(k_0/k)^2]$$

(8)

in the vicinity of the stability threshold. Here

$$\alpha = \frac{\sqrt{2\pi} c \Omega_i (1 - \beta_{\|e} \Delta_e)(1 - \Delta_i)^{3/2}}{\sqrt{\beta_{\perp i}} \omega_{0i} \Delta_i^2 (2 - \Delta_i)^2}$$

$$k_0 = \omega_{0i}/c\sqrt{2\beta_{\perp i}(1 - \Delta_i)}.$$

If $k_*^2 \ll k_0^2$, then the growth rate approaches the maximum value γ_m at

$$k_m \simeq k_*(1 - k_*^2/2k_0^2). \tag{9}$$

The relative width of the instability band is of the order

$$\delta k/k_m \simeq \beta_{\perp i}\Delta_i^2/(1 - \beta_{\|e}\Delta_e). \tag{10}$$

For typical parameters $\Omega_i \sim 0.5$ s^{-1}, $\omega_{0i} \sim 4 \times 10^3$ s^{-1}, $\beta_{\|i} \sim 0.5$, $\beta_{\|e} \sim 1.5$, $\Delta_i \sim 0.5$, $\Delta_e \sim 0.2$ we have $\gamma_m \sim 2 \times 10^{-4}$ s^{-1}, $k_m \sim 10^{-7}$ cm^{-1} and $\delta k/k_m \sim 0.1$.

Exercises 6.5.3.

Find the angular dependence of the growth rate of the cyclotron instability of the interplanetary plasma.

Solution 6.5.3.

Let us consider the wave propagation at an arbitrary but fairly small angle θ to the interplanetary magnetic field. At $k_\perp^2 \ll \Omega^2/w_\perp^2$ it is sufficient to retain only the terms with $s = 0, \pm 1, \pm 2$ in the dielectric permeability tensor (6.5). Let us also consider that ε_{zz} is much larger than all the other components of $\varepsilon_{\alpha\beta}$. Then for quasi-parallel propagation and $\theta^2 \ll 1$ equation (6.6) is reduced to the form

$$n^2 = (\varepsilon_{xx} \mp i\varepsilon_{xy})(1 + \theta^2/2) + (\varepsilon_{yy} - \varepsilon_{xx})/2. \tag{1}$$

Substituting the expressions for $\varepsilon_{\alpha\beta}$ into (1), we search for the growth rate angular dependence using the method of successive approximations over θ. In the zero approximation, we obtain the solution of the preceding problem. At $k_z \sim \omega_{0i}/c$, $\omega \sim \Omega_i$, $\beta \sim 1$ the main contribution to the first nonvanishing approximation is made by the Cherenkov term ($s = 0$). Near the stability threshold we have

$$\gamma(k_z, \theta) = \alpha(k_* - k_z)\exp[-(k_0/k_z)^2] - \mu\theta^2$$

where

$$\mu = \frac{\sqrt{\pi}\,\Omega_i \Delta_i [\beta_{\perp i}(1 - \beta_{\|e}\Delta_e)]^{1/2}}{\sqrt{2}(2 - \Delta_i)}$$

$$\times \left\{ \sqrt{\frac{m_e}{m_i}} \left(\frac{T_{\|i}}{T_{\|e}}\right)^{3/2} \left(\frac{T_{\perp e}}{T_{\perp i}}\right) + \exp\left[-\frac{(1 - \beta_{\|e}\Delta_e)}{2\beta_{\perp i}}\right] \right\}.$$

Since $\mu > 0$, the waves propagating strictly along the magnetic field ($\theta = 0$) are enhanced more rapidly. The angular width of the growth rate is of the order $\delta\theta \sim 0.2$ at $\beta_{\|e}\Delta_e \sim 0.2$, $\beta_{\perp i} \sim 0.5$, $\Delta_i \sim 0.5$.

Exercise 6.5.4.

Estimate the amplification coefficient of MHD waves by the beam of the reflected protons in the upstream region.

Solution 6.5.4.

The configuration of the upstream region is shown in figure 6.10 for the typical orientation of the interplanetary magnetic field. The reflected protons form two populations: 'reflected' and 'diffusive' particles, differing in the distribution of particles over velocities and in space. The first population represents a narrow beam, directed along the field lines from the front with the following parameters: the modulus of the velocity exceeds that of the solar wind velocity by several times (in the laboratory reference system), the effective temperature is equal to several millions of degrees, and the anisotropy is characterized by the ratio $T_\perp/T_\parallel \sim 2$–3. The particles are observed on the field lines crossing the quasi-perpendicular sector of the bow shock, i.e. adjoining points of contact in figure 6.10. The parameters of the second population are: the directed velocity is somewhat smaller than the wind velocity, the temperature is several tens of millions of degrees, angular distribution is almost isotropic, and distribution by energies is nonmonotonic. The particles are observed on the field lines crossing the quasi-parallel sector of the bow shock. The concentration of particles in both populations is of the order of $N' \sim 0.1$ cm^{-3}. (Note that in the major flow of the solar plasma $N \sim 5$ cm^{-3}.) We observe a more or less smooth transition from one population to another.

The complicated structure of the upstream region hampers adequate formulation of the problem at a sufficient level of rigour. Here we outline a general skeleton of considerations.

The solar wind velocity is larger than the Alfvén velocity approximately by an order of magnitude. Therefore, MHD waves may arrive at the Earth only when they are carried down by the solar plasma flow directly onto the forefront of the magnetosphere. On their way these waves cross the upstream region, which acts as a selective amplifier. The selection of MHD waves, always present in the interplanetary medium, happens as follows: the waves, that have $k_\perp = 0$ and $k_\parallel = k_r$, where $k_r \simeq \Omega_p/u$ is the resonant wavenumber, and u is the beam velocity, experience maximum amplification.

At $k_\perp = 0$ the dispersion equation for the waves in the upstream region is derived from equation (1) in exercise 6.5.2 by means of a formal substitution in the right-hand side $\omega \to \omega - k_\parallel u$. It is also supposed that the distribution function of the reflected particles has the form (6.52) with the substitution $v_z \to v_z - u$. The particles of the background plasma will be described in the MHD approximation, taking the wavelength to be sufficiently large. In addition, for simplicity the anisotropy of the background plasma will be neglected. For further simplification we consider the beam to be moderately dense in the sense that

$$\frac{c_A^2 w_\parallel}{u^3} \ll \frac{N'}{N} \ll \left(\frac{w_\parallel}{u}\right)^3$$

where $w_\parallel$ is the thermal scatter of beam particles over longitudinal velocities[9]. Then the local coefficient of amplification reaches the maximum value

$$\text{Im}k \simeq \left(\frac{N'u}{Nw_\parallel}\right)^{1/2} \frac{\Omega_p}{V}$$

in the case when the real part of the wavenumber is equal to the resonance value

$$\text{Re}k \simeq \Omega_p/u.$$

Here and above u represents the beam velocity in the attendant reference system.

The given formulae refer to magnetosonic waves. The amplification of the Alfvén waves is exponentially small. Additional analysis testifies to the fact that in the attendant reference system the magnetosonic waves propagate in a direction similar to that of the beam, i.e. away from the Earth, and the waves have clockwise polarization—as it should be. However, when passing to the laboratory system there occurs an inversion of polarization, i.e. in the reference system connected with the Earth the waves have counter-clockwise polarization (and propagate earthwards).

The total amplification has the order $Q \sim \text{Im}kl$, where l is the distance between the front boundary of the upstream region and the bow shock. Evidently, the value Q may reach 40 dB.

The unwieldy proof to the effect that enhancement is maximum at $k_\perp = 0$ is omitted.

Bibliography

For section 6.1

The kinetic equation with the self-consistent field (6.1) was formulated by Vlasov (1938). Chew *et al* (1956) deduced the equations of magnetohydrodynamics for a plasma with anisotropic pressure (see (6.4)). Sagdeev and Shafranov (1960) deduced the dispersion equation which is widely used when analysing the cyclotron instability of the magnetospheric and interplanetary plasma.

Of the works concerning the problem of the interpretation of geomagnetic pulsations, the papers by Gintzburg (1961a,b) deserve special mention. Few people make reference to them, but in their time these papers influenced many

---

[9] The given inequalities are evidently yielded in the upstream region. If the first one is violated, the system passes into the kinetic regime, and if the second one is violated, then it passes to the hydrodynamic regime of the wave amplification. Here we have the intermediate amplification regime.

scientists and played a special part in the formation of the modern concept of the origin of geomagnetic pulsations. The theory of Pc1 (Cornwall 1965), the theory of IPDPs (Fukunishi 1969) and the theory of Pc3 (Guglielmi 1974) originated from these papers.

Extensive monographic literature is dedicated to the different aspects of plasma kinetics. The general theory of instability is presented by Lifshitz and Pitaevsky (1979). Mikhailovskii (1974) and Hasegawa (1975) gave a systematic account of practically all aspects of the linear theory of the stability of homogeneous and inhomogeneous plasma.

The drift instabilities arising in inhomogeneous plasma play an essential role. Above we have limited ourselves to a short mention of instabilities of that kind. In order partly to fill this gap we present here the corresponding references. Of general guidance, besides the book by Mikhailovskii (1974), we point out the book by Petviashvili and Pokhotelov (1992) where the reader will find information on linear and nonlinear drift waves, and on flute and ballooning instabilities in the magnetosphere of the Earth. Hasegawa (1969) discovered a so-called drift-mirror instability of the hot plasma, when investigating the mechanism of excitation of hydromagnetic waves at the periphery of the magnetosphere within the framework of a comparatively simple model (see also Lanzerotti *et al* 1969). The papers by Mikhailovskii *et al* (1976) and Marchenko *et al* (1988) are dedicated to the generalization and development of the theory of drift-mirror and field-swelling instabilities. Mikhailovskii and Pokhotelov (1975a,b, 1976, 1977), Kozhevnikov *et al* (1976), Nezlina *et al* (1984) investigated excitation of the drift Alfvén waves by fast particles in the geomagnetic trap. Meerson *et al* (1978) deduced a general expression for the instability growth rate at an arbitrary ratio between the parameters β and a/R, where a is the characteristic dimension of the plasma inhomogeneity and R is the radius of curvature of the magnetic field. The influence of high-frequency noise on the stability of low-frequency Alfvén waves is examined by Gokhberg *et al* (1981). The influence of bounce effects on evolution of the instability is analysed by Kaladze *et al* (1976). Ivanov and Pokhotelov (1987) generalized the well-known Kadomtzev criterion and analysed the flute instability when there is plasma convection from the tail of the magnetosphere.

Pokhotelov and Pilipenko (1976), Pokhotelov *et al* (1985, 1986a), and Woch *et al* (1988) developed the theory of low-frequency waves, driven by the curvature of the magnetic field lines in an anisotropic plasma. These authors had discovered that the field curvature effect is responsible for a new instability of the coupled compressional Alfvén and drift modes similar to the drift-mirror instability. The comparison of theoretical predictions with the properties of the so-called storm-time Pc5 pulsations shows that many features of observed data can be reasonably explained in terms of the developed theory. By using plasma and magnetic measurements *in situ* Pokhotelov *et al* (1986b) and Woch *et al* (1990) could show that a drift-mirror mode with small field-aligned wavelength and a large radial perturbation component is driven unstable if the effects of a strongly curved magnetic field are taken into account.

Southwood and Kivelson (1993) review the physical mechanism of linear mirror instability with special emphasis on the problem of our understanding of the observations of low-frequency signals in space.

So the drift excitation of the Alfvén oscillations of the magnetosphere causes the appearance of long-period pulsations, in particular, of the storm-time Pc5 pulsations (Engebretson and Cahil 1981, Pokhotelov *et al* 1986a,b, Takahashi *et al* 1987, 1990a, Woch *et al* 1990, Tian *et al* 1991, Grant *et al* 1992). Flute and ballooning instabilities are evidently responsible for the excitation of Pi2 (Ivanov and Pokhotelov 1987, Ivanov *et al* 1992). These and other instabilities of the inhomogeneous plasma significantly influence the character of magnetosphere–ionosphere connections (e.g. Hasegawa and Maclennan 1990).

For section 6.2

Russell *et al* (1970) discovered specific noises in the equatorial vicinity of the plasmapause. Figure 6.5 is borrowed from this paper with kind permission of the authors. Guglielmi *et al* (1975) advanced a hypothesis on the origin of the 'strange' form of the noise spectrum and suggested an appropriate theory. The observations made by Gurnett (1976) leave no doubt that we are actually dealing with the excitation of waves at high harmonics of the gyrofrequency of ions. Figures 6.6 and 6.7 are borrowed from the paper by Gurnett (1976) with kind permission of the author.

Although the estimates of the distribution of energetic particles in the evening sector of the magnetosphere (6.31)–(6.35) are original to a certain extent, nevertheless, they are based completely on the results of classical works by Kavanagh *et al* (1968) and Chen (1970). The particle traces constructed by Chen (1970) are reproduced in figure 6.4 with the kind permission of the author. Concerning the formula (6.31) see the book by Alfvén and Fälthammar (1963). The article by Smith and Hoffman (1974) was the source of information on energetic protons during the magnetic storm on 24 February 1972.

For section 6.3

Gintzburg (1961b) was the first to realize that the cyclotron interaction of the MHD waves with energetic ions of the radiation belt is an effective mechanism, which leads to the generation of geomagnetic pulsations. Cornwall (1965, 1966) put forward a theory on Pc1 generation which is based upon the concept of the cyclotron instability of protons of the outer radiation belt. The papers by Liemohn (1967), Tverskoy (1967), Criswell (1969), Feygin and Yakimenko (1969) and Gendrin (1975) are dedicated to the development of this theory. In these papers longitudinal inhomogeneity of the geomagnetic trap is taken into account, evolution of wave packets in the active part of the radiation belt is considered, and theoretical conclusions are compared with the observational data.

Feygin *et al* (1970b) made a direct estimate of the growth rate of cyclotron instability with the application of the satellite information on the distribution of energetic protons in the radiation belt. It turned out that with the enhancement of the proton flux considerable intensification of Pc1 takes place. Troitskaya *et al* (1969) made an indirect estimate of the data on the enhancement and damping of oscillations in the isolated Pc1 series.

Feygin and Yakimenko (1970, 1971) and Gendrin *et al* (1971) proposed an interesting mechanism for the formation of Pc1 dispersion anomalies. By taking into account the quasilinear effects, these authors constructed the theory of cyclotron instability of the radiation belt, within the framework of which they succeeded in imitating pearl pulsations with both normal ($D > 0$) and anomalous ($D \leq 0$) dispersion.

Guglielmi (1971) made an attempt to explain the discreteness of pearl pulsations by self-modulation of the growth rate within the framework of the phenomenological theory, an essential element of which is the 'duct capacity' of the generation region, i.e. rapid transition of resonant particles across the wave field. Polyakov *et al* (1983) explained the effect of discrete self-modulation of the decrement. Self-modulation arises owing to the fact that the ionosphere plays the role of a nonlinear mirror. The coefficient of reflection R from this mirror changes with energetic particle precipitation into the ionosphere through the loss cone, and the intensity of precipitation is controlled by pearl pulsations.

In the attempts to perceive discreteness and anomalous dispersion, one has to exceed the limits of the linear theory and bring in the concepts of quasilinear relaxation, nonlinear oscillations of the resonant particles in the wave field, bounce-resonances, etc. Thus, sometimes authors succeed in explaining some other mysterious properties of Pc1 (Bud'ko *et al* 1972, Gendrin *et al* 1971, Gokhberg *et al* 1972, Roux *et al* 1973, Feygin and Kurchashov 1975, Gendrin 1975, 1981, Khabazin *et al* 1979, Pokhotelov and Khabazin 1979, Kurchashov *et al* 1987), for example the fine structure of the Pc1 spectrum, which is explained by the so-called side-band instability (Bud'ko *et al* 1972).

However, there are still a number of interesting and difficult questions within the linear theory of Pc1. They arise, in particular, in connection with the fact that Pc1 are excited and propagate in the multicomponent plasma. The presence of ions of He^+ and O^+ complicate the phenomenon and enrich the problems (Kalisher *et al* 1982, Mauk 1982, Perraut *et al* 1984, Fraser and Sentman 1991, Hu *et al* 1991, Andersson *et al* 1992).

For section 6.4

During a magnetic storm, Duffus *et al* (1958) observed 'pearls', the frequency of which increased from 2 to 3 Hz. We cannot exclude that this was a fragment of an IPDP. Troitskaya and co-workers carried out fundamental research, the result of which was the discovery of IPDPs, as a geophysical phenomenon unknown before, and a description of the most important characteristics of IPDPs (Troitskaya and Melnikova 1959, Troitskaya 1961, 1964, 1967, Troitskaya and Maltseva 1968). Although decades have passed since the discovery of IPDPs and hundreds of articles have been written on this remarkable phenomenon, it still attracts the close attention of geophysicists. We noticed some correlation between the number of annual publications on the IPDP problem and the Wolfe numbers characterizing solar activity, although a 'solar' control of the interest in the IPDP hardly follows from this observation.

The period 1961–63 is characterized by the 'geographic conquests' of IPDPs and gradual recognition of the significance of the discovery (Duncan 1961, Chrzanowski *et al* 1961, Jacobs and Watanabe 1962, Kozlowski 1963). The first attempts at obtaining theoretical insight into IPDPs also occurred during this period. This initial stage was completed in 1964, when Troitskaya (1964) summed up the achievements of the study of IPDPs and outlined further research.

After that there ensued a decline in the interest in IPDPs. In 1965 a review by Barrington and Fejer (1965) was published, in which the authors mentioned IPDPs. Judging by the literature, nothing pointed to the steady rise of interest for the IPDP in the years to come and the 'appearance' of the maximum number of publications in 1968–72 approximately coincided with the peak of solar activity.

Analysis of the literature and, in particular, analysis of the cross-references allows us to suppose that the intensification of the research was indirectly caused by the observations at the Sogra–Kerguelen conjugate points within the framework of Soviet–French scientific cooperation (Gendrin and Troitskaya 1965, Gendrin *et al* 1967). Not least was the fact that first successful attempts to understand the physical mechanism of excitation of IPDPs refer to the beginning of this period.

Particular interest in IPDPs is evidently connected with their marked temporal evolution, distinguishing IPDPs from other oscillations of the magnetosphere. The efforts were concentrated on constructing a self-consistent model, within the framework of which the most characteristic property of IPDPs, namely the nonstationarity of the spectrum, would be fundamentally connected with the excitation process itself. Gendrin *et al* (1967) and Heacock (1967) independently proposed the mechanism of cyclotron instability of energetic protons drifting radially into the magnetosphere with the enhancement of the electric field of a large-scale convection (see also Knaflich and Kenney 1967, Lacourly 1969, Fukunishi 1973).

The idea of the spreading of the proton cloud during azimuthal drift as being the cause of increase of the oscillation frequency belongs to Fukunishi (1969). As a test to verify his theory Fukunishi used the dependence df/dt on the length of the interval between the time of particle injection and the time of appearance of IPDPs. Maltseva *et al* (1970) detected the effect of the 'western frequency drift', also testifying in favour of the Fukunishi hypothesis (1969). When presenting Fukunishi's idea (1969) we followed the version suggested by

Guglielmi and Zolotukhina (1975, 1978, 1983).

The Fukunishi injection mechanism accounts for many items, but not all. We know of a variety of IPDP that is not directly connected with the impulsive particle injection (so-called 'convection-type IPDP') and which also appears during magnetic storms. However, the oscillation frequency increase is obviously associated with the radial drift of the emitters into the magnetosphere. On the relative role of the azimuthal and radial drift of the emitters see the papers by Kangas *et al* (1974), Heacock *et al* (1976), Arnoldy *et al* (1979), Kangas (1982).

Bossen *et al* (1976), Fraser and McPherron (1982), Maltseva *et al* (1982) registered IPDPs at the geosynchronous orbit. Horita *et al* (1978, 1979) and Søraas *et al* (1980) employed satellite information on energetic ions in the magnetosphere when analysing ground IPDP observations.

A special variety of IPDP was discovered by Fukunishi and Toya (1981) from the observational data in the Antarctic. These oscillations appear in the morning hours and not in the evening, as it is usually the case in mid-latitudes. The frequency range of the 'morning IPDP' is 0.1–0.6 Hz, and the frequency drift velocity is 0.2–1 Hz h^{-1}. The dynamic spectrum consists of discrete irregular elements. In contrast to ordinary ones, the 'morning IPDPs' appear most frequently under weak and moderate magnetic activity.

Kangas *et al* (1976, 1984, 1988), Bösinger *et al* (1981) and Pikkarainen (1987, 1989) presented a thorough investigation of IPDPs, based on the data from the Finnish north–south magnetometer and riometer chain, which was supplemented by the electric field data from the EISCAT incoherent scatter radar.

For section 6.5

The hypothesis of the extramagnetospheric origin of Pc3 was first put forward by Guglielmi (1974). It suggests itself if we take account of the discovery of MHD waves in the upstream region (Fairfield 1969) and the totality of the morphological properties of Pc3 (frequency range, permanence, 24 hour variation, etc). Coincidence of the Pc3 frequency on the Earth and the MHD wave frequency in front of the magnetosphere could be a strong argument in favour of this hypothesis. For various reasons instead of comparing these two frequencies, the frequency f of pulsations on the ground was compared with the value B of the interplanetary magnetic field. The works by Troitskaya *et al* (1971), Guglielmi and Bol'shakova (1973), Guglielmi *et al* (1973), Guglielmi and Troitskaya (1973), Guglielmi (1974), Russell and Flemimg (1976), Verö and Hollo (1978), Greenstadt *et al* (1979), Green *et al* (1983), Odera (1984), Yumoto *et al* (1984, 1985), Verö *et al* (1985), Yumoto (1985), Miletits *et al* (1988, 1990), Slawinski *et al* (1988), Engebretson *et al* (1990), Wolfe *et al* (1990), Olson *et al* (1991) are dedicated to this aspect of the experimental research. The measurement results of the coefficient of connection between f and B are partly summed up in the paper by Guglielmi (1988b). The conceptual problem of the Pc3 theory, mentioned in this text is also formulated. Guglielmi *et al* (1989b) modified the method of $f-B$ diagrams to prove the dual origin of Pc2 and Pc4. The regular Chapman Conference of the American Geophysical Union (Chapman Conference 1992) held recently was dedicated to the problem of the origin of Pc3 pulsations.

Russell and Hoppe (1981) and also Green *et al* (1983) measured the ratio f/B by wave observation directly in the upstream region and found out that this ratio in the limits of measurement errors coincides with the value g known from ground observations of Pc3. For all the importance of these two works it should be admitted that the result was not unexpected. The same cannot be said of the outstanding paper by Hoppe and Russell (1982). The value of $g \simeq 5.8$ mHz nT^{-1} found by these authors on the basis of the observations in the vicinity of Mercury, Venus, the Earth and Jupiter unexpectedly coincided within the limits of measurement accuracy with the coefficient found by the observations of Pc3 on the Earth's surface (see also the review by Russell and Hoppe 1983). In contrast to the opinion expressed sometimes, we consider that it could not have been foreseen (and no one had predicted it).

The article by D'Angelo (1975) opened the long list of critical publications on this problem. The remarks, in many respects, stimulated additional analysis and promoted further improvement of argumentation. The answer to the question raised by D'Angelo (1975) was given by Guglielmi (1976). Webb and Orr (1976), Greenstadt and Olsen (1979), Wolfe (1980) and Wolfe *et al* (1980), stressing the dependence of the oscillation amplitude on the solar wind velocity and on the orientation of the interplanetary magnetic field, all arrived at the conclusion that Pc3 are excited at the magnetopause owing to the Kelvin–Helmholtz instability. The fading effect of the Pc3 amplitude, detected by Bol'shakova and Troitskaya (1968), is explained by Nishida (1978) through the stabilization of the Kelvin–Helmholtz instability at some orientation of the interplanetary magnetic field. Meanwhile the fading is absolutely transparent in the framework of the theory of extramagnetospheric generation.

All this confirms the idea that, unless account is taken of the spectral dependencies, the amplitude analysis does not lead us anywhere. The amplitude is formed as a result of far more complicated and intricate processes than the spectrum. Therefore, common sense suggests that a test in the form of the $f-B$ diagram to check the hypothesis on the origin of Pc3 is the best possible course of action.

Plyasova–Bakunina *et al* (1978) and Golikov *et al* (1980) made a revision of the Guglielmi theory (1974) on account of the fact that 'It is difficult to understand how waves observed upstream from the bow shock can propagate through the bow shock, magnetosheath and magnetopause'. To overcome difficulties the authors appealed to the Kelvin–Helmholtz theory. Russell *et al* (1983) criticized the concept put forward by Golikov *et al* (1980) and adduced arguments in favour of the Guglielmi theory (1974).

A special issue of the *Journal of Geophysical Research* (1981 volume A **86**, number 6) and the review by Russell and Hoppe (1983) are dedicated to remarkable discoveries made from observations by spacecraft in the upstream region. In addition, we shall just point out a few more references. Asbridge *et al* (1968) detected a flux of reflected protons. Gosling *et al* (1978) and Paschmann *et al* (1981) investigated spatial and energetic distribution of protons in detail and found out two populations of these particles ('reflected' and 'diffusive'). The fundamentals of the respective theory were laid down by Sonnerup (1969) and Paschman *et al* (1980). In the range 0.01–0.1 Hz permanent magnetic pulsations were observed by Greenstadt *et al* (1968), Fairfield (1969), Greenstadt (1981), Hoppe and Russell (1983). Bonifazi *et al* (1980) detected the deceleration of the solar wind in the region occupied by these pulsations. Discrete wave packets in the range 0.1–1 Hz were discovered and closely examined by Russell *et al* (1971), Hoppe and Russell (1980), Hoppe *et al* (1981) and Le *et al* (1989).

Serpentine emission was observed by Guglielmi and Dovbnya (1974) at the Vostok station and by Morris and Cole (1987) at the Davis station. Both stations are situated in the Antarctic. In the Arctic this emission was observed by Rusakov from a drifting ice-floe (see Guglielmi 1979). The depth of the ocean under the observational point exceeded the skin-layer thickness. This factor made it possible to check up the absence of SE in the vertical component of the magnetic field and thus parry the criticism that the SE is just local interference.

DSs in the polar caps are described in the works by Matveeva *et al* (1976, 1978), Bondarenko and Guglielmi (1976), Cole *et al* (1982), Fraser-Smith (1982), Morris *et al* (1982). The test for the verification of the hypothesis of the extramagnetospheric origin of DSs was proposed by Guglielmi (1985a).

Exercise 6.5.1 is based on the ideas and estimates suggested by Parker (1963). When compiling exercises 6.5.2–6.5.4 we used the results obtained by many scientists. The dispersion equation (1) in exercise 6.5.2 belongs to Sagdeev and Shafranov (1960). The calculation techniques of the growth rate near the instability boundary was borrowed from Mikhailovskii (1974). We learned of the fire-hose instability as applied to the solar wind from the works by Parker (1963) and Kennel and Sagdeev (1967). The cyclotron instability of the solar wind is investigated in more detail in the works by Kennel and Scarf (1968) and Hollweg and Völk (1970).

Chapter 7

Nonlinearity

The exposition of the wave theory in the preceding chapters of this monograph was restricted by the linear approximation with just slight exceptions. When we analyse the excitation of waves induced by external sources, this idealization is justified in many cases. But it cannot be considered an adequate representation of reality when analysing the self-excitation of waves. Exponential growth of the amplitude rapidly withdraws the oscillation system from the region of applicability of the linear theory. After a period of time of the order of γ^{-1}, starting with the moment of transition to the unstable state, nonlinear processes are switched on which saturate the amplitude of oscillation. In a nonlinear regime the superposition principle is no longer valid. The properties of waves depend on the amplitude, and the medium in which the waves propagate is modified.

We shall distinguish between dissipative and conservative nonlinearities. Formally, in the case of a dissipative nonlinearity the modification of the anti-Hermitian part of the dielectric permeability tensor takes place under the influence of waves. In the magnetosphere the dissipative nonlinearity is manifested, for example, when resonant interaction of geomagnetic pulsations occurs with energetic particles, which constitute a relatively small admixture to the background (cold) plasma. This applies to a wide circle of problems connected with so-called quasilinear relaxation of magnetospheric plasma, leakage of energetic particles from the radiation belt, etc. We, however, will focus our attention on the conservative (nondissipative) nonlinearity, when a modification of the Hermitian part of the dielectric permeability tensor takes place. In the magnetosphere this nonlinearity is manifested in the perturbation of the background plasma and geomagnetic field, as well as in the excitation of solitons and vortices, i.e. specific self-consistent wave structures.

We consider these nonlinear effects to be small and take them into account by introducing corresponding corrections into the dielectric permeability tensor, into the dispersion ratio, or directly into the dynamic equations of the linear theory.

7.1 Ponderomotive forces

Here we shall consider the action of an electromagnetic field on a medium. We shall be interested in the time-averaged ponderomotive forces which influence a medium immersed in the quickly oscillating field. Let us represent this field in the form

$$E = E_0(x, t)e^{-i\omega t} \qquad b = b_0(x, t)e^{-i\omega t} \tag{7.1}$$

where E_0, b_0 vary in time slower than $\exp(-i\omega t)$, i.e. the field is a quasi-monochromatic one. We introduce the designation f for the force which acts on the unit of volume averaged over the period $2\pi/\omega$. The fairly general expression for f is

$$f = f^A + f^{WK} + f^M + f^P + f^{KP} \tag{7.2}$$

where

$$f^A = \frac{1}{8\pi c}\mathrm{Re}\frac{\partial}{\partial t}\left[(D - E) \times b^*\right]$$

$$f^{WK} = \frac{\omega}{8\pi c}\mathrm{Re}(F \times b^*)$$

$$f^M = \frac{1}{16\pi}(\varepsilon_{\alpha\beta} - \delta_{\alpha\beta})\nabla E_\alpha^* E_\beta$$

$$f^P = \frac{1}{16\pi}E_\alpha^* E_\beta \frac{\partial \varepsilon_{\alpha\beta}}{\partial B_\gamma}\nabla B_\gamma$$

$$f^{KP} = \frac{1}{16\pi}B \times \mathrm{curl}\left(\frac{\partial \varepsilon_{\alpha\beta}}{\partial B}E_\alpha^* E_\beta\right).$$

Here $B(x)$ is the stationary external magnetic field, and

$$D_\alpha = \varepsilon_{\alpha\beta}E_\beta \qquad F_\alpha = \frac{\partial \varepsilon_{\alpha\beta}}{\partial \omega}\frac{\partial E_{0\beta}}{\partial t}.$$

As before the asterisk denotes the complex conjugate.

The applicability of (7.2) is restricted by a number of conditions. The medium is considered stationary, immobile and transparent. The tensor $\varepsilon_{\alpha\beta}$ is Hermitian ($\varepsilon_{\alpha\beta}^* = \varepsilon_{\beta\alpha}$). Spatial dispersion is absent, i.e. $\epsilon_{\alpha\beta}$ does not depend on the wave vector k. However, these limitations may be relaxed somewhat.

Let us clarify the physical sense of the separate terms on the right-hand side of (7.2). The first term, or f^A, represents the Abraham force, which is connected in some way with the action of the Lorentz force on the displacement current. The characteristic combination $(D - E)$ in this expression can be understood if we note that in the broad sense the force is equal to the change of the impulse per unit time, and the Abraham force is acting only on the medium. Thus, one has to extract the change of the field impulse from the change of the whole impulse[1]. If the amplitude of the field oscillations is constant (i.e. the oscillations are purely monochromatic) then $f^A = 0$.

---

[1] The density of the field impulse equals $(4\pi c)^{-1}E \times b$ and after averaging over the period $2\pi/\omega$ we obtain $(8\pi c)^{-1}\mathrm{Re}\,E \times b^*$.

The second term $\boldsymbol{f}^{WK}$ is an addition to the Abraham force, and was found by Washimi and Karpman (1976). The force $\boldsymbol{f}^{WK}$ differs from zero only in dispersive media when $\varepsilon_{\alpha\beta}$ is frequency-dependent. Its appearance is connected with the fact that, due to dispersion, the variation in time of the electric induction of the quasi-monochromatic field equals approximately

$$\frac{\partial D_\alpha}{\partial t} \approx -\mathrm{i}\omega\varepsilon_{\alpha\beta}(\omega)E_\beta + \frac{\partial(\omega\varepsilon_{\alpha\beta})}{\partial\omega}\frac{\partial E_{0\beta}}{\partial t}\mathrm{e}^{-\mathrm{i}\omega t}.$$

The term $\boldsymbol{f}^M$ is known as the Miller force. In a certain sense this term is analogous to that known in mechanics as the Kapitza force which acts on a material point in a fast-oscillating field. The simple form of $\boldsymbol{f}^M$, presented in (7.2), is obtained under the condition $\rho\,\partial\varepsilon_{\alpha\beta}/\partial\rho = \varepsilon_{\alpha\beta} - \delta_{\alpha\beta}$.

The Pitaevsky force $\boldsymbol{f}^P$ appears because of the induced magnetic moment with density

$$M = \frac{1}{16\pi}\frac{\partial\varepsilon_{\alpha\beta}}{\partial B}E_\alpha^* E_\beta \tag{7.3}$$

under the action of the wave on the medium.

Finally, the origin of the last term, $\boldsymbol{f}^{KP}$, is connected with the Lorentz interaction between the external magnetic field and the current that is induced by the wave in the medium (Klima and Petrjilka force).

The ponderomotive force given by equation (7.2), when acting on the medium, modifies the dielectric permeability, and thus affects the electromagnetic field that gave raise to this force. This is one of the possible mechanisms of self-action of the electromagnetic field. It plays an important role in the formation of self-consistent nonlinear waves. Expression (7.2) may be applied when solving corresponding problems.

A simplified set-up of the problem is also possible: we can try to find the perturbation of the medium under the action of the ponderomotive force, considering the electromagnetic field to be given. Let us use (7.2) to consider problems of this kind in application to the Earth's magnetosphere.

7.1.1 Equatorial plasma condensation

As a first example, we consider the ponderomotive forces induced by the ion cyclotron waves travelling along the geomagnetic field lines. We shall show that ponderomotive forces can push the plasma up to the equatorial plane and this leads to an appreciable thickening at the tops of the geomagnetic field lines[2].

We shall avoid the difficulties connected with the Abraham force by assuming the wave amplitude to be constant in time. In the case of geomagnetic pulsations of Pc1-2 type, this is valid if we average square combinations of the field components not over the oscillation period but over a much longer

---

[2] Standing MHD waves compress or rarify the plasma at the equator of the oscillating magnetic shell if a node or antinode is located there.

period, for example over the repetition period of signals. Then the longitudinal component of the ponderomotive force (7.2) is

$$f_\parallel = \frac{1}{16\pi} \left\{ (\varepsilon_{\alpha\beta} - \delta_{\alpha\beta}) \nabla_\parallel E_\alpha^* E_\beta + E_\alpha^* E_\beta \frac{\partial \varepsilon_{\alpha\beta}}{\partial B_\gamma} \nabla_\parallel B_\gamma \right\}. \tag{7.4}$$

For relatively short Pc1-2 waves the WKB approximation can be used, and we take it as a solution for the field $E(x, t)$ with the following reservation.

In reality the Pc1-2 pulsations are reflected from the ionosphere and form oppositely moving wave beams in the magnetosphere. This may lead to the formation of the interference picture, where the field amplitude oscillates in space, even if averaging over time has been carried out. It is desirable to ignore these small amplitude oscillations, if they appear. For that we just have to accomplish an additional averaging along the field lines over the spatial period of oscillations. But if we take distances from the Earth that are much longer than the wavelength, then it is not even necessary. The nonmonochromatic character of Pc1-2 leads to fluctuations of the interference picture, and averaging over a relatively long time interval will automatically lead to the necessary spatial averaging (along the field line)[3].

In any case, we shall neglect the interference effects when calculating $f_\parallel$. Then the amplitude variation of the wave, when travelling along the geomagnetic field line, will be determined by the law of conservation of the wave energy flow in each of the oppositely moving wave beams only. The effect of the ponderomotive force, which we are going to demonstrate, does not depend on whether the wave propagates parallel or anti-parallel to the direction of the geomagnetic field. So, for simplicity we shall refer to one of the two oppositely moving wave beams.

We focus on the circularly polarized waves with counter-clockwise rotation of the electric field vector. Expression (7.4) then reduces to

$$f_\parallel = \frac{1}{8\pi} \left[(n^2 - 1) \nabla_\parallel E^2 + E^2 \frac{\partial n^2}{\partial B} \nabla_\parallel B \right] \tag{7.5}$$

where n is the refractive index for the waves of the counter-clockwise rotation (upper sign in (4.4)) and E is the amplitude of the electric field oscillations. We consider a collimated wave beam localized in the magnetic tube. In the WKB approximation $E \propto (B/n)^{1/2}$, where B and n are taken at the axis of the tube. Substituting this into (7.5) and taking into account (4.5) we obtain

$$f_\parallel = -\frac{n^2 E^2}{16\pi} \left[\left(\frac{\omega}{\Omega - \omega} \right) \nabla_\parallel \ln B + \nabla_\parallel \ln \rho \right]. \tag{7.6}$$

Here Ω is the ion gyrofrequency and it is assumed that $\omega < \Omega$ and $n^2 \gg 1$.

---

[3] These considerations break down if we speak of the field at relatively short distances from the Earth, since at least one element of the interference structure here is fixed by the boundary condition on the ionosphere.

Moving away from the equator along the field lines, the magnetic field B as well as the plasma density ρ increase. (The latter is the unperturbed state, see below.) Hence, the ponderomotive force (7.6) moves the plasma up towards the equatorial plane. This force is counter-balanced by the parallel component of the gravity force $\rho g_\parallel$, but the latter approaches zero in the vicinity of the equator and the plasma 'shovelling' is efficient when the wave amplitude is sufficiently strong.

For a qualitative estimate, we consider the balance of forces

$$\nabla_\parallel p = \rho g_\parallel + f_\parallel \tag{7.7}$$

in the close vicinity of the equator. Here $p = c_s^2 \rho$ is the plasma pressure and c_s is the velocity of isothermic sound. The gravitational acceleration g is directed radially towards the centre of the Earth. The equation for the field lines is described by $r = R_E L \cos^2 \varphi$, where φ is the geomagnetic latitude, R_E is the Earth's radius and L is the McIllwain parameter. The radial component of the unit vector τ which is tangent to this line equals

$$\tau_r = \frac{2 \sin \varphi}{(1 + 3 \sin^2 \varphi)^{1/2}}.$$

Then, the longitudinal component of the gravity acceleration $g_\parallel = (\tau \cdot g)$ in the vicinity of the equator ($\varphi^2 \ll 1$) approximately equals

$$g_\parallel \approx 2 g_0 L^{-2} \varphi$$

where $g_0 = 980$ cm s^{-1}. With the same accuracy the magnetic field equals

$$B \approx B_0 L^{-3} \left(1 + \tfrac{9}{2} \varphi^2 \right).$$

Here $B_0 = 0.31$ G. The longitudinal differentiation operator is obviously

$$\nabla_\parallel \approx \frac{1}{R_E L} \frac{\partial}{\partial \varphi}.$$

Substituting all this into (4.7) and then into (7.6) and (7.7) we get

$$\alpha \frac{\partial \rho}{\partial \varphi} = \beta \rho \varphi \tag{7.8}$$

where

$$\alpha = c_s^2 + \frac{\Omega}{\Omega - \omega} \left[\frac{cE}{2B} \right]^2$$

$$\beta = 2 g_0 \frac{R_E}{L} - \frac{\Omega \omega}{(\Omega - \omega)^2} \left[\frac{3cE}{2B} \right]^2.$$

Here the values E, B and Ω have to be evaluated at $\varphi = 0$. Hence it follows, that a maximum of plasma density exists when

$$E > \frac{2\sqrt{2}}{3} \frac{B}{c} \left(\frac{g_0 R_E}{L} \right)^{1/2} \left(\frac{\Omega}{\omega} \right)^{1/2} \left(1 - \frac{\omega}{\Omega} \right). \tag{7.9}$$

If criterion (7.9) is not satisfied but nevertheless the amplitude E is comparable with the value

$$2B(c_s/c)(1 - \omega/\Omega)^{1/2} \tag{7.10}$$

then the density ρ has a minimum at the equator as usual. However, the ponderomotive force significantly increases the value of this minimum relative to the unperturbed state. These conclusions are, of course, only valid when the ponderomotive force acts during a sufficiently long time in order to establish the stationary state.

Let us make a numerical estimate, assuming that $L = 5.5$ and that the wave frequency is 1 Hz. In this case $\omega/\Omega = 0.35$, which is quite typical for Pc1 pulsations. Then the critical amplitude according to (7.9) is equal to 6.5×10^{-6} V cm^{-1}. This is quite a moderate value. Taking into account the relation $b = nE$ we find the critical amplitude of the magnetic field oscillations $b = 0.65$ nT at the typical value $n = 300$.

Far away from the equator, i.e. at high latitudes where the geomagnetic field lines are nearly radial and, consequently, the operator $\nabla_\parallel$ in (7.7) may be substituted for $\partial/\partial r$, similar consideration leads to the following equation of the balance of forces

$$\frac{1}{\rho} \frac{d\rho}{dr} \left(c_s^2 + \frac{\alpha}{\sqrt{\rho}} \right) = -\frac{GM}{r^2}. \tag{7.11}$$

Here G is the gravitational constant, M is the Earth's mass, $\alpha = b_0^2/(16\pi\sqrt{\rho_0})$, and b_0 and ρ_0 are the amplitude of the magnetic field oscillations and the plasma density at a certain level $r = r_0$. In the derivation of (7.11) from (7.6) and (7.7) we have assumed that $\omega \ll \Omega$.

In order to estimate the effect of the ponderomotive force, we integrate (7.11) and formally make the limiting transition $r \to \infty$. The result is

$$c_s^2 \ln \frac{\rho_0}{\rho_\infty} + \frac{b_0^2}{8\pi\rho_0} \left[\left(\frac{\rho_0}{\rho_\infty} \right)^{1/2} - 1 \right] = \frac{GM}{r_0}. \tag{7.12}$$

Let us consider the obvious case $c_s^2 \ll GM/r_0$. Then the density ρ_∞ is much smaller then ρ_0, if $b_0 = 0$. If $b_0 \neq 0$, then ρ_∞ is comparable to ρ_0 at realistic oscillation amplitudes. It is easy to confirm that the characteristic value of b_0, at which the intense plasma stretching due to the ponderomotive force takes place, is

$$(8\pi GM\rho_0/r_0)^{1/2}. \tag{7.13}$$

For example, at $r_0 = 1.5\,R_E$ and $\rho_0 = 2 \times 10^{-21}$ g cm^{-1} the value (7.13) is equal to 14.5 nT. In the auroral zone above the ionosphere the amplitude of the Alfvén waves sometimes exceeds this value considerably. Waves with such amplitudes can create dense layers and fibres streaming up along the geomagnetic lines.

7.1.2 Polar wind acceleration

The polar wind denotes the outflow of the plasma from the ionosphere into the tail of the magnetosphere along the 'open' field lines of the geomagnetic field. The origin of the polar wind is in many ways similar to the origin of the solar wind.

The general analogy between the polar and solar winds suggests the idea of translating certain concepts about the acceleration mechanisms of the solar wind, which have been accumulated in the course of long-term research works, into the case of the polar wind. The most important concept of this kind is that the MHD waves that propagate away from the Sun into the interplanetary space make a noticeable contribution to the acceleration of the solar wind due to the action of the ponderomotive force on the plasma of the solar corona. Above the polar caps in the magnetosphere of the Earth there exist intense MHD oscillations and they are capable of producing a similar effect.

Let us make a qualitative analysis within the framework of a simple Parker's model, adding the ponderomotive force to it

$$\frac{1}{v}\frac{dv}{dr}(v^2 - c_s^2) = \frac{3c_s^2}{r} - \frac{GM}{r^2} + \frac{f_{\parallel}}{\rho}. \tag{7.14}$$

Here v is the velocity of the plasma movement, $c_s = (2T/m_i)^{1/2}$ is the sound velocity, T is the temperature, m_i is the ion mass, G is the gravitational constant, M is the mass of the Earth, ρ is the plasma density and r is the distance from the Earth's centre. The polytropic index is assumed to be equal to unity for simplicity, i.e. an isothermal plasma is being considered.

Without accounting for the ponderomotive force ($f_{\parallel} = 0$), the critical point r_c of the supersonic transition is defined by the expression

$$r_c = \frac{GM}{3c_s^2}. \tag{7.15}$$

Obviously the influence of the ponderomotive force shifts the critical point. We find the shift value Δr. Assuming the shift to be small we find from (7.14) using the perturbation method

$$\Delta r = -\frac{r_c^3 f_{\parallel}}{GM\rho}. \tag{7.16}$$

Here $f_{\parallel}$ and ρ should be taken at the point r_c.

From (7.16) it is evident that at $f_{\parallel} > 0$ the point of the supersonic transition is shifted towards the Earth. In this sense one may speak about polar wind acceleration.

For the ion cyclotron waves, $f_{\parallel}$ is given by (7.6). Supersonic transition may occur at altitudes of 2000–5000 km. At these altitudes $\omega \ll \Omega$ within the range of geomagnetic pulsations and consequently the first term on the right-hand side of (7.6) may be neglected in comparison with the second term

$$f_{\parallel} = -\frac{n^2 E^2}{16\pi\rho} \frac{d\rho}{dr}.$$

It is suggested that the plasma flow is conserved in the magnetic field tubes. Thereby $\rho v/B = $ constant and

$$\frac{1}{\rho}\frac{d\rho}{dr} = -\frac{3}{r} - \frac{1}{v}\frac{dv}{dr}.$$

From (7.11) it follows that $dv/dr = c_s/r_c$ at $r = r_c$. So we have

$$\frac{1}{\rho}\frac{d\rho}{dr} = -\frac{4}{r_c}$$

and finally

$$f_{\parallel} = S/c_A r_c. \tag{7.17}$$

Here the expression for the Poynting vector modulus

$$S = \frac{c}{4\pi} E^2 n$$

and the formula $n = c/c_A$ for the refractive index in the range $\omega \ll \Omega$ are used.

Let us substitute (7.17) into (7.16) and find the relative shift of the supersonic transition point

$$\frac{\Delta r}{r_c} = -\frac{S}{3c_A p}$$

or

$$\frac{\Delta r}{r_c} = -\frac{b^2}{12\pi p} \tag{7.18}$$

where $p = c_s^2 \rho$ is the plasma pressure at $r = r_c$. The relative shift of the critical point increases with increasing r_c, since the pressure p decreases rapidly when moving away from the Earth. Furthermore, the critical point always shifts towards the Earth, independent of whether the waves propagate away from or towards the Earth. Finally, we should expect a gusty polar wind, since the intensity of geomagnetic pulsations is quite variable.

The experimental status of the theory of polar wind acceleration, as well as of the theory of equatorial plasma condensation, is not clear at present. Real confirmations have not been found. Nevertheless, let us proceed with our discourse and follow the life of the polar wind. Here it passes away from the Earth along the magnetospheric tail as if along a tube. A portion of plasma flow lines, following the lines of the geomagnetic field, deviates towards the equatorial plane.

We recall, that the geomagnetic tail is divided into northern and southern halves by the so-called neutral sheet, which is immersed in a thicker layer of relatively dense plasma. This plasma layer is formed as an after-effect of a number of factors; but can at least some of its properties be explained by the influence of the ponderomotive force?

The geomagnetic field decreases when approaching the neutral sheet, and if the intensity of MHD waves increases in the plasma layer, then the plasma 'shovelling' may begin towards the neutral sheet of the wave by the same mechanism, which is supposed to lead to the equatorial plasma condensation in the magnetosphere. This, in its turn, will cause the enhancement of the drift current along the neutral layer, which stimulates the tearing instability with well-known consequences.

Exercise

Exercise 7.1.1.

An ion cyclotron wave propagates towards a decreasing external magnetic field. While approaching the resonance ($\omega \to \Omega$) the ponderomotive force increases. Find the maximum value of the ponderomotive force in the vicinity of the resonance.

Solution 7.1.1.

In the close vicinity of the resonance the second term on the right-hand side of (7.6) is small compared with the first one and may be neglected

$$f_{\parallel} = -\frac{S}{4c} \frac{\Omega^{1/2}\omega_{0i}}{(\Omega - \omega)^{3/2}} \frac{\nabla_{\parallel} B}{B}. \tag{1}$$

Here $S = (c/4\pi)nE^2$ is an average density of the wave's energy flow. The growth $|f_{\parallel}|$ at $\omega \to \Omega$ is not infinite, because when approaching the resonance the wave amplitude begins to attenuate owing to the cyclotron damping. The minimum value of the difference $\Delta\omega = \Omega - \omega$ is estimated as $\Delta\omega \simeq kv_T$, where v_T is the thermal velocity of ions and $k = (\omega/c)n$ is the wavenumber. Using the expression for the refractive index

$$n = \frac{\omega_{0i}}{[\Omega(\Omega - \omega)]^{1/2}}$$

we find

$$(\Delta\omega)_{\min} \approx (v_T/c_A)^{2/3}\Omega$$

where $c_A = c(\Omega/\omega_{0i})$. Substituting this into (1) we obtain

$$|f_\|\|_{max} \approx \frac{S}{4v_T} \frac{|\nabla_\| B|}{B}.$$

The force is directed towards decreasing magnetic field, i.e. upwards in the case of the Earth.

To adapt this exercise to the ion whistlers discussed in Chapter 5, two considerations should be taken into account: (1) the nonstationarity of the ion whistler; (2) absorption of waves and heating of the plasma in the resonant region. In any case the ion whistler pushes the plasma upwards. In fact, the wave impulse of the whistler is directed upwards. Since the whistler is absorbed completely, its impulse is transferred to the medium completely.

7.2   Solitons

In the preceding section we examined the influence of the ponderomotive force on the medium in the approximation of the given wave field. But the ponderomotive force, modifying the medium, disturbs the field of the wave creating this force. Thus the effect of self-action of the wave appears[4]. It has interesting and various consequences.

One of the results of self-action is the appearance of solitons, i.e. nonlinear solitary waves propagating without distortion of their form due to the fact that the dispersive spreading is exactly balanced by the nonlinearity (figure 7.1).

We shall be interested mainly in applied aspects of the theory, particularly, in the problem of searching for MHD solitons among the geomagnetic pulsations. The interest in solitons is justified by the consideration that wave structures of this kind emerge naturally as a result of decay of more or less arbitrary initial perturbations or as a result of modulation instability.

7.2.1   The nonlinear Schrödinger equation

Let us consider the so-called envelope solitons, i.e. nonlinear wave packets. In linear theory the wave packets in a dispersive medium spread in the course of propagation. Also, the amplitude of the wave packet is a free parameter. It does not depend on the packet's width and carrier frequency. As was mentioned above, the soliton is free of dispersion spreading and the wave profile is not distorted. Another important feature of the soliton is that its amplitude is not an independent parameter. The certain connection between the amplitude, the packet width and carrier frequency may be used as a distinctive indication in the search for solitons in the near-Earth plasma.

---

[4] Here for shorter presentation we mean the ponderomotive force, but one should bear in mind that the inverse Faraday effect also leads to self-action. The essence of this consists in magnetizing the medium by the electromagnetic field (see formula (7.3) and exercise 7.2.1).

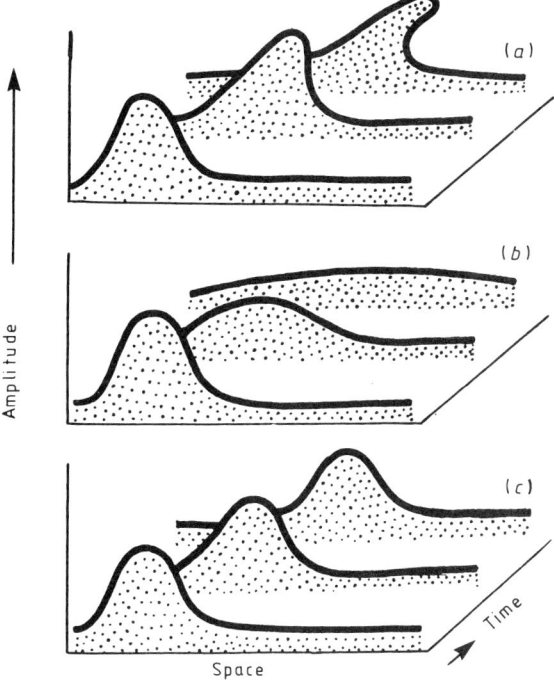

Figure 7.1. Propagation of waves of different types: (a) a nonlinear wave in a nondispersive medium becomes more and more steep; (b) a linear wave in a dispersive medium undergoes dispersive spreading; (c) a nonlinear wave in a dispersive medium may form a soliton structure and propagate without distortion of its shape.

We shall present a simplified derivation of the nonlinear Schrödinger equation, which contains solitary solutions. At the heuristic level the standard procedure of derivation of the nonlinear equation involves expanding the dispersion relation $\omega(k)$ in the vicinity of k_0 into a series and taking into account the nonlinear frequency shift $\delta\omega = \alpha|\psi|^2$, where α is the coefficient of nonlinearity and ψ is the amplitude of the wave.

We consider the magnetoactive plasma assuming the z axis to be along the direction of the external magnetic field. We focus on a one-dimensional case where all the values depend only on z and t. Let the wave energy be localized in a narrow vicinity of the wavenumber k_0

$$E_x(z, t) = \varphi(z, t) \exp(ik_0 z - i\omega_{k_0} t)$$
$$E_y(z, t) = \mp i\varphi(z, t) \exp(ik_0 z - i\omega_{k_0} t). \tag{7.19}$$

Thus $\varphi(z, t)$ varies in time and space more slowly and smoothly than the oscillations inside the packet. The upper (lower) sign in (7.19) corresponds to the wave of the left (right) circular polarization, as usual. For simplicity we shall consider the wave packet to be formed of the waves of one out of the two polarizations—either left or right.

The Fourier components of the functions $\varphi(z, t)$ and $E_x \equiv E(z, t)$ are connected by the relation

$$E_{\omega k} = \varphi_{\nu \kappa} \tag{7.20}$$

where $\nu = \omega - \omega_{k_0}$, $\kappa = k - k_0$). The spectral components E_y and φ are connected similarly with each other.

In the linear approximation, the wave equation for the spectral components $E_{\omega k}$ has the form

$$(\omega - \omega_k) E_{\omega k} = 0. \tag{7.21}$$

Consequently, for $\varphi_{\nu \kappa}$ from (7.20) we have

$$(\nu - \omega_k + \omega_{k_0}) \varphi_{\nu \kappa} = 0. \tag{7.22}$$

Now we shall take into account that the wave spectrum is localized in the small vicinity of k_0 and expand ω_k into a series in the vicinity of this wavenumber

$$\omega_k = \omega_{k_0} + v_g \kappa + \tfrac{1}{2} v_g' \kappa^2. \tag{7.23}$$

Here $v_g = \partial v_g / \partial k$, $v_g' = \partial v_g / \partial k$, and both parameters are calculated at $k = k_0$. In the following stage we introduce the term $\alpha |\psi|^2$, which imitates the nonlinear frequency shift, into the right-hand side of (7.23). Here we have introduced a dimensionless amplitude $\psi = b/B$, where $b = \sqrt{\varepsilon}\varphi$, $\varepsilon = \omega_{0i}^2 / \Omega_i^2$ and $\varepsilon \gg 1$. After that we substitute (7.23) into (7.22), accomplish the inverse Fourier transformation and obtain the nonlinear Schrödinger equation

$$i \left(\frac{\partial \psi}{\partial t} + v_g \frac{\partial \psi}{\partial z} \right) + \frac{v_g'}{2} \frac{\partial^2 \psi}{\partial z^2} - \alpha |\psi|^2 |\psi| = 0. \tag{7.24}$$

The first two terms in (7.24) describe the packet transition as a whole with the group velocity in the positive z direction. The third term describes the linear effect of a packet spreading in a dispersive medium. Then the fourth term describes the nonlinear effect of self-action of the wave packet.

If the Lighthill criteria

$$v_g' \alpha < 0 \tag{7.25}$$

is satisfied then equation (7.24) has a solution of the form of a solitary wave

$$|\psi| = |\psi_0| \operatorname{sech}[(z - v_g t)/\zeta]. \tag{7.26}$$

The width of the solitary wave ζ is connected with the maximum amplitude $|\psi_0|$ by the ratio

$$|\psi_0|^2 \zeta^2 = -v_g'/\alpha. \tag{7.27}$$

Except for the solution that describes single soliton states, equation (7.26) generally contains multi-soliton solutions. They describe the interaction of two, three and more solitons (figure 7.2).

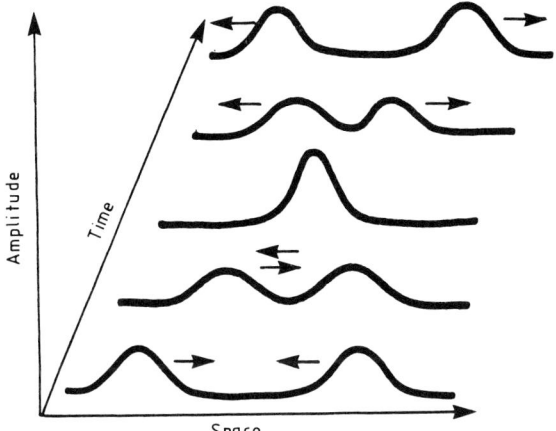

Amplitude

Time

Space

Figure 7.2. The interaction of two solitons: first, they move towards each other and then, after interaction, they restore their initial form and move in different directions.

The coefficients of the linear terms of equation (7.24) are not difficult to find if the explicit dependence $\omega(k)$ is known. For example, in the MHD limit, i.e. for $\omega \ll \Omega_i$, we have

$$v_g = c_A \qquad v'_g = \mp c_A / k_{0i}. \qquad (7.28)$$

Much more difficult to evaluate is the coefficient of nonlinearity α. This is a delicate problem. From the standpoint of the physics of geomagnetic pulsations it is a problem of extraordinary importance since the possibility of the existence of solitons depends on the sign of α (see criteria (7.27)). The complexity lies in the following.

We consider one-dimensional solitons. A one-dimensional approximation allows us to enter the domain of the fundamental concepts of soliton physics without any particular mathematical difficulty, and hence it is useful. Moreover, the linear one-dimensional waves, propagating strictly along or strictly across the field lines of the external magnetic field, satisfactorily approximate certain kinds of geoelectromagnetic waves. It is natural to try to conserve this assumption in the nonlinear theory too. However, it is obvious even from general considerations of existence and stability that in the case of solitons the dimension plays a much more important role than it does in the case of linear wave packets.

We shall return to this problem in section 7.3, but now we proceed with the discussion of one-dimensional solitons.

In order to find α in the MHD range we shall write the dispersion relation in the form $\omega = k(c_A + v_z)$. Here the small dispersion term is neglected; but the Doppler shift frequency, connected with the movement of plasma at velocity

v_z under the action of the ponderomotive force, is taken into account. In a cold plasma v_z is found from the equation of motion

$$\rho \dot{v}_z = f_z. \tag{7.29}$$

In the case of solitary waves when v_z, f_z and other slowly changing values depend on t and z only in the combination $z - c_A t$, the longitudinal component of the ponderomotive force (7.2) equals

$$f_z = \frac{B^2}{8\pi c_A} \frac{\partial |\psi|^2}{\partial t}. \tag{7.30}$$

Substituting (7.30) into (7.29) we find

$$v_z = (c_A/2)|\psi|^2. \tag{7.31}$$

Under the action of the ponderomotive force not only the velocity is disturbed but also the other macroscopic parameters of the medium which are taken into account in the dispersion relation. In the first approximation these perturbations, averaged over the fast oscillations, are in square dependence on the wave amplitude. The corresponding correction to the frequency (7.23) is

$$\delta \omega = \sum \frac{\partial \omega}{\partial \xi_i} \delta \xi_i = \alpha |\psi|^2. \tag{7.32}$$

In the case under consideration there are three parameters $\xi_1 = v_z$, $\xi_2 = \rho$, $\xi_3 = B$. We have already examined the perturbation of velocity. From the continuity equation we find the connection between the density perturbations and that of the velocity: $\delta \rho / \rho = v_z / c_A$. As regards to the perturbations of the magnetic field B, at strictly longitudinal propagation they do not appear at all due to the equation $\nabla \cdot B = 0$. Taking this into account we find from (7.31) and (7.32)

$$\alpha = c_A k_0 / 4. \tag{7.33}$$

So the coefficient of nonlinearity is positive. It means that according to the Lighthill criterion (7.20) the existence of the Alfvén solitons is possible while for solitons it is impossible[5]. This conclusion (equation (7.33)) is in accordance with the rigorous treatment in the framework of the perturbation theory. However, formula (7.33) and the conclusion we have made cannot be considered accurate enough if we are interested in the search for real MHD solitons in the near-Earth plasma. The sign of α may, for various reasons, be opposite to that indicated in (7.33).

Let us start with the finite pressure effect. In the isotropic plasma the sign of α changes from positive at $c_s < v_g$ to negative at $c_s > v_g$ (see exercise

[5] It will be recalled that we are considering one-mode solitons so this conclusion relates to this approximation only. It does not exclude the existence of the soliton containing Alfvén and magnetosonic waves simultaneously.

7.2.2). This feature is the result of the so-called inversion of actions, which is well known in gas dynamics. In an anisotropic plasma the situation is even more complicated. First, $|\alpha|$ increases when approaching the threshold of the fire-hose instability. Second, it turns out that while calculating α the anisotropy should be taken into account even if plasma is isotropic in the unperturbed state. The point is that the anisotropy, appearing under the influence of the wave, essentially affects nonlinear properties of the plasma, and at high pressure it makes the main contribution to α.

The sign of α may be changed even for $c_s = 0$ if the wave is not strictly one-dimensional. For strictly field-aligned propagation, the plane wave, stretched infinitely in the transverse direction, does not perturb the longitudinal component of the magnetic field. Owing to the high symmetry, this configuration appears to be strongly degenerated in the following sense.

If we take the real collimated quasi plane wave, then because of the inverse Faraday effect there appears a perturbation of the magnetic field, which is square with respect to the wave amplitude. The perturbation of the magnetic field δB makes a contribution to a nonlinear correction to the frequency (7.32). This may lead to a change of the sign of α. For example, if the wave is localized inside a cylinder of arbitrary large radius, then in a cold plasma

$$\delta B = -B|\psi|^2. \tag{7.34}$$

Substituting (7.34) into (7.32) and taking into account the respective modification of $\delta \rho$, we find

$$a = -c_A k_0/4. \tag{7.35}$$

The final analysis gives an idea of the difficulties of the problem of one-dimensional approximation in the description of MHD solitons. Comparing equations (7.25), (7.28), (7.33) and (7.35) one may assume the existence of one-dimensional solitons formed either by Alfvén or magnetosonic waves. But even if criterion (7.20) is satisfied, the one-dimensional solitons may be unstable under transverse deformations.

Without any doubt the development of the theory will lead to better understanding of the existence and structure of quasi one-dimensional MHD solitons. Besides this one should try to search now for solitons among the geomagnetic pulsations according to the above-given general features of the solitons.

The example of pulsations, to which we may apply the theory of solitons, is presented in figure 4.7. We see the series of 'pearls'. When considering the magnetogram from an observatory it may seem that a whole string of wave packets arrives at the observation point—sometimes up to one hundred and even more. But we already know that in reality it corresponds to only one or two packets. The wave packet propagates in the magnetosphere along the field line, reaches the Earth, reflects, moves back, again reaches the Earth and again reflects. Since the process repeats again, there appear to be a multitude of packets. We note by the way that figure 4.7 suggests an interaction between the two solitons.

7.2.2   The search for solitons

The problem of extraction and study of solitons in the magnetosphere naturally enters the realm of physical problems of the near-Earth environment. Specific theoretical problems, connected with solitons, may find a proper experimental basis in the observational data of the wave activity of the magnetosphere. In literature devoted to MHD solitons, the problem of seeking such objects among the geomagnetic pulsations still only has a modest place.

In seeking the solitons, it is reasonable to use relation (7.27) which gives the connection between the soliton width ζ and its amplitude $|\psi_0|$. The rational application of the given criterion may lead to a rather simple but straightforward strategy for the investigation. Of course it is out of the question to use this criterion for the separate events. One ought to examine not individual events but the whole series, united by the community of signs.

Some other methods are used in the search for solitons among the natural electromagnetic emissions. For example, the soliton can be identified by a typical form (see (7.26)). However, following this method, it is rather difficult to distinguish the soliton from a linear wave packet with a bell-shaped amplitude envelope.

Another method is based on the fact that generally the soliton velocity depends on its amplitude. However, such dependence exists not only for solitons but also for other nonlinear wave packets. Both methods are less efficient than the method based on (7.27) but they can be used as additional tools for seeking solitons.

The form of the amplitude envelope of a separate Pc1 packet is similar in general to the form of the curve described by (7.26). In addition the velocity of Pc1 displacement along the field lines depends on the amplitude of the signal. It is noticeable at the mounting stage of the Pc1 series. To observe the effect it is reasonable to choose 'clean' Pc1 series, in which the intervals between the wave packets are without parasitic signals. In figure 6.8 we see four echo-signals of rising intensity. Every signal appears at the observation point after the double pass through the radiation belt, where it is amplified. As the intensity increases, the period of signal repetition decreases. By natural assumption that the 'optical' path length is not changed in time, a decrease of the repetition period signifies an increase of the efficient propagation velocity as the amplitude increases.

The soliton-like features of Pc1 signals suggest a method for testing these signals by means of (7.27). At a qualitative level the ratio (7.27) is accomplished. Let us compare, for example, the first and the second signals in figure 4.7. The width of the weak signal exceeds the width of the strong one twofold at half the maximum amplitude.

For the quantitative analysis let us rewrite (7.27), taking into account (7.28), in the form

$$b = \frac{2m_p c}{e\tau} \left(\frac{\Omega}{\omega}\right)^{1/2} \tag{7.36}$$

where e is the elementary charge, m_p is the proton mass, $\tau = \zeta/c_A$ is the duration of the soliton and ω is the carrier frequency. Equation (7.36) corresponds to the Alfvén or magnetosonic waves in terms of the dependence on two nonlinearity coefficients, either (7.33) or (7.35). The result of processing several Pc1 series indicates an increase of the amplitude with an increase of the parameter $\xi = 1/\tau\sqrt{\omega}$ (figure 7.3).

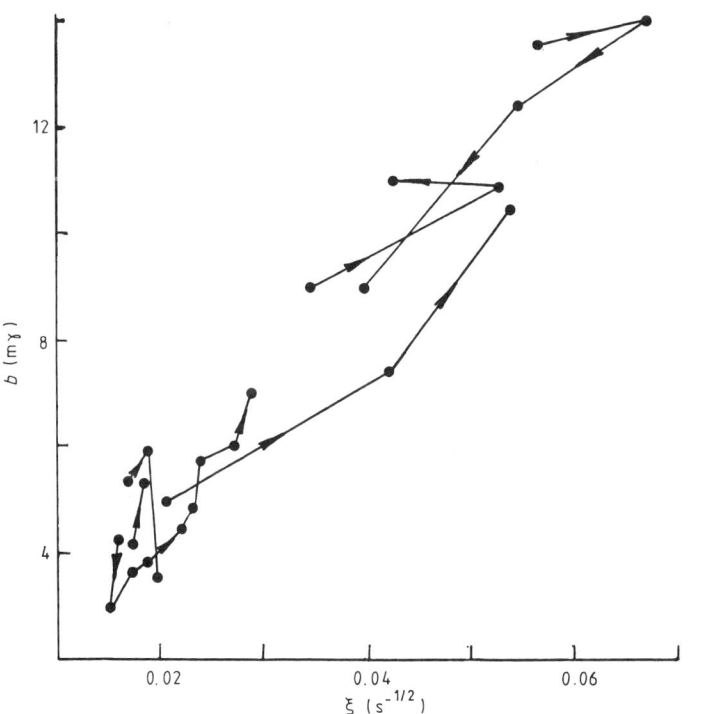

Figure 7.3. The dependence of Pc1 amplitude on the parameter ξ. The segments connect the points which correspond to the adjacent packets. The arrows signify the direction of time. The plot was based on the observations of the Sogra Observatory. 'Clean' pearl series were selected which consist of signals distinctly separated from one another.

The result is promising, but the problem remains unclear. The Pc1 signals are excited as Alfvén waves in the radiation belt. For these waves the Lighthill criterion (7.25) is accomplished if $\alpha > 0$. But the value (7.35) seems to be more real than (7.33). This contradiction cannot be eliminated in the framework of the simple theory explored above.

Nevertheless no set of signatures nor even the test (7.27) by itself guarantees that some MHD signal should be considered as a soliton of type (7.26). The general idea concerning the origin of the signals turns out to be more significant. For example, let us examine figure 7.4 in which a pair of Pi2 signals is shown. It turns out that the weaker signal is wider as it corresponds to solitons. Indeed, the analysis of some tens of such events shows that the width of the Pi2 signal is smaller for higher amplitude. A ratio of type (7.27) is accomplished. However, it is hard to imagine that Pi2 pulsations, the wavelength of which is comparable to the dimensions of the magnetosphere, may be in the form of a travelling wave packet.

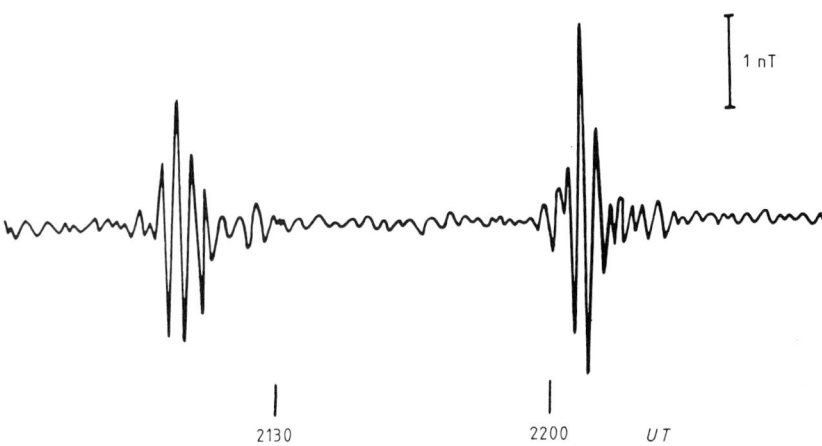

1 nT

2130   2200   *UT*

Figure 7.4. Oscillogram of Pi2 pulsations (Borok Observatory, 27 January 1976). Nonlinearity of oscillations may be revealed by means of statistical analysis of a large number of similar pairs of adjacent impulses.

7.2.3 Solitons in an active medium

Up to now we have discussed the MHD solitons in a nonlinear dispersive but conservative medium. This limitation allows us to formulate simple criterion for seeking solitons in the geomagnetic pulsations. The question appearing in this connection is as follows. How can the criterion (7.25) and test (7.27) be modified if one takes into account the fact that pulsations are excited as a result of a plasma instability and, consequently, the conservative idealization is not appropriate? It is natural to expect that close to the threshold of the stability (7.22), linking the amplitude of soliton and its width, will keep its form.

The account of nonconservatism, which is needed for the above reason, is interesting in one more respect. Either the amplitude or the width of the soliton in relation (7.27) is arbitrary, i.e. it is the free parameter which depends on initial conditions. The carrier frequency is also arbitrary. Another state of affairs is in the nonconservative system. In this case the formed soliton can exist without any free parameters: the carrier frequency, the amplitude and the width are determined completely by the parameters of the medium.

Finally, if we seriously take into account the hypothesis of Pc1 soliton structure, then the account of dissipative effects allows us to eliminate one significant objection. In the case of conservative idealization the growth rate of modulation instability, leading to soliton creation, is proportional to the square amplitude of oscillations and is so small that the characteristic time of the instability far exceeds the usual duration of Pc1. But if we deal with an unstable plasma, i.e. with an active (amplifying) medium, then the nonlinear evolution takes place during a period of the order γ^{-1}, where γ is the linear growth rate. This time does not depend on the amplitude and it can be significantly less than the characteristic time of the modulation instability in the passive medium.

Let us use (6.56) for the growth rate of the Alfvén waves due to the ion cyclotron instability of the radiation belt. For the longitudinal propagation, the growth rate $\gamma(k)$ has a maximum at the resonant value of the wavenumber $k_m = \Omega/w_\parallel$, where $w_\parallel = \sqrt{2T_\parallel/m_p}$ is the longitudinal thermal velocity of energetic protons. We expand $\gamma(k)$ in the vicinity of this maximum

$$\gamma(k) = \gamma_m - \eta(k - k_m)^2. \tag{7.37}$$

Here $\eta = 2\gamma_m/k_m^2$, γ_m is the maximum value of the growth rate.

Now we shall take into account the absorption. There are several different mechanisms of dissipation. The rigorous calculation of the explicit form for this value is quite complicated. So let us introduce the effective parameter decrement δ, which includes all kinds of losses. Let δ in the vicinity of k_m be independent of k. Then the net growth rate will keep the form (7.37), where we have to replace γ_m by $\bar{\gamma}_m = \gamma_m - \delta$. Let us present this in the form

$$\gamma(k) = \bar{\gamma}_m \left[1 - 4(k - k_m)^2/(\Delta k)^2\right] \tag{7.38}$$

where $\Delta k = 2\sqrt{\bar{\gamma}_m/\eta}$ is the width of the instability range. Close to the threshold of stability, i.e. for $\bar{\gamma}_m \ll \gamma_m$, we have $\Delta k \ll k_m$.

The evolution of small perturbations looks as follows.   The plasma instability leads to the fact that Alfvén waves whose frequency is near to $\omega_m = c_A k_m$ increase with growth rate $\bar{\gamma}_m$. As the amplitude increases the plasma nonlinearity begins to play an important role, and the modulation instability is developed. As a result of this instability wave spectral broadening appears, which leads to energy pumping from the region of amplification $|k - k_m| < \Delta k/2$ into the region of dissipation $|k - k_m| > \Delta k/2$. The wave amplitude can be stabilized by this process.

The equation describing the wave modulation in an unstable plasma is obtained from (7.24) if, according to (7.22), (7.23) and (7.38), we add to the right-hand side of (7.24) the term

$$i\bar{\gamma}_m \left[\psi + \frac{4}{(\Delta k)^2} \frac{\partial^2 \psi}{\partial z^2} \right]. \tag{7.39}$$

This term describes the plasma instability and the dispersion of the growth rate.
Let us introduce the new variables

$$\phi = \left[\frac{w_\parallel}{2c_A} \right]^{1/2} \psi \qquad \tau = \left[\frac{c_A}{w_\parallel} \right]^2 \frac{\Omega t}{2} \qquad \xi = k_m(z - c_A t)$$

and rewrite the equation for Alfvén wave modulation in dimensionless form

$$\phi_\tau + i\phi_{\xi\xi} + i|\phi|^2\phi = \nu\phi - \mu\phi_{\xi\xi}. \tag{7.40}$$

Here $\nu = 2\bar{\gamma}_m w_\parallel^2/c_A^2\Omega$, $\mu = 4\gamma_m w_\parallel^2/c_A^2\Omega$.
We are seeking a solution to (7.40) in the form

$$\phi(\xi, \tau) = \phi_0 \operatorname{sech}(\xi/\xi_0) \exp\left(-i \int_0^\tau \kappa \, d\tau'\right). \tag{7.41}$$

It is well known that for $\nu = \mu = 0$ this solution is rigorous, and the amplitude
ϕ_0, the width ξ_0 and phase κ of the soliton are constant. They are connected with
each other by the relations $\phi_0 = \sqrt{2}/\xi_0$, $\kappa = \phi_0^2/2$ (see e.g. (7.26) and (7.27)).
Let us suppose that these connections are kept at nonzero, but sufficiently small μ
and ν. However, now ϕ_0, ξ_0 and κ will be slowly changing in time. Substituting
(7.41) into (7.40) we find the equation describing the soliton amplitude variation

$$d \ln \phi_0/d\tau = 2\nu - \mu\phi_0^2/3$$

Its stationary stable solution has the form

$$\phi_0 = \sqrt{6\nu/\mu}. \tag{7.42}$$

The width of the stationary soliton is

$$\xi_0 = \sqrt{\mu/3\nu}. \tag{7.43}$$

It is evident that the relation between the amplitude and width of the soliton is

$$\phi_0\xi_0 = \sqrt{2} \tag{7.44}$$

which is quite similar to that in the conservative medium.
So the account of nonconservatism makes the theory physically more
attainable and it opens up new possibilities for applications. The soliton features
in the active medium do not depend on the initial conditions at all. They are
determined completely by the features of the medium. The carrier frequency is

determined by the cyclotron resonance condition $\omega = \Omega(c_A/w_\parallel)$. The amplitude and the width of the soliton are connected by the usual relation, but each of these values is determined separately by the width $\Delta\omega$ of the instability band

$$b = \frac{m_p c}{e} \left(\frac{3 w_\parallel}{c_A} \right)^{1/2} \Delta\omega \tag{7.45}$$

$$\zeta = \frac{2 c_A}{\sqrt{3} \Delta\omega}$$

where $\Delta\omega = c_A \Delta k$.

However, to what extent does this theory allows us to improve our knowledge of 'pearls'? These signals are sometimes observed for several hours. During this period the wave packet covers a distance of millions of kilometres, none the less in general it keeps its individuality even though it propagates in a dispersive medium. In other words, the wave packet resembles the soliton.

But it not only resembles the soliton it really is the soliton. Magnetospheric plasma on the ray path of the 'pearls' represents the active medium. The wave packets arise spontaneously and are amplified within it. Its amplitude increases rapidly, and nonlinearity begins to play a role. The further evolution of the packet is not yet quite clear. Probably, the dispersion is compensated. But it may be that the amplifying properties of the medium are modified under the action of waves, and this leads to the specific modulation of the wave field, resulting in the effect of 'pearl pulsations'. The alternative point of view is that it is not the amplifying coefficient that is modulated but the damping decrement. In any case the process ends with the appearance of the Alfvén auto-soliton, which looks like a series of 'pearls' on the magnetogram. (The prefix 'auto' emphasizes the spontaneous character of soliton appearance).

Exercises

Exercise 7.2.1.

Evaluate the magnetic momentum induced in a plasma by a transverse circularly polarized electromagnetic wave propagating along the external magnetic field.

Solution 7.2.1.

Let a homogeneous magnetic field be directed along the z axis. The electric field of the wave has the following components

$$E_x = E \exp(ikz - i\omega t)$$

$$E_y = \mp iE \exp(ikz - i\omega t).$$

The upper (lower) sign refers to the left (right) circular polarization. The magnetic moment of a single charged particle, appearing in the wave field, is

$$\mu = (e/2c)\overline{x \times v}$$

where $v = \dot{x}$, $x(t)$ is the trajectory of a particle and the bar denotes averaging over time. First we consider a cold plasma. In this case the equation of motion (3.41) may be linearized, i.e. $\dot{v} = F + v \times \Omega$, where $F = (e/m)E$, $\Omega = eB/mc$. Integration gives

$$v_x = \frac{iF}{\omega \mp \Omega} \qquad v_y = \mp i v_x \qquad v_z = 0. \qquad (1)$$

Taking into account that $\overline{x \times v} = (1/4)x^* \times v + \text{c.c.}$, $x = (i/\omega)v$ and $M = \sum N\mu$ we obtain the magnetic moment per unit volume

$$M_z = \mp \frac{E^2}{8\pi c \omega} \sum \frac{e}{m} \frac{\omega_p^2}{(\omega \mp \Omega)^2}. \qquad (2)$$

We may easily verify that (2) coincides with (7.3) if the tensor of dielectric permeability of a cold plasma (3.46) is used.

However, the Pitaevsky formula (7.3) is inadequate for media with spatial dispersion, for example in a hot plasma. Therefore, we proceed with the straightforward calculation.

In a hot plasma we must distinguish nonresonant particles, interacting adiabatically with the wave, and resonant particles trapped in the wave field. The magnetic moment of nonresonant particle is obtained in the following way. Let us pass to a reference system, travelling along B at velocity v_z. For nonresonant particles the formulae (1) are correct in this system. Turning back to the initial system, connected with a plasma, we should make a substitution $\omega \to \omega - kv_z$, $E \to E - (1/c)v \times b$, where $b = (c/\omega)E \times k$ is the magnetic field of the wave. As a result we get an explicit dependence μ_z on v_z

$$\mu_z = \mp \frac{e^3 E^2 (\omega - kv_z)}{2cm^2\omega^2(\omega \mp \Omega - kv_z)^2}. \qquad (3)$$

Hence we see that when approaching the resonant condition $v_z = (\omega \mp \Omega)/k$ and the magnetic moment of the particle increases. However, in the immediate vicinity of the resonance, equation (3) breaks down, since the nonlinear character of the particle motion in the field of the wave is not taken into account. We make an assumption that the contribution of an anomalous magnetic moment of resonant particles may be neglected.

The magnetic moment per unit volume may be expressed in the form

$$M_z = \sum \int_{-\infty}^{\infty} \mu_z(v_z) f(v_z) \, dv_z \qquad (4)$$

where $f(v_z)$ is the distribution function of charged particles, normalized so that $\int_{-\infty}^{\infty} f(v_z)dv_z = N$, N is the particle's number density. We choose the Maxwell distribution and assume that $|\omega - \Omega| \gg kw_\parallel$, where $w_\parallel = \sqrt{2T_\parallel/m}$, $T_\parallel$ is the longitudinal temperature. This condition is sufficient to neglect the contribution of resonant particles to M_z. Substituting (3) into (4) and integrating, we find the value of plasma magnetization in the field of the circularly polarized wave

$$M_z = \mp \frac{E^2}{8\pi B} \sum \frac{\omega_0^2 \Omega}{\omega(\omega \mp \Omega)^2} \left[1 + \frac{k^2 w_\parallel^2(\omega \mp 2\Omega)}{2\omega(\omega \mp \Omega)^2} \right].$$

Magnetization of the medium by the circularly polarized wave is termed the inverse Faraday effect. It is assumed that the effect has a square dependence on the wave amplitude and in fact this is so, but not always. In the plasma which contains a noticeable admixture of resonant particles $M_z \propto E^{3/2}$, i.e. an anomalous inverse Faraday effect appears.

Exercise 7.2.2.

Define the nonlinearity coefficient in a nonlinear dispersion relation taking into account the effect of finite pressure in the case of longitudinal propagation of transverse electromagnetic waves.

Solution 7.2.2.

Let us assume the pressure to be isotropic. Then at $k_\perp = 0$ the pressure does not influence the linear dispersion relation for the transverse waves. The corresponding coefficient is

$$\alpha = \frac{v_g k}{4} \frac{c_A^2}{v_g^2 - c_s^2}. \tag{1}$$

To derive this from (7.32) it is sufficient to take into account the term ∇p in the equation of motion (7.29).

We see that α changes its sign from being positive at $c_s < v_g$ to negative at $c_s > v_g$. Another important feature is that $|\alpha|$ is the larger, the closer the group velocity is to the sound velocity[6]. Both features are a consequence of the inversion of the action law known from gas dynamics.

The account of anisotropy of the plasma pressure introduces corrections into (1). First, $|\alpha|$ increases when approaching the threshold of the fire-hose instability. Second, it turns out that when calculating α the anisotropy should be taken into account even in the case when plasma is isotropic in a steady state.

---

[6] To avoid misunderstanding, we note that in fact the increase of $|\alpha|$ while v_g approaches c_s will not continue infinitely. The applicability of (1) is limited by the condition of smallness of $\delta\omega$ as compared with ω. At $\omega \ll \Omega$ it gives the following limitation: $|(c_A - c_s)/c_A| \gg (1/8)(b/B)^2$.

7.3   Vortices

If one does not take into account the difference in signs of dispersion (7.28), then for the case of parallel and quasi-parallel propagation $(k_\perp \ll k_\parallel)$ of Alfvén and magnetosonic waves we have similar behaviour. Both of them are approximately described by equations similar to (7.24). A completely different story arises for the case of perpendicular or quasi-perpendicular propagation $(k_\perp \gg k_\parallel)$. A fast magnetosonic wave propagates freely in an arbitrary direction, relative to the external magnetic field. Due to the existence of the magnetosonic waveguide, quasi-one-dimensional propagation $(k_\perp \gg k_\parallel)$ is easily realized in the magnetosphere (see section 4.3).

A different story arises for Alfvén waves. The linearization of the MHD equations leads to the dispersion relation of Alfvén waves $\omega = c_A k_\parallel$. This means that for any $k_\perp$ the energy of the wave is transported strictly along the ambient magnetic field. The perpendicular structure of the wave field is completely arbitrary.

Such an arbitrariness is unsatisfactory from the theoretical point of view. The account of nonlinear and dispersion effects in the description of Alfvén waves restricts arbitrariness in selecting the perpendicular structure of the field and removes the paradox.

But even this does not represent the principal point. The spontaneous amplification of Alfvén waves may arise in unstable plasma. It is caused, for example, by the drift-dissipative instability of the inhomogeneous plasma. In this case nonlinearity becomes inevitable. The analysis testifies to the fact that the combined action of nonlinearity and dispersion results in the condensation of Alfvén waves into vortices, highly stretched in the longitudinal direction. Moreover, vortex structures of various forms were observed in the Earth's magnetosphere (e.g. figure 7.5 shows an oscillogram of the Alfvén vortices excited in the auroral zone).

The derivation of equations necessary for the description of Alfvén vortices demanded, at the proper time, an analysis of the basic notions and initial premises. One could use the usual MHD equations. However, they also contain information on magnetosonic waves, which means that they are too complicated for the description of Alfvén waves. It is reasonable to make a reduction of the MHD equations, i.e. to exclude unimportant elements, using *a priori* notions on the field structure of the Alfvén waves. On the other hand, the MHD equations are not complete in the sense that they do not take into account the transverse dispersion of the Alfvén waves. So, they should be supplemented by the necessary terms which cannot be treated in the framework of the one-fluid approximation.

We present the basic equations of the theory of Alfvén vortices in the Petviashvili–Pokhotelov form

$$\frac{d\Gamma}{dt} = -\frac{\partial J}{\partial z} \tag{7.46}$$

IC-B-1300   02 MARCH 1982

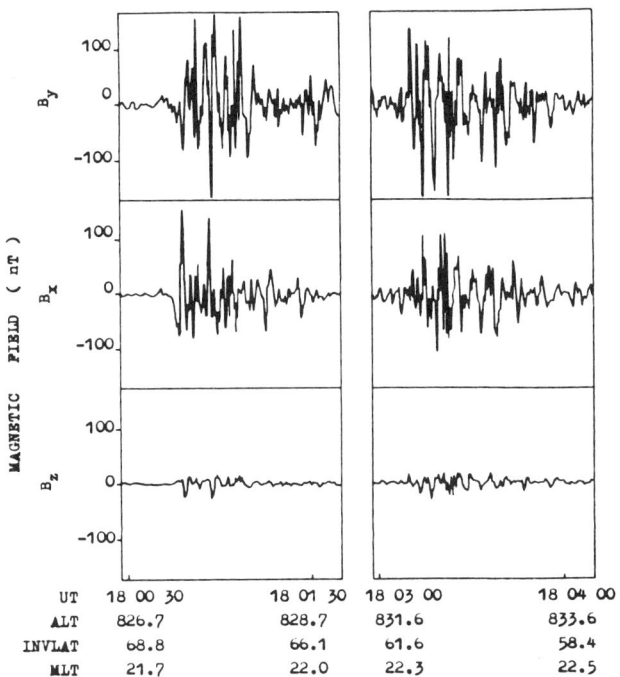

Figure 7.5. Oscillograms of the magnetic field oscillations in the frequency range 0.1–8 Hz along the *IC-Bulgaria 1300* satellite orbit (Chmyrev *et al* 1988). These intense oscillations do not reach ground level. They have the structure of Alfvén vortices.

$$\frac{\partial A}{\partial t} - \epsilon \frac{\mathrm{d}J}{\mathrm{d}t} = \frac{\mathrm{d}}{\mathrm{d}z}(N - \Phi) \qquad (7.47)$$

$$\frac{\mathrm{d}N}{\mathrm{d}t} = -\frac{\mathrm{d}J}{\mathrm{d}z}. \qquad (7.48)$$

Here $\Gamma = \Delta\Phi$ is the vorticity, $J = \Delta A$ is the field-aligned current, Φ is a scalar potential, A is the component of vector potential, $N = (n/n_0) - 1$ is the normalized plasma density perturbation, n_0 is an unperturbed plasma density, $\epsilon = (m_e/m_i)(c_A/c_s)^2$, $m_{e(i)}$ is the electron (ion) mass, $c_A = \Omega/k_{0i}$ is the Alfvén velocity, $c_s = (T_e/m_i)^{1/2}$ is the sound velocity, $k_{0i} = \omega_{0i}/c$, ω_{0i} is the ion plasma frequency and T_e is the electron temperature. Differential operators in (7.46)–(7.48) have the form

$$\frac{\mathrm{d}}{\mathrm{d}t} = \frac{\partial}{\partial t} + \{\Phi, \ldots\} \qquad \frac{\mathrm{d}}{\mathrm{d}z} = \frac{\partial}{\partial z} - \{A, \ldots\}$$

$$\Delta = \frac{\partial^2}{\partial x^2} + \frac{\partial^2}{\partial y^2} \qquad \{A, \Phi\} = \frac{\partial A}{\partial x}\frac{\partial \Phi}{\partial y} - \frac{\partial A}{\partial y}\frac{\partial \Phi}{\partial x}.$$

All the values are given in the dimensionless form. In order to restore the dimension one should divide (x, y), z, t by k_s, k_{0i}, Ω and multiply Γ, J, Φ, A by Ω, k_{0i} B_z, T_e/e, $(T_e/e)(c/c_A)$ respectively. Here $k_s = \Omega/c_s$, e is the elementary charge and B_z is an unperturbed magnetic field. Finally, the transverse components of the magnetic field are expressed in the following way through the z component of the vector potential

$$B_\perp = [\nabla A, \zeta]$$

where ζ is the unit vector along the z axis.

The following inequalities were used for the derivation of (7.46)–(7.48)

$$\partial/\partial t \ll \Omega \qquad \nabla_\| \ll \nabla_\perp \qquad B_\perp \ll B_z \qquad c_s \ll c_A \qquad \epsilon \ll 1.$$

It was also assumed that $B_z = $ constant, $n_0 = $ constant. Some of these limitations may be omitted without any significant modification of the theory.

Before we proceed to the description of vortex solutions of equations (7.46)–(7.48) let us point out the origin of each term and clarify its physical sense. Equation (7.46) is derived from the equation of perpendicular ion motion and the current continuity equation. In doing so the ion inertia is taken into account but the velocity of parallel ion motion is considered to be zero. The effects of ion pressure are disregarded. Equation (7.47) is obtained from the equation of parallel electron motion. The electron inertia and electron pressure are taken into account for the analysis of parallel but not perpendicular electron motion. The velocity of perpendicular motion is considered to be equal to electron drift velocity. Equation (7.48) represents the electron continuity equation.

Let us explain the physical sense of each term in (7.46)–(7.48). All nonlinear terms are contained in Jacobians of the type $\{A, \Phi\}$. Their origin is connected with the operators $(v_\perp \cdot \nabla)$ and $(B_\perp \cdot \nabla)$. The first term represents the plasma transport at $E \times B$ velocity and the second one gives the bending of the magnetic field lines. Neglecting these terms we obtain the following dispersion equation

$$\frac{\omega}{k_\|} = \left(\frac{1 + k_\perp^2}{1 + \epsilon k_\perp^2} \right)^{1/2}. \tag{7.49}$$

Now we may clarify the origin of the transverse dispersion of the Alfvén waves. It is connected with the effects of ion pressure and inertia (the numerator and the denominator on the right-hand side of (7.49) respectively). If we omit both the nonlinearity and the dispersion, then we obtain $\omega = k_\|$. So the perturbation propagates along the z axis at the Alfvén velocity, if we proceed to dimensional values. The transverse structure of the wave field in this case is arbitrary. Equations (7.46) and (7.47) may be reduced to Strauss equations if one neglects the dispersion effects (i.e. to substitute $\epsilon = N = 0$). These equations also follow from the Kadomtsev and Pogutse equations, obtained in the one-fluid approximation. Finally, for $A \ll \Phi$ and $N = 0$ the system (7.46)–(7.48) reduces to the Hasegawa and Mima equation.

Therefore, equations (7.46)–(7.48) contain all the physical information and are fairly universal. In particular, they imitate the Alfvén waves in a weakly inhomogeneous plasma. Below we shall consider plasma to be inhomogeneous in the x direction. Let us substitute $N \to \kappa x$ in the vicinity of the $x = 0$ plane. For simplicity we consider $\kappa =$ constant. We note that in this case the term $\kappa k_y / \omega$ should be added to the numerator and denominator of the expression in the brackets of (7.49). After that the dispersion relation will describe the drift-Alfvén waves.

7.3.1 Classification of Alfvén vortices

The existence of Alfvén vortices is confirmed by the straightforward construction of solutions of equations (7.46)–(7.48). This is a rather delicate and laborious procedure. We limit ourselves to the description of some classes of solutions known to date. The classification will simplify the interpretation of observational data in terms of the physical origin of the phenomena.

We consider four classes of wave structures: monopole and dipole vortices, vortex chains and vortex cells. Dipole vortices and vortex chains are obtained as the solutions for the equations (7.46)–(7.48) in the form presented here. To obtain the solutions in the form of monopole vortices a small scalar nonlinear term should be added to equation (7.47). Convective cells represent the solutions of equations (7.46)–(7.48) with an additional condition that $\partial/\partial_z = 0$.

Besides these solutions there are also other types of vortex structures (flute and ballooning vortices, magnetic islands, etc). However, their description would require a more significant modification of equations (7.46)–(7.48).

The solutions in the vicinity of the $x = 0$ plane should be found within the class of functions depending only on x and $\eta = y - ut + \alpha z$. Here u is the vortex velocity and α is the inclination of the vortex relative to the z axis. The monopole vortex is axially symmetric, i.e. all its parameters depend only on $r = (x^2 + \eta^2)^{1/2}$. As was mentioned above, such a solution exists if (7.47) is supplemented by a small term (which is of the order of m_e/m_i). The need to investigate monopole vortices arose in connection with the observations of such wave structures in space plasma.

All variables in dipole vortices depend on r and $\theta = \tan^{-1}(\eta/x)$. The example of a dipole vortex is presented in figure 7.6. We see a pair of closely packed vortices with positive and negative vorticity. An essential feature of the structure is its strong localization i.e. all perturbations decrease exponentially with increasing r. The energy of the dipole vortex in inhomogeneous plasma, i.e. when $\kappa \neq 0$, is negative. In the presence of dissipation, the energy of the vortex decreases and hence the intensity of vortex motion increases. This process may be treated by adding the term $\nu \Delta A$, where ν is the magnetic viscosity, to the right-hand side of (7.47).

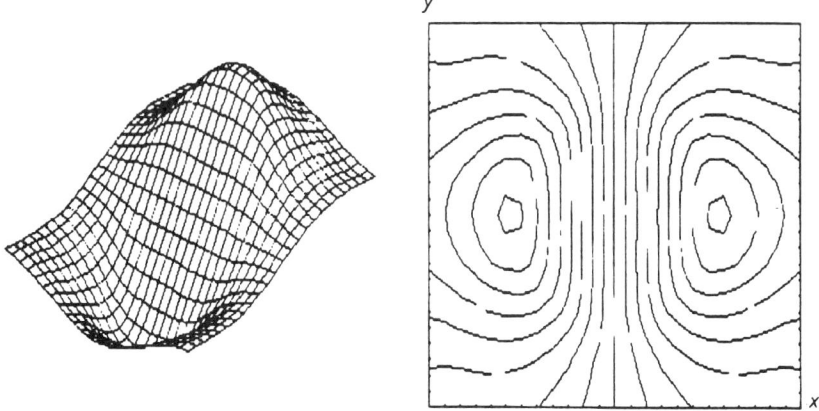

Figure 7.6. Dipole drift-Alfvén vortex. Plot of the potential relief (left). Lines of the level of this potential in the plane perpendicular to the vortex axis (right).

Equations (7.46)–(7.48) have solutions in the form of vortex chains, similar to the Kelvin–Stewart solutions in incompressible fluid. The relief of the chain of Alfvén vortices is displayed in figure 7.7. The axis of symmetry here coincides with the y axis. The chain may be elongated along this axis in both directions up to infinity. Then the general picture of the Alfvén wave train, localized exponentially along x and travelling along y at velocity u, will be clear. The potentials A and ϕ have qualitatively identical shape, as shown in figure 7.7. The distribution of the electromagnetic field as well as of the field-aligned current and vorticity may be found by similar differentiation of A and ϕ.

Finally, we shall consider convective cells, which represent vortices, to be homogeneous along the ambient magnetic field. The corresponding basic equations may be derived from equations (7.46)–(7.48) with the assumption $\partial/\partial z = 0$.

Let us consider the vector potential $A = A_0 \ln(\cosh \beta y)$ to be given. It corresponds to the field-aligned current

$$J = \Delta A = A_0 \beta^2 \cosh^{-2} \beta y. \tag{7.50}$$

Now we superimpose the axially symmetric perturbation in the form $\phi = \phi_0 \exp(-\gamma r^2)$, where $r = (x^2 + y^2)^{1/2}$, and consider the evolution of the system. The result of numerical simulation of equations (7.46)–(7.48) is presented in figure 7.8. The time interval between the adjacent plots is 48 units in dimensionless time. A structure of the 'curl' type is seen to appear with time. If we change the sign of A_0, then the direction of vorticity in the structure will reverse. This is displayed in figure 7.9, where the result of evolution is presented at a moment of time $t = 200$.

The result of numerical simulation for a superimposed perturbation of the

y

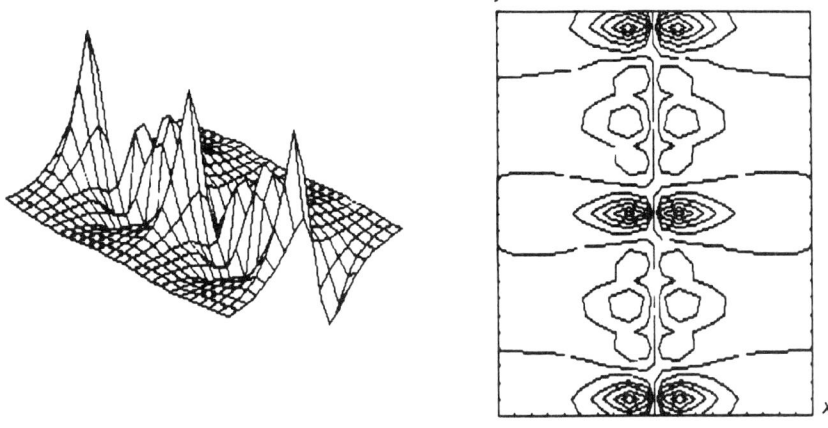

x

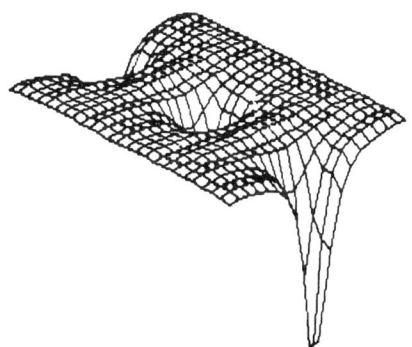

Figure 7.7. A vortex chain. Plot of the relief (left) and isolines (right) of potential (upper panel) and density (lower panel).

form $\phi = \phi_0 \exp[-\gamma(x - y)^2]$ is given in figure 7.10. It shows the fold that is fairly common in observations of the polar aurora.

Let us consider the background to be consistent not only with the current layer (7.50) but also with the scalar potential $\phi = \phi_0 \ln(\cosh \beta y)$. We analyse the evolution of the localized initial perturbation $\delta\phi = \phi_1 \exp(-\gamma r^2)$. Fragments of the evolution are shown in figure 7.11. For $\phi_1 > 0$ we have the formation of an odd chain, and the case $\phi_1 < 0$ corresponds to the even structure. Further evolution results in merging of adjacent vortices.

Some new features appear when perturbation of the current is superimposed. Let, for example, the background distribution have the form (7.50) and $\phi =$

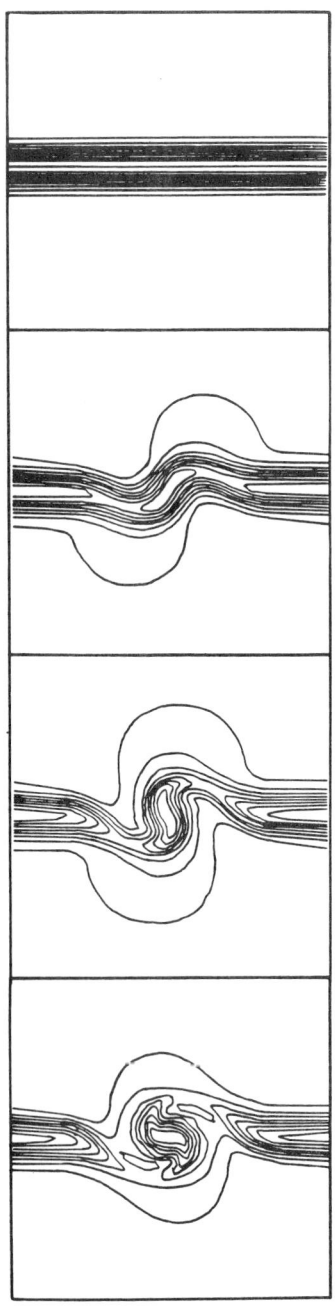

Figure 7.8. Formation of the curl-like structure. The initial perturbation of the potential is assumed to be axially symmetric.

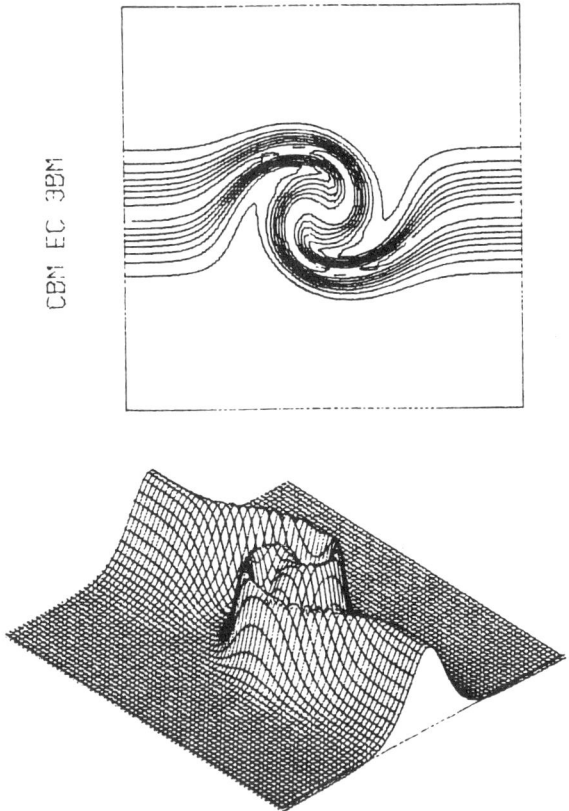

Figure 7.9. Isolines (upper panel) and relief (lower panel) of the field-aligned current in a curl-like structure.

$\phi_0 \tanh \beta y$ and the initial distribution have the form $\delta A = A_1 \exp(-\beta r^2)$. The result of the calculation is presented in figure 7.12 (left panel). The time step between different fragments is $\Delta t = 5$. One may observe the rapid growth of the field and the current in the vortex chain which results in the degeneration effect. The latter is distinctly displayed in figure 7.12 (right panel), where the background distribution of the scalar potential was given in the form $\phi = \phi_0 \exp(-\gamma y^2)$.

7.3.2   Observations of Alfvén vortices

Thin vortex tubes, the perpendicular dimension of which is much smaller than the thickness of the atmosphere, cannot be observed on the ground. In the magnetosphere they may be observed only when the satellite crosses the vortex tube. Fortunately, the generation of vortex structures represents an ordinary phenomenon in the auroral zone, so that they may be easily observed on a polar

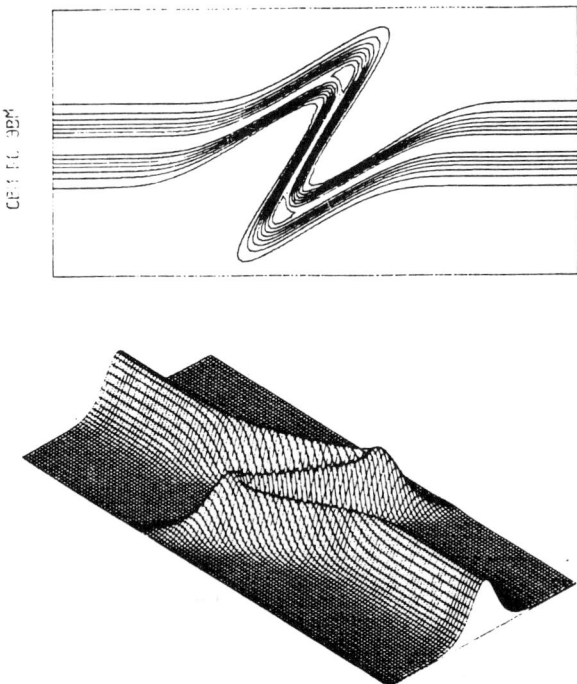

Figure 7.10. Isolines (upper panel) and relief (lower panel) of the field-aligned current. A plot of the intermediate stage of the evolution of the fold-type structure is presented.

orbiting satellite.

The satellite *IC-Bulgaria-1300* was flying at an altitude of 850 km. The inclination of the orbit was 81.3°. Six components of the electromagnetic field in the frequency range 0.1–8 Hz were being measured. An example of records in the evening sector during the magnetic storm on 2 March 1982 is presented in figure 7.5. The first signal appeared at the moment when the satellite crossed the northern boundary of the auroral oval, and the second signal appeared when it passed the southern edge. We see intense oscillations of transverse components of the geomagnetic field. The amplitude reaches 170 nT peak to peak which has never been observed on the Earth's surface in this frequency range. The electric field components were oscillating in a similar manner. Their amplitude reached 100 mV m^{-1}. The spectral maximum occurred at a frequency of 0.5 Hz.

Localized oscillations of such intensity should undoubtedly be members of the class of nonlinear oscillations. The fact that the oscillations do not penetrate down to the Earth's surface indirectly testifies to transverse localization. The localization, in its turn, suggests that we are dealing with Alfvén waves. This idea is confirmed by two supplementary arguments: (1) the longitudinal component of magnetic field oscillations is much smaller than the transverse;

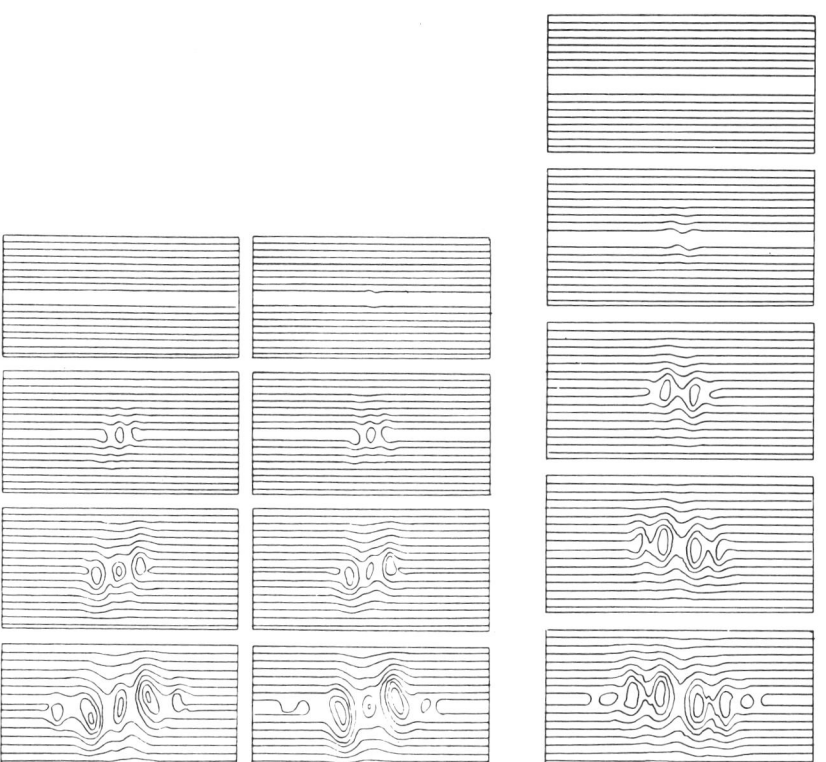

Figure 7.11. Formation of the odd (left) and even (right) vortex chains. The time is going from top to bottom.

(2) the relationship of transverse components of magnetic and electric fields is close to the expected value of the refractive index for the Alfvén waves at the height of the satellite.

So, there are grounds to suppose that we are dealing with nonlinear small-scale Alfvén waves. The theory, mentioned above, predicts that such waves have vortex structure. But how can this structure be revealed in the experiment? The simplest way is to construct the distribution of oscillation polarizations along the satellite orbit. Without any doubt, we can see the traces of vortex chains in figure 7.13. Both counter-clockwise and clockwise rotation is observed in various fragments of these chains. For example, the counter-clockwise rotation in the cells of a strong field within the time interval from the 21st to the 26th second is then replaced by the clockwise rotation. The inversion of rotation also took place at 53 min.

Additional analysis of oscillation hodograms revealed not only the vortex chains in the auroral plasma but dipole and monopole Alfvén vortices as well. The transverse dimension of a single vortex is of the order of several kilometers.

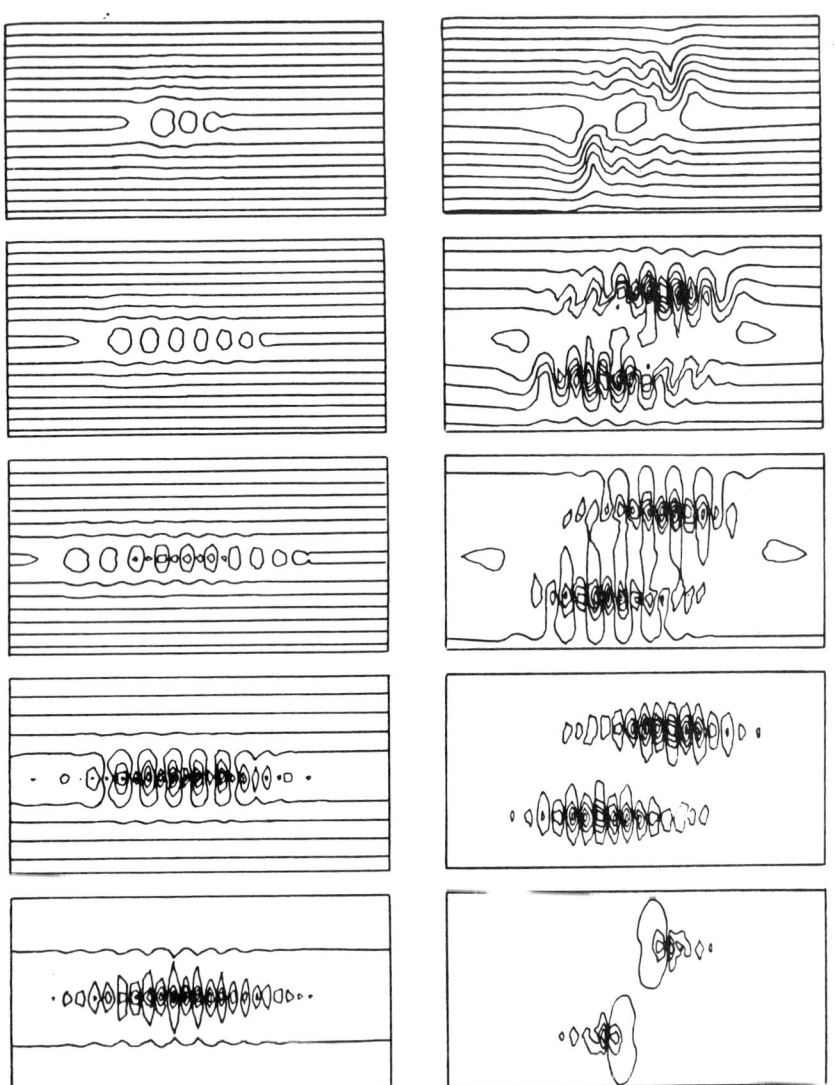

Figure 7.12. Formation of the single (left) and double (right) vortex chains.

Such small thickness allows us to consider them as vortex fibres or filaments. Due to their localization near the outer and inner edges of the plasma sheet one may conclude that they are excited through the development of a drift-dissipative instability.

This is evidence of excellent agreement between theory and observations, but it must not mislead us. The problem of Alfvén vortices in the auroral plasma is far from being completed. On the contrary, the possibility of formulating new

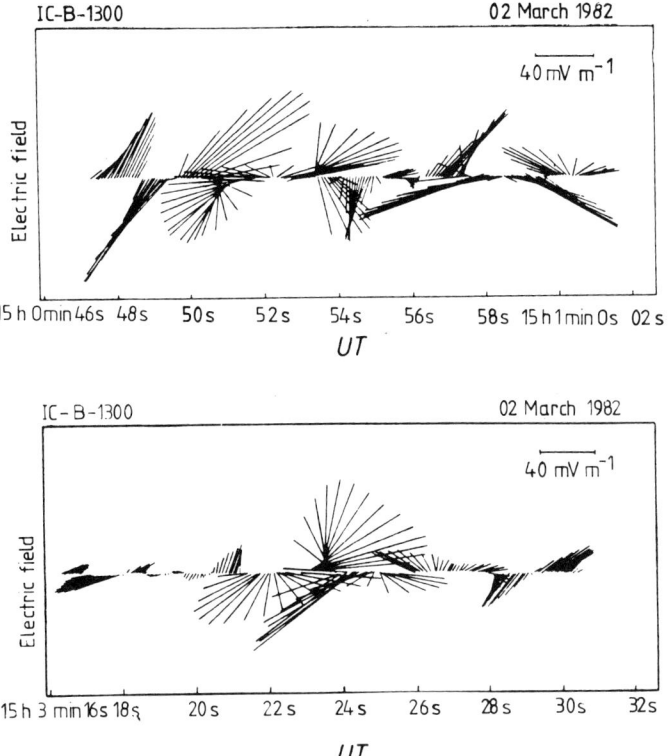

Figure 7.13. Variations of the horizontal projection of electric field oscillations along the satellite orbit in the first (upper panel) and second (lower panel) intervals of figure 7.5.

problems and finding new methods of studying nonlinear wave motions in the magnetosphere emerges. Part of the problem is connected with the specification of the mechanism for the generation of vortex filaments. One of the unsolved problems is the location of the source regions. These structures may be excited at the edges of the plasma sheet or in the ionospheric resonator.

It is rather early to draw a final conclusion in such a complicated problem, but let us examine figure 7.14 in detail. A signal in the form of intense electron flux and very low frequency (VLF) emission has appeared at 25 s. The general feature of the phenomenon suggests that we are observing a precursor which forestalls the intense Alfvén vortex by few seconds. Presumably the precursor has arisen in the region of vortex generation. Then due to time delay this region is located at a distance of 5–6 R_E, i.e. in the plasma sheet and not at ionospheric altitudes.

A wide circle of interesting problems is connected with the analysis of the influence of nonlinear Alfvén waves on the ionospheric and magnetospheric plasma. These problems have been attracting the attention of researchers for

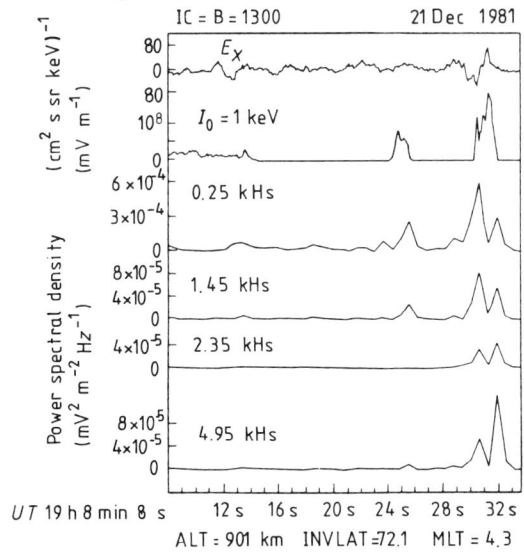

Figure 7.14. Distribution of the electric field oscillations (upper panel), auroral electron fluxes (second upper panel) and spectral density of VLF emissions in different frequency ranges (four lower panels) along the satellite orbit. One has to indicate the rise of the electron flux at 25 s forestalling the intense electric signal at a few seconds.

a long time. Advances in the theory of Alfvén vortices give hope for future progress in this area.

In conclusion let us cast a general look at the perspectives and development tendencies of the physics of geoelectromagnetic waves.

At present observations are the main source of knowledge regarding geoelectromagnetic waves. But those who try to estimate the achievements in this field by the number of new facts will be mistaken. The most important and vital results usually appear when new ideas are applied to the known facts.

Our hopes and plans for the future are connected with the formation and investigation of qualitatively new models of oscillation processes. A reasonable model allows us to translate the observational data into substantial statements concerning the physical nature of the phenomema. Modelling is capable of producing a compact and clear descriptive picture of geomagnetic pulsations. It gives us the opportunity to shorten an enormous logical gap between the morphology of pulsations and the theory of waves. In fact, the language of traditional morphology is poor for the description of a fairly wide circle of phenomena. As a result, a considerable portion of observational data is cut off from the most well-developed branches of the theory.

All this engenders many interesting problems. One group of them is profoundly studied and well known; but other problems are still being worked out or have not even been formulated. Vague but alluring perspectives might stem from these problems. While admitting that there exists a large spread of viewpoints regarding further research, here we shall just put the matter as we see it ourselves.

At one time eigen and forced oscillations, running and standing waves, wave packets and other comparatively simple objects of the linear theory of oscillations and waves served as prototypes for geomagnetic pulsations. Now this set has been enriched by solitons, vortices and similar nonlinear wave structures.

In all this one sees the desire to choose well-regulated objects, possessing certain stability and reproducibility. We will retain this idea since it reflects the real processes of self-organization of wave fields in the magnetosphere, i.e. the spontaneous appearance of spatio-temporal structures out of chaotic noises in an unstable nonlinear dispersive medium. At the same time, from the way we see it, an opposite tendency has been appeared—interest in fluctuations and critical phenomena, strange attractors, etc, has arisen. In other words, the interest in stochastic and chaotic oscillations has been countered. These two tendencies, supplementing each other, will lead to the construction of a well-composed and fairly complete picture of oscillations of the magnetosphere.

Bibliography

A systematic presentation of the theory of waves in linear and nonlinear media is contained in the fundamental monograph by Whitham (1974). The monograph by Kadomtsev (1976) may serve as an introduction to nonlinear plasma physics. The monographs by Hasegawa (1975) and Petviashvili and Pokhotelov (1992) are dedicated to the nonlinear waves in the magnetosphere.

The first research works on the resonant nonlinear interaction of hydromagnetic waves with a single group of energetic particles in the magnetosphere were carried out by Dragt (1961) and Wentzel (1961), Davis and Chang (1962), Chang and Pearlstein (1965) and Watanabe (1966). Kennel and Petschek (1966) put forward the concept of self-restriction of energetic protons in the radiation belts owing to excitation of Pc1-2 pulsations, pitch-angle diffusion and the escape of particles into the loss cone. Based on this most important concept, known in the literature as the 'limit of stability of the radiation belt', Cornwall *et al* (1970) suggested an explanation for the correlation of radial distribution of the ring-current protons with the location and displacement of the plasmapause.

Nonresonant effects of the geomagnetic pulsations are presented by a wide set of processes of energetic electron flow modulations in the radiation belt, pulsating forms of aurora, quasi-periodic VLF emissions, modulation of the Earth–ionosphere radio waveguide, etc.

The modification of energetic ions and electrons that form a relatively small admixture to the magnetospheric plasma is widely known and has been reported in the literature (e.g. Thorne 1974, Nambu 1974, Kimura 1974).

For section 7.1

Equation (7.2) is one of the prominent results of phenomenological electrodynamics. Pitaevsky (1961) started the derivation of this expression and Washimi and Karpman (1976) completed it. Some terms in this equation were even derived before as a result of microscopic modelling. The origin of other terms goes back to Helmholtz and Abraham. Much effort had to be made trying and retrying to understand the physical meaning contained in (7.2) in order to remove the paradoxes, etc (see, for example, Ginzburg 1975).

Allan *et al* (1991) and Allan (1992) discovered the nonmonotonic distribution of plasma density in the neighbourhood of a node of the standing Alfvén wave. The hypothesis on the equatorial plasma condensation and polar wind acceleration under the action of ion cyclotron waves was suggested by Guglielmi (1992c) and Guglielmi *et al* (1993). The theory of the polar wind was constructed by Banks and Holzer (1969) by analogy with the solar wind theory proposed by Parker (1963). The acceleration of the solar wind under the action of MHD waves of solar origin was studied by Belcher (1971), McKenzie (1991) and others. On neutral and plasma layers of the geomagnetic tail, on the tearing instability and its consequences see Nishida (1978).

Hultqvist *et al* (1988), Lundin (1988), Lundin and Hultqvist (1989) observed on board the *Viking* satellite that perturbations in the ion distribution above the high-latitude ionosphere correlate with intense Pc1 pulsations.

For section 7.2

Extensive literature is dedicated to solitons. In addition to the books by Whitham (1974), Kadomtsev (1976), Petviashvili and Pokhotelov (1992), we point out, for example, the book by Zakharov *et al* (1980). One-dimensional nonlinear waves were analysed by Bhatnagar (1979). *Solitons* (Bullough and Caudrey 1980) contains profound and diverse information, in particular, the history of the discovery of the soliton and a list of the published papers by John Scott Russell, who was the first to discover and estimate this unusual form of waves back in 1834.

The nonlinear Schrödinger equation (7.19) with coefficients (7.28) and (7.31) that describes the low-frequency circularly polarized waves in a cold magnetoactive plasma was derived by Mjølhus (1976) and Mio *et al* (1976). (In these two works a more general, so-called derivative nonlinear Schrödinger equation was proposed.) Hasegawa (1970) discovered that in a finite pressure plasma the sign of the nonlinearity coefficient depends on the sign of the difference $c_s - v_g$ (exercise 7.2.2).

One-dimensional MHD solitons in a weakly unstable plasma were examined by Hasegawa and Mima (1976), Guglielmi (1980) and Guglielmi and Repin (1981). Magnetosonic solitons in the plane plasma waveguide were analysed by Guglielmi (1979) and Churilov and Shukhman (1983). Nakaryakov and

Fainshtein (1991) investigated the nonlinear generation of the second harmonic in such a waveguide.

The search for solitons among the geomagnetic pulsations was undertaken by Guglielmi (1973, 1979) and Guglielmi and Repin (1978). The idea of the auto-modulation growth rate as the cause of the appearance of discrete series of 'pearls' was suggested by Guglielmi (1971). Polyakov *et al* (1983) suggested a theory of the discreteness of 'pearls' based on the idea of the auto-modulation of decrement.

For section 7.3

The idea of the reduction of the MHD equations and the first system of reduced nonlinear equations was suggested by Kadomtsev and Pogutse (1973). Equations (7.46)–(7.48) that we proposed as a basis for the analysis of Alfvén vortices were derived by Petviashvili and Pokhotelov (1985). These equations, on the one hand, generalize the equations of Kadomtsev and Pogutse (1973), Hasegawa and Mima (1978) and, on the other hand, they admit interesting generalizations themselves (see the monograph by Petviashvili and Pokhotelov 1992).

The nonlinear condensation of the Alfvén waves into vortex filaments at the drift-dissipative instability of an inhomogeneous plasma was considered by Kaladze *et al* (1987). When classifying vortex solutions we followed the works by Petviashvili and Pokhotelov (1977, 1985, 1986), Petviashvili *et al* (1982, 1986), Kaladze *et al* (1986), Chmyrev *et al* (1988, 1991, 1992), Ivanov and Pokhotelov (1988) and Streltzov *et al* (1990). In particular, the construction of the solution in the form of the dipole vortex was described by Petviashvili and Pokhotelov (1985). The odd and even chains of vortices in figure 7.11 were constructed by Streltzov *et al* (1990).

Figures 7.5 and 7.13 are borrowed from the paper by Chmyrev *et al* (1988). Other examples of vortex structure observation in the magnetosphere are presented by Chmyrev *et al* (1991).

Chapter 8

Fluctuational and critical phenomena

Our description of the geomagnetic phenomena, though certainly unable to exhaust the whole complex of problems, comes to a finale. In the next chapter, the applied aspects of the theory will be described. In this chapter we intend to abandon the determinism which prevailed in all the previous chapters. Now we want to take into account the stochasticity intrinsic to the geoelectromagnetic waves and try to do so at the phenomenological level.

8.1 Phenomenological modelling

By analogy with the wave phenomena in optics, acoustics and radiophysics, we would naturally expect that fluctuations of the field of geomagnetic pulsations would contain information about the excitation and propagation mechanisms and—a particularly important point—on the effects of nonlinearity. One can cite only a few papers that have allowed for the stochasticity of geomagnetic pulsations. To a large extent, this situation is due to the difficulty of experimental study of the corresponding effects in the range of geomagnetic pulsations. With the measurement apparatus available today, it is not possible to take up many interesting problems, e.g. that of fluctuations in the angle of arrival of the emission.

There is yet another cause of the delay in stochastic studies of pulsations. This is the inadequate level of development of the deterministic theory. Dynamic problems, if they are posed more or less appropriately, i.e. if the nonlinearity of pulsations and the complex structure of the magnetosphere are taken into account, are not amenable to solution. For the time being we are thus blocked from taking the customary path to analysing stochastic systems, which is essentially one of replacing the numerical functions in the corresponding deterministic model by random functions and evaluating the probability for some state or other of the system.

With these comments in mind, we should refer to the study of fluctuation and critical phenomena presented in this chapter. Only the 'coarse' parameters of the pulsations, i.e. parameters that could be measured reliably (amplitude, group delay, etc) were selected for the study. In an effort to avoid the second difficulty listed above, the deterministic models of the pulsations were replaced by extremely simplified phenomenological models, which were used to formulate some simple problems for choosing between alternative possibilities.

The basis of our approach is the presentation of the magnetosphere as a black box. The features of this box should be determined from statistical features of the signals from its output. This means that we are interested not only in the statistical features of the geomagnetic pulsations themselves but we also use it as a source of information on the dynamics of the black box. Problems of that kind are referred to as the class of inverse problems of the statistical theory of oscillations.

8.1.1 Autogenerator or filter?

Let us consider the simple example which gave us the initial idea about the methods of phenomenological modelling.

Geomagnetic pulsations are quite frequently quasi-monochromatic oscillations. Two types of models are being discussed in the geophysical literature in an effort to explain these oscillations. We shall call these the 'self-oscillation' and 'filtration' models. In models of the first type the pulsations are supposed to arise from a plasma instability, i.e. at a bifurcation from an equilibrium state of the focus type, with a transition to a nonlinear regime and the formation of a nontrivial attractor (e.g. limiting cycle). In theories based on models of the second type it is assumed that the magnetosphere contains selective filters or amplifiers which pass narrow bands of the spectrum of noise which penetrates into the magnetosphere from the solar wind.

The carrier frequency and other spectral properties of the pulsations are simulated equally well by models of the two types. Nevertheless, a choice between the two types of model can be made by studying fluctuations of the amplitude of pulsations. In the case of an autogenerator, a Gaussian amplitude distribution will be observed at the output, while in the case of a selective filter there will be a Rayleigh distribution.

Experimentally, one constructs an empirical distribution of the fluctuations of the pulsation amplitude and compares them with Rayleigh and Gaussian distributions. In the case of a reliable approximate agreement with one of them, a conclusion may be drawn about the type of oscillatory system that generated the pulsations. Illustration of such an approach may be found in the papers indicated in the bibliography for this chapter.

8.1.2 Fluctuations near threshold

In the papers devoted to the interpretation of pulsations, the Landau theory is frequently used in direct or indirect form. The basic statements of this theory in application to magnetosphere are the following.

There is a critical parameter I_c (e.g. anisotropic flux of energetic particles). The physical sense of I_c is that for $I < I_c$ pulsations are not excited, and for $I > I_c$ self-excitation of pulsations takes place. Let us introduce the value $\mathcal{E} = A^2$ for the square of the amplitude of pulsations averaged over the characteristic period of fluctuations. The equation

$$\dot{\mathcal{E}} = 2\gamma\mathcal{E} \tag{8.1}$$

describes the evolution of pulsations. In the general case $\gamma = \gamma(I, \mathcal{E})$. Near the threshold area ($I \simeq I_c$) the form of this dependence is taken from phenomenological consideration

$$\gamma(I, \mathcal{E}) = \gamma_0(I) - \alpha\mathcal{E}$$
$$\gamma_0(I) = \eta(I - I_c). \tag{8.2}$$

The theory contains three parameters I_c, α, η. Proceeding from microscopic linear theory of instability of the magnetosphere, it is easy to ascertain the physical sense of I_c and η. The parameter α takes the nonlinearity into account. We consider $\alpha > 0$. This corresponds to the soft regime of self-excitation. We will not need the real values of these parameters here.

At $I < I_c$ in the steady state there are no pulsations ($\mathcal{E} = 0$). When crossing the threshold ($I > I_c$) the magnetosphere (or rather some oscillatory subsystem of the magnetosphere) exhibits a phase transition of the second kind. The amplitude of pulsations $A = \sqrt{\mathcal{E}}$ in this case is proportional to the square root of supercriticality

$$A = \text{constant}\sqrt{I - I_c}. \tag{8.3}$$

Then the correlation interval τ of the amplitude fluctuations is in inverse dependence on supercriticality

$$\tau = \text{constant}/|I - I_c|. \tag{8.4}$$

Now let us raise the right-hand and the left-hand sides of equation (8.3) to the second power and multiply them by the right-hand and left-hand sides of (8.4) respectively. As a result we obtain the expression

$$A^2\tau = \text{constant} \tag{8.5}$$

which contains values that may be easily measured and verified experimentally.

The form of (8.5) resembles the criterion (7.44) for solitons. Although these formulae correspond to two different physical objects, this similarity exhibits some heuristic content.

In numerous papers devoted to Pc1, PiC, Pc4, etc, the phenomenological theory (8.1) and (8.2) is used in the hidden or explicit form. But if this theory really reflects significant features of these types of pulsations then the criterion

(8.5) has to be fulfilled. In accordance with this criterion the square of the pulsation amplitude is in inverse proportion to the interval of correlation. It is rather useful and interesting to check the criterion (8.5) with the experiment.

Now suppose $\alpha < 0$ (the system with rigid self-excitation). Then wave hysteresis is possible. It would also be useful to find this experimentally. At the phenomenological level the phenomenon is described as follows.

A stabilizing term $-\beta \mathcal{E}^2$, where $\beta > 0$, should be added to the right-hand side of (8.1). In this case the amplitude increases abruptly from zero up to the value

$$A_0 = (|\alpha|/\beta)^{1/2} \tag{8.6}$$

with the increase of I and its transition through the critical value I_c. If I starts decreasing after that, the oscillations will not disappear during the inverse transition through I_c. Disruption of oscillations will occur at the value of the governing parameter

$$I_c' = I_c - |\alpha|^2/4\beta\eta$$

i.e. at $I < I_c$. The amplitude value of oscillations before they disappear is

$$A_1 = (|\alpha|/2\beta)^{1/2}. \tag{8.7}$$

Hence we may derive a simple strategy to search for hysteresis wave phenomena in the magnetosphere.

The value I can hardly be detected by experiment. Frequently the sense of the control parameter is completely unknown, so one should watch the amplitude of oscillations attentively and seek oscillatory regimes where the amplitude increases abruptly from zero to some finite value A_0, after which it experiences a smooth variation, and at some value A_1 the oscillations disappear abruptly. The realization of the criterion

$$A_0/A_1 = \sqrt{2}$$

serves as an indication of the rigid self-excitation regime, i.e. the amplitude is approximately one and a half times smaller at the moment of disappearance than at the moment of the rise of oscillations.

8.1.3 A black box without an input

The black box idea always arises where the object under study is inaccessible to direct observation. In dealing with such an entity we advance hypotheses on its internal structure, i.e. we construct models as structural and functional approximations of the object. The hypotheses are usually tested against experimental results in an 'input–output' approach. The object is approximated by a purely dynamic model, i.e. the fluctuation phenomena which occur in any real system are totally ignored. For example, the hypothesis on the extramagnetospheric origin of Pc3 pulsations was studied using such a scheme.

The value of the interplanetary magnetic field was controlled at the 'input'. The carrier frequency of pulsations was measured at the 'output' (see section 6.5).

The incorporation of fluctuations in the model and the use of the methods of the statistical theory of oscillations opens up the possibility of obtaining information from the output signal alone. The diagnostic model of a 'black box without an input' is based on the idea that the distribution of fluctuations of the amplitude and phase in oscillatory systems of various types may differ from each other essentially. It is thus possible to draw certain conclusions about the internal structure and operation of the system using the output signal. The methods pointed out above provide a very simple realization of this idea. And here we shall discuss a more complex example, in which the correlation properties of the pulsations are used to seek dynamic equations describing certain aspects of the oscillation process. In other words, the inverse problem of the statistical theory of oscillations is being solved.

The correlation method for studying an uncontrolled self-oscillation system on the basis of its signals is as follows. Dynamic equations for modelling pulsations are sought in a class of models with a single degree of freedom. The perturbation of the geomagnetic field is taken to be the signal at the output and it is assumed that the instantaneous state of oscillation of the system is characterized by a point in the phase plane. The motion of the imaging point is described by a system of two first-order differential equations.

In addition to the dynamic characteristics (which are to be sought) the system contains fluctuational δ-correlated terms. It is assumed that if fluctuations are ignored the system has a limiting cycle, i.e. an asymptotically stable closed orbit in the phase plane. Fluctuations lead to normal and tangential excursions from the limiting cycle. The dynamic parameters of the system are determined through a correlation analysis of the excursions of the trajectory from the average limiting cycle.

By taking this route one can find the form of the limiting cycle simulating pulsations of magnetospheric origin, the rigidity and anisochronicity of the system, the nonlinear distortion factor, the amplitude dispersion and the phase diffusion. Consequently, the output signal alone, which has the form of a sinusoidal segment, contains much nontrivial information on the magnetosphere. The realization of this procedure will be presented in section 8.2 using the example of Pc4 pulsations.

A narrow-band filter, with noise fed in at its input, may be taken as an alternative. However, the diagnostic scheme of the 'black box without an input' allows a much wider set of alternatives. For example, we may take an oscillator with friction under the action of the resonant external force in the presence of a Langevin source. This formal model corresponds to the concept of the extramagnetospheric origin of Pc3 pulsations, supplemented with the ideas on local field-line resonances. The waves coming behind the shock front penetrate into the magnetosphere and act on the resonator as a periodical force. The additional Langevin force imitates the noises. From the corresponding Fokker–

Planck equation for the distribution functions of the Pc3 amplitude and phase we obtain the solution which allows the formulation of the criterion for verification of the model.

Exercise

Exercise 8.1.1.

Find the change pattern of the magnetic moment of unit volume of magnetospheric plasma at the transition of the cyclotron instability threshold.

Solution 8.1.1.

The magnetic moment of the unit volume is formed by the moment M_0, which is caused by the thermal motion of plasma particles and the magnetization of the plasma in the field of the circularly polarized wave (the inverse Faraday effect). Taking into account (7.3) we have

$$M = M_0 + \frac{\partial n^2}{\partial B} \frac{E^2}{8\pi}.$$

Below threshold $(I < I_c)$ the amplitude $E = 0$; at $I > I_c$ the amplitude E increases as the square root of supercriticality (see (8.3)). Hence it is clear that M is continuous when crossing the threshold while the derivative $\mathrm{d}M/\mathrm{d}I$ experiences a finite jump at $I = I_c$. In the case of cyclotron instability, I indicates a flux of energetic particles.

8.2 An auto-oscillatory model of pulsations

The theory of geomagnetic pulsations based on the consistent application of plasma electrodynamics is aimed at the construction of a model of an oscillatory magnetospheric system that more or less corresponds to the intuitive concept of the bifurcation and appearance of the limiting cycle or another nontrivial attractor in the phase space of the magnetosphere. The difficulties that arise here are well known. Owing to the lack of information on the magnetosphere, these difficulties arise as soon as the problem is posed. But the problem, even if formulated correctly, turns out to be mathematically involved. Exact solutions are unknown, as a rule. Approximate solutions obtained through linearization, for example, offer but a limited opportunity to test the theory by a comparison with the signals observed in reality.

  All this justifies the attempt to apply another approach; its essence is presented in section 8.1. The electrodynamic model of oscillations is not formulated in the explicit form. Instead dynamic equations that describe some aspects of the oscillatory process are sought by the statistical properties of the observed signals, i.e. the inverse problem of the statistical theory of oscillations is being solved.

8.2.1 Selection of the pulsation type

From the variety of pulsations we select Pc4, being guided by the following considerations.

Pulsations are excited when the level of magnetic activity is low. This is a favourable factor for the use of the method, one of the applicability conditions of which is the low order of slow accidental changes of the external parameters. Oscillations last long, sometimes for hours which allows us to single out the intervals with a quasi-stationary regime and store the necessary information for the statistical analysis. The form of oscillations is rather simple (figure 8.1). There is no complicated amplitude or frequency modulation. The spectral composition is rather poor. This gives us grounds to suppose that the appearance of Pc4 signifies excitation of a relatively small number of degrees of freedom of the magnetosphere and to a certain extent justifies attempts to use the finite-dimension approximation of the oscillations.

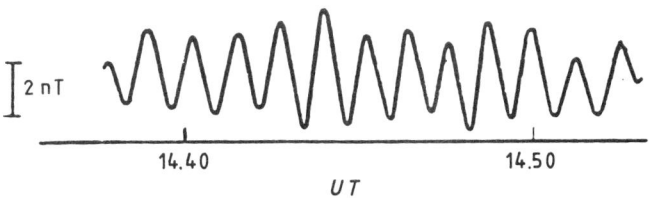

Figure 8.1. Oscillogram of Pc4 pulsations registered at a mid-latitude observatory on 8 January 1971. Nonlinearity, which is hardly observed by eye, becomes evident if one constructs the phase diagram of the oscillations.

Finally, the fact that the morphology of Pc4 has been studied carefully is also of importance. However, a satisfactory theory of Pc4 has not yet been constructed, so it is advisable to investigate Pc4 using the inverse problem of the theory of oscillations to estimate dynamic parameters, hoping that it will facilitate the construction of an adequate physical model.

We start with the general morphological characteristics of pulsations. According to the definition, Pc4 pulsations are quasi-periodic oscillations in the range of 45–150 s. Observations indicate that oscillations of one and the same kind are excited in a somewhat wider range—approximately from 40 s to 170 s. Pc4 are observed on the day side during the recovery phase of the storm when the level of magnetic activity is relatively low. The characteristic period is 100 s. Once begun, the oscillations last for many hours with short interruptions. At mid-latitudes the amplitude is of the order of 2 nT and at high latitudes 10 nT.

Pc4 form a fairly homogeneous class of oscillations, although modifications, morphological varieties, etc, do occur. A special variety is formed by so-called 'giant pulsations' or Pg. These are quite regular pulsations of a beautiful form with a very large amplitude (dozens of nT). They appear in the morning sector within the narrow band stretching along the geomagnetic parallel near the equatorial boundary of the auroral zone. It is appropriate to mention here that initially Pc4 were detected exactly as 'giant pulsations'.

The daily variation in the number of occurrences of Pc4 has an ample maximum in the vicinity of the noon meridian. The daily variation in the period of oscillations is observed at the synchronous orbit: it increases from 45 s to 160 s in the interval 07–21 LT and then oscillations with a large period appear in the evening sector under extremely calm conditions ($K_p \leq 1$).

The seasonal variation is characterized by a maximum number of occurrences in December–February and a minimum in April–May. At the conjugate points the oscillations start and finish simultaneously, the periods are identical and the amplitudes are of the same order. There is detailed information on the spatial distribution of polarization, on the connection of Pc4 with the parameters of the magnetosphere and the solar wind, etc. Corresponding references may be found in the bibliography to this chapter.

For investigation using the 'black box without an input' scheme, ten intervals of records of geomagnetic pulsations, made at the mid-latitudinal Borok Observatory, containing Pc4 with quasi-stationary modulation of the amplitude, were selected. The primary processing consisted in digitization of the oscillograms with the step, providing 20 points over the period. The change limits of the local time, visible period, interval value, amplitudes of H and D components and K_p index from one case to another are as follows.

LT (h)	T (s)	Δt (min)	H (nT)	D (nT)	K_p
10–17	60–80	10–12	1.3–3.5	0.3–1	0–0$_+$

Then the question was raised: can the selected oscillations be considered to be a result of the action of the autonomous generator and not, for example, the result of noise filtration? A hypothetical answer was given by analysis of the empirical distribution of the oscillation amplitudes. The solid line in figure 8.2 indicates the distribution of relative deviations of the Pc4 amplitude from the average value. With a probability of 0.8 according to the Peerson criterion χ^2, this distribution approaches that of Gauss (shown by the dashed line). The Rayleigh distribution poorly approximates the empirical one (the probability is less than 10^{-2}). Hence it follows that the Pc4 pulsations selected for the analysis are the result of self-oscillations, rather than a result of narrow-band filtration.

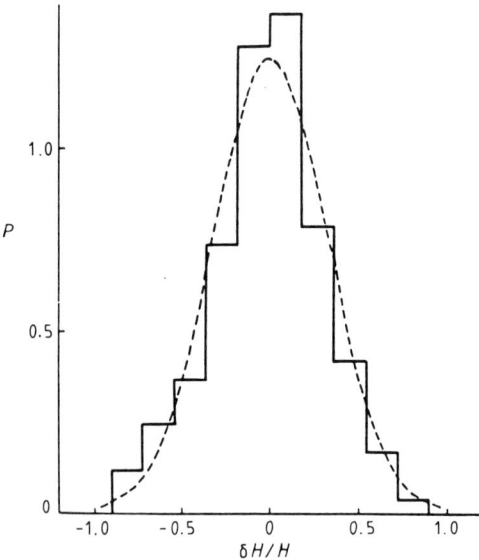

Figure 8.2. Empirical distribution of Pc4 amplitude fluctuations. Vertical axis—density of the probability, horizontal axis—normalized deviation of the amplitude from the averaged value. The dashed line indicates the Gaussian distribution.

8.2.2 Limiting cycle

Let us try to describe Pc4 pulsations by means of the dynamic model with a single degree of freedom. We choose the perturbation of the H-component of the magnetic field as the signal at the output $x(t)$. Let us assume that the instantaneous state of the system is characterized by the point on the phase plane (x, y) where $y = dx/dt$. The presentation of the behaviour of the real oscillatory system in two-dimensional phase space is a strong idealization and it should be borne in mind.

The motion of the imaging point on the phase plane is described by two differential equations of the first order. Following the general idea, we introduce fluctuational δ-correlated terms in the system. Suppose that regardless of fluctuations the system has a limiting cycle, i.e. an asymptotically stable closed orbit (linear attractor). Under the action of fluctuations the imaging point will deviate from the limiting cycle. Figuratively speaking, under the action of fluctuations the imaging point 'roams over' the phase plane, inspecting the vector field of the dynamic system in the vicinity of the limiting cycle. This, in fact, offers an opportunity to realize the idea of the 'black box without an input', i.e. to investigate the dynamics of the system by the output signal alone.

It is convenient instead of (x, y) and t to use the local coordinates (n, γ) and

θ, where n and γ are the normal and tangential excursions from the limiting cycle respectively, and $\theta = t + \gamma$. Then in the linear with respect to n approximation the equations of the auto-oscillation system have the form

$$dn/d\theta = -N(\theta)n + F_n(\theta)$$
$$d\gamma/d\theta = \kappa(\theta)n + F_\gamma(\theta)$$

(8.8)

where F_n and F_γ are stationary fluctuations which make the system deviate from the limiting cycle.

The problem is to determine the rigidity $N(\theta)$ and non-isochronicity $\kappa(\theta)$ using the data on the observed signal. The form of the limiting cycle and its position on the phase plane should be found in advance. The problem of defining the dispersion of the amplitude and diffusion of the phase may be included.

First of all one has to construct the limiting cycle for every individual Pc4. To this end it is necessary to pass to dimensionless variables

$$x_1 = (x - \bar{x})/(\bar{x} - \bar{x}_{min}) \qquad y_1 = \Delta x_1 T / 2\pi \Delta t$$

where $\bar{x}$ is the average value of x, $\bar{x}_{min}$ is the average value of the minima of x, T is the period of motion over the limiting cycle and Δt is the digitization step of the Pc4 oscillogram. Points $x(t_i)$ and $y(t_i)$ are plotted on the phase plane with an interval of digitization (here and below the index '1' is omitted). Then they are connected by a smooth curve. The averaged limiting cycle is obtained by the method of successive iterations.

Figure 8.3 illustrates this procedure using the event presented in figure 8.1. The number of turns of the imaging points $s = 10$, and the period of motion over the average cycle is $T = 82$ s. The motion over the cycle is uneven, i.e. the arc $y > 0$ is run through in 37 s, and the arc $y < 0$ in 45 s.

Owing to the small number of turns, the accuracy in determining the average limiting cycle and other characteristics of the system in individual cases is not great. In order to improve the accuracy, below we make an average over the events. Beforehand, the values containing time in their dimensions are normalized to average the period of motion over the cycle.

The average limiting cycles obtained over ten events are presented in figure 8.4. The interval between the lines that the imaging point runs over is $T/12$. One can see the irregularity of the movement over the cycle.

The form of the cycle varies slightly from one event to another. This allows us to conclude that we have obtained the generalized phase portrait of Pc4. Its main feature is connected with the asymmetry relative to the x axis. The shift of the figure centre upwards testifies to the anharmonicity of the oscillations.

Let us introduce the measure of anharmonicity (a sort of nonlinear distortion factor) using the formula $\Delta = 2(y_+ - y_-)/(y_+ + y_-)$, where y_+ and y_- are the values of y in the upper and the lower points of the limiting cycle. On average $\Delta = 0.7 \pm 0.15$. The visible value of the distortion factor points out the significant nonlinearity of Pc4. As $\Delta > 0$ the oscillations have a sawtooth form, the teeth of which are shifted to the left.

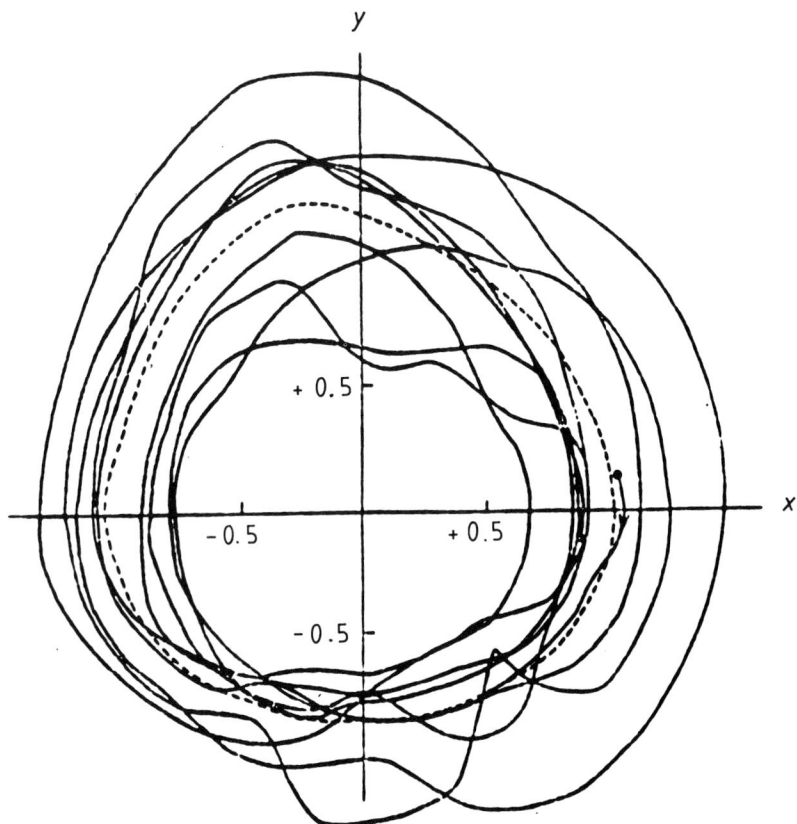

Figure 8.3. Trajectory of the imaging point on the phase plane.

So, we have obtained an interesting result. The sawtooth nature of Pc4 was first discovered by analysing the phase diagram of the oscillations. The result proved to be fairly stable. It can be easily reproduced over the whole complex of other cases of Pc4.

8.2.3 Rigidity and nonisochronicity

The dynamic parameters of N and κ are determined by correlation analysis of the trajectory deviations from the average limiting cycle using the equations

$$N(\theta) = -\langle \dot{n}(\theta)n(\theta - \tau)\rangle / \langle n(\theta)n(\theta - \tau)\rangle$$
$$\kappa(\theta) = \langle \dot{\gamma}(\theta)\gamma(\theta - \tau)\rangle / \langle \gamma(\theta)\gamma(\theta - \tau)\rangle. \tag{8.9}$$

Here the dot indicates the derivative over θ, the angular brackets indicate averaging over the ensemble and τ is chosen so that $\tau > \tau_F$, where τ_F is

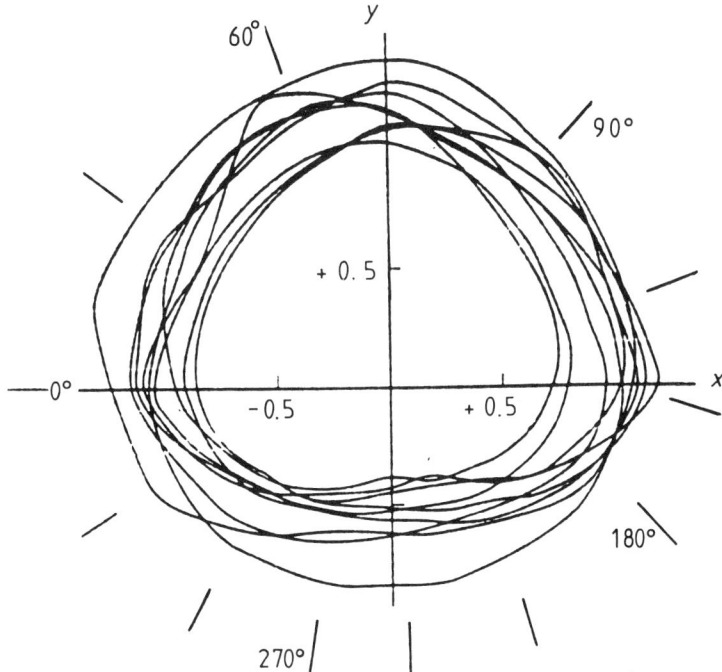

Figure 8.4. Generalized phase diagram of Pc4 pulsations. The asymmetry of trajectories of the imaging point relative to the x axis indicates the nonlinearity (saw-like structure) of the oscillations.

the correlation interval of fluctuations. The time is measured in units of T.

The rigidity N characterizes the relaxation of the perturbations of the phase trajectory which are normal to the limiting cycle. From the condition of orbital stability of the average trajectory it follows that $\bar{N} > 0$ (the bar signifies averaging over θ). This inequality is yielded for all the events. The type of autogenerator is qualitatively estimated by the value of $\bar{N}$: at $\bar{N} \ll 1$ it is of Thomson type, and at $\bar{N} \gg 1$ it is of the relaxation type. For Pc4 pulsations we have $\bar{N} = 2.7 \pm 1.2$.

The phase dependence of the limiting cycle rigidity $N(\theta)$, averaged over all the events, is shown in figure 8.5. We note that the rigidity increases at $\theta \approx 90°$. This corresponds to the beginning of the transition from the 'fast' motion along the cycle to the 'slow' motion (e.g. figure 8.4).

Anisochronicity κ characterizes the variation of the velocity of the imaging point when it departs from the limiting cycle. For Pc4 the parameter κ is statistically indistinguishable from zero, i.e. the oscillations are isochronous.

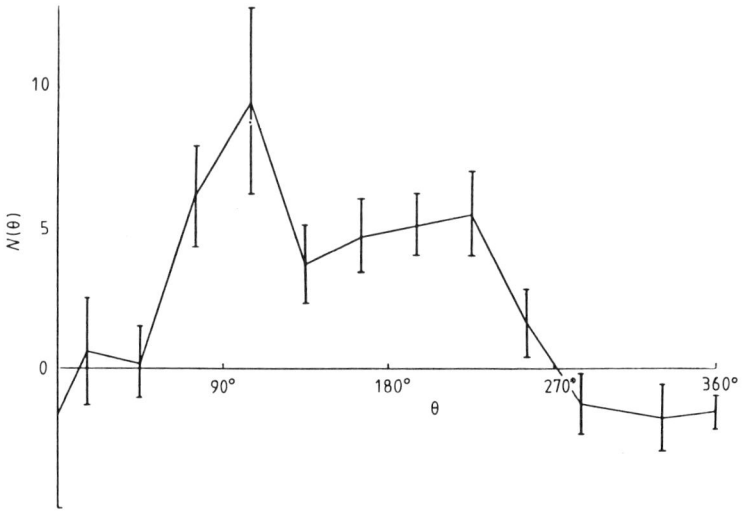

Figure 8.5. Dependence of rigidity of the Pc4 limiting cycle on the oscillation phase. Solid line—the result of averaging over ten events, vertical bars—scatter of the rigidity values.

8.2.4   Amplitude and phase fluctuations

The normal deviations from the cycle have a Gaussian distribution with periodic dispersion D_n. The dispersion depends on the rigidity in the following way

$$D_n(\theta) = A \int_{\theta-T}^{\theta} \exp\left[2 \int_{\theta-T}^{\xi} N(\eta)d\eta\right] d\xi \Big/ \{\exp\left[2 \int_{\theta-T}^{\theta} N(\xi)d\xi\right] - 1\}. \quad (8.10)$$

Here it is supposed that $\langle F_n(\theta+\tau)F_n(\theta)\rangle = A\delta(\tau)$, where $A = $ constant. If the empirical dependence $N(\theta)$, presented in figure 8.5, is substituted into (8.10), we obtain curve 1 in figure 8.6. On the other hand $D_n(\theta)$ may be determined by direct analysis of phase trajectories (figure 8.6, curve 2). Good agreement of these two independent results serves as an additional argument in favour of the auto-oscillation model of Pc4.

It follows from (8.8) that in isochronous motion the dispersion of the tangent deviations from the average cycle increases linearly with time

$$\langle \gamma^2(t)\rangle = 2D_\gamma t. \quad (8.11)$$

Here the diffusion coefficient D_γ signifies the diffusion of the phase over time to be greater than the interval correlation. If we put this time equal to T, then

$$D_\gamma = \langle \gamma^2(T)\rangle/2T.$$

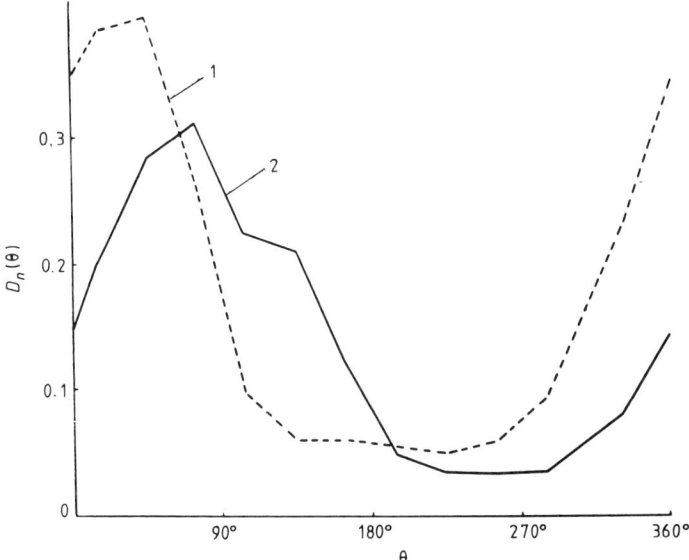

Figure 8.6. Dispersion of the normal deviations from the limiting cycle calculated by two methods (see text).

No reliable deviations from the diffusion law were detected. This testifies to the fact that we actually deal with random perturbations of the auto-oscillation system, and not with additive noises.

The average parameters of Pc4 are as follows.

Δ	N	κ	D_n	D_γ
0.7 ± 0.15	2.7 ± 1.2	0.01	0.15 ± 0.08	0.02 ± 0.005

8.2.5 The search for stochastic attractors

We have put forward the result of Pc4 investigations using the 'black box without an input' scheme known from medical diagnostics. It is based on the assumption that Pc4 may be simulated by the behaviour of a nonlinear dynamic system with a single degree of freedom. At first glance this assumption contradicts the MHD description of the magnetosphere as a continuous medium, i.e. a medium with an infinite number of degrees of freedom. This contradiction will lessen if we apply the finite-dimension approximation of the motion of a continuous medium. However, it cannot be removed completely for the following reason.

We may suppose that the appearance of quasi-periodical Pc4 oscillations signifies the perturbation of a small number of degrees of freedom ν in the magnetosphere. However, it is possible to consider $\nu = 1$ only as the test step. In fact, already at $\nu = \frac{3}{2}$ the dynamic system as it is, regardless of external fluctuations, may possess stochastic behaviour. This is not characteristic of the systems with $\nu = 1$.

Up to now we have not succeeded in revealing the features of spontaneous stochastic behaviour of Pc4 experimentally. The main difficulty lies in obtaining the intervals of specific sufficiently long realizations. The Pc1 series is a more probable candidate in the search for strange attractors. The realizations of hundreds or even thousands of periods of oscillations may be used here. We considered that a search of this kind would be fruitful. It would be strange if the phase space of the magnetosphere was not in fact packed with a great number of strange attractors.

At the same time we should admit that even the simple model of an autonomous generator in the Pc4 case is a reasonable approximation to reality. It not only gives the new phenomenological representation of Pc4 in the form of a set of dynamic and statistical parameters but allows us to reveal nontrivial features (saw-toothed oscillations) and regulations (the connection of the amplitude dispersion with the rigidity and the phase of cycle).

8.3   Phenomenology of magnetic storms

It has been mentioned and explained above that it is not obligatory to know the origin of small-scale fluctuations acting on oscillatory systems of the magnetosphere when modelling these systems. In fact, there is no reliable information on the rapidly varying processes responsible for the stochastic behaviour, for example, of geomagnetic pulsations Pc3 or Pc4. At the same time these pulsations themselves may be considered as rapidly varying fluctuations with respect to slow large-scale processes, for example, as regards magnetic storms. Such a representation is most useful from the heuristic standpoint. It enables us to enrich the traditional set of problems for the phenomenological theory of magnetic storms. In particular, below we shall prove that Pc3 should evidently be regarded as multiplicative noise that affects the origin of a magnetic storm.

8.3.1 D_{st}-variation model

A magnetic storm is a global perturbation of the magnetosphere, accompanied by strong enhancement of the ring current in the radiation belt. Let us choose an appropriate model for the evolution of the ring current which is responsible for the main and recovery phases of the storm.

The complex structure of the magnetosphere and the complex behaviour of its constituent structural elements hinder a 'microscopic' modelling of a magnetic storm, i.e. a systematic description based on the equations of plasma physics. The microscopic approach provides an understanding of the fragments of the overall picture, but if we proceed from first principles alone, then it will be practically impossible to draw an overall picture of, for example, a D_{st} variation, which is the most important manifestation of a magnetic storm. As in other similar cases, this justifies the use of phenomenological modelling.

The construction of the phenomenological model to describe D_{st} means choosing the simplest possible evolution equation. It is desirable that the choice be motivated by physical and geophysical considerations. The parameters of the equation must be found from observations. For example, the familiar RBM model (see Burton *et al* 1975) in its original version has the form

$$\dot{D} = q(t) - \alpha D \tag{8.12}$$

where

$$D = D_0 - D_{st} + aV\rho^{1/2}$$

$$q = \begin{cases} 0 & E < E_0 \\ \mu(E - E_0) & E \geq E_0 \end{cases}$$

$$E = -(V/c)B_z.$$

Here V and ρ are the velocity and density of the solar wind, B_z is a component of the interplanetary magnetic field, D_{st} is the storm-time variation of the geomagnetic field, and D is connected with D_{st} in such a way that it is equal (with the negative sign) to the magnetic disturbance due to the ring current. The right-hand side of (8.12) imitates the sources and sinks which form the ring current of a magnetic storm. The parameters D_0, a, E_0, μ, and α of the model are sought from observations.

The model (8.12) and its modifications are used for the short-term forecast of the magnetic storms. Then the problems of diagnostics come to the foreground, since in many cases major forecast mistakes ensue from the incorrect evaluation of the current state. In this case the state of the interplanetary medium in front of the magnetosphere should be evaluated in advance. For this purpose satellite measurements of the solar wind and interplanetary magnetic field parameters are used. According to the experimental data we find $q(t)$ with the advance $\Delta t = \Delta r/V$, where Δr is the distance from the satellite to the magnetopause. Then we calculate D and make a forecast of D_{st} for some hours ahead. Figure 8.7 gives an example of realization of the described procedures. The solid line is the D_{st} variation, the crosses are predictions based on the RBM model and the circles are predictions based on one of its modifications.

The observations of the pulsations' behaviour during magnetic storms allowed accumulation of a number of empirical forecast methods. For example, when the external sources feeding the ring current are switched off or weaken abruptly and the recovery phase begins, i.e. the final phase of the storm, the researcher is faced with the question of how long this phase will continue. The RBM model gives an estimate of the decay time of the ring current to be $\sim 1/\alpha$. It appears that this estimate may be considerably improved on the basis of the information about the current state of the magnetosphere carried in geomagnetic pulsations. Distinct symptoms of the termination of the storm were found experimentally. Thus the reduced activity of Pc2, Pi2 pulsations and the activization of Pc1 indicate a short duration of the storm. On the other hand,

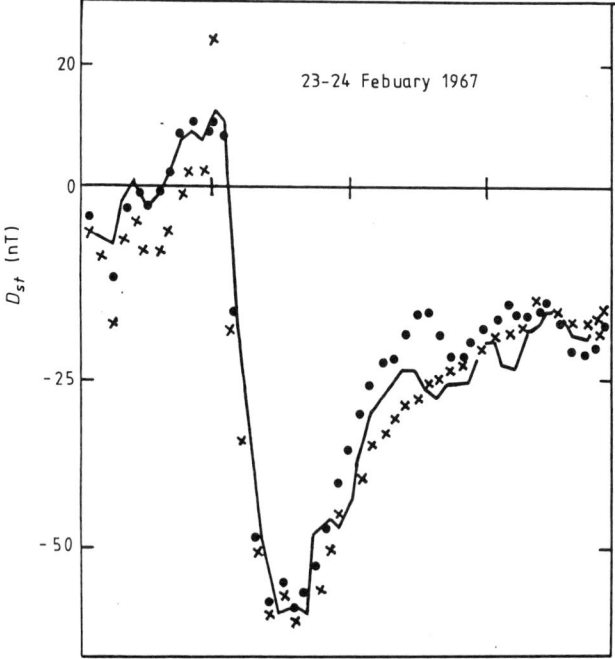

Figure 8.7. Example of D_{st} variation during the magnetic storm of 23–24 February 1984 (solid line). Crosses and circles indicate the result of the forecast. The plot is taken from the paper by Baumjohann (1986) with kind permission of the author.

the absence of Pc1 and high activity of Pc2 and Pi2 are signs of a long-lasting perturbation.

8.3.2 Stochastic equivalent of the RBM model

In models of the RBM type, a D_{st} variation is treated as the output signal of a dynamic system

$$\dot{D} = -\partial V / \partial D. \tag{8.13}$$

Here $V(D, \ldots)$ is some potential function and $D(t, \ldots)$ is the state function defined as the solution of equation (8.13), where the dots are the set of the parameters of the system. Parameters may vary with time so in general the system (8.13) is not autonomous. If we suppose that

$$V = \alpha D^2 / 2 - q D \tag{8.14}$$

then we obtain (8.12).

The dynamic model (8.13) performs a phenomenological reduction of an indefinitely large number of degrees of freedom of the magnetosphere. The degrees of freedom that were not taken into account will experimentally create a scatter which converts the deterministic function $D(t)$ into a random function. As in other, similar cases, here we may try to simulate the disregarded degrees of freedom by random forces. For this we substitute (8.13) with the Langevin equation

$$\dot{D} = -\partial V/\partial D + \xi(t) \tag{8.15}$$

where now, in contrast to (8.13), the state of the system $D(t)$ is the random function and $\xi(t)$ is the random force (Langevin source) with zero mathematical expectation and δ-correlation

$$\langle \xi(t) \rangle = 0 \qquad \langle \xi(t)\xi(t') \rangle = 2N\delta(t - t'). \tag{8.16}$$

Then from (8.15) and (8.16) and using (8.14) at $q = 0$ we obtain

$$\langle D(t) \rangle = 0 \qquad \langle D(t)^2 \rangle = N/\alpha$$
$$\langle D(t)D(t') \rangle = (N/\alpha) \exp(-\alpha |t - t'|). \tag{8.17}$$

Substituting (8.12)–(8.14) for (8.15)–(8.17) we introduce in the model the additional (sixth) parameter N—the intensity of the Langevin source. The parameter N imitates the action of the rapidly varying processes in the magnetosphere or/and solar wind on the ring current. But how can we control such processes in the experiment and how can the numerical values of N be obtained? Without answering these questions, the stochastic differential equation (8.15) has no meaning with respect to D_{st}. Since there are no final prescriptions for the construction of phenomenological models, additional efforts are necessary in this direction. Now we may reasonably use the data on the intensity of geomagnetic pulsations for an estimation of N.

Let us introduce the distribution function $F(D, t)$ and write the kinetic equation for it

$$\frac{\partial F}{\partial t} + \frac{\partial}{\partial D} \left[(q - \alpha D)F \right] = N \frac{\partial^2 F}{\partial D^2} \tag{8.18}$$

which is the Fokker–Planck equation. It is deduced in the usual way from (8.14)–(8.16). Here for simplicity, we assume that the intensity N of a random force does not depend on the state of the system D. The stationary solution of (8.18) is

$$F(D) = \text{constant} \times \exp \left[N^{-1}(qD - \alpha D^2/2) \right]$$

has the peak $(N/\alpha)^{1/2}$ width at $D = q/\alpha$.

The advantage of the stochastic model D over the deterministic model is as follows. On the basis of (8.12)–(8.14), the short-term forecast, mentioned above, will be expressed by the point estimate of D_{st} without indicating the confidence interval. The stochastic generalization of the RBM model allows us to seek the confidence interval of the forecast D (of the order of $(N/\alpha)^{1/2}$).

8.3.3   Sources and sinks

A modification of an equation like (8.12) and the introduction of additional parameters in the model are widely used for modelling D_{st}. In passing from (8.12) to (8.18) we also introduced a new parameter, N. If we wish to go further in this direction, however, it is useful to choose some guiding principle. The theory of critical phenomena appears to be the most appropriate guidance here. Really the form of the dependence of the source function q on the controlling parameter E in the RBM model suggests that we are dealing with a phase transition at a certain critical value $E = E_0$ (figure 8.8). The role of phase transitions in the formation of the sinks (the second term on the right-hand side of (8.12)) is less obvious. We shall confirm that in this case the theory of critical phenomena also gives grounds for one or another modification of the RBM model.

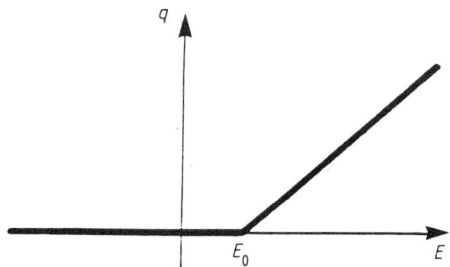

Figure 8.8.   Dependence of the D_{st} source on the azimuthal component of the interplanetary electric field in the RBM model (schematic presentation).

We consider the dynamic system

$$\dot{q} = -\frac{\partial W}{\partial q}$$

so that we can reach an understanding, at the phenomenological level, of how $q(E)$ dependence of the type shown in figure 8.8 arises. The question before us is reduced to the choice of the form of the potential $W(q, E)$. The postulate

$$W \propto \mu(E_0 - E)\frac{q^2}{2} + \frac{q^3}{3} \qquad (8.19)$$

under the additional discrimination $q \geq 0$ is sufficient to give us an expression for $q(E)$ in the RBM model, since the stable critical points of (8.19) are $q = 0$ at $E < E_0$ and $q = \mu(E - E_0)$ at $E \geq E_0$.

However, what besides the empirical correspondence governs the choice of specific expression (8.12) for the potential W? Burton *et al* (1975) in their pioneering work used the dependence $q \propto E^2$ as an alternative for the dependence $q \propto E$ at $E > E_0$ and showed that it poorly approximates the experimental points. However, the choice of $q \propto E^2$ for the comparison clearly had no basis. A dynamic approach indicates a more appropriate alternative.

For comparison with (8.19) we adopt the Ginzburg–Landau potential

$$W \propto \eta(E_0 - E)\frac{q^2}{2} + \frac{q^4}{4}. \tag{8.20}$$

Then $q = 0$ for $E < E_0$ as before, but for $E > E_0$ we have

$$q = [\eta(E - E_0)]^{1/2}. \tag{8.21}$$

This dependence is shown by the solid line in figure 8.9. Also shown here are the experimental points from the paper by Burton *et al* (1975). The reasonably good agreement between theory and observations is evidence that expression (8.21) is at least no worse than $q(E)$ in the RBM model.

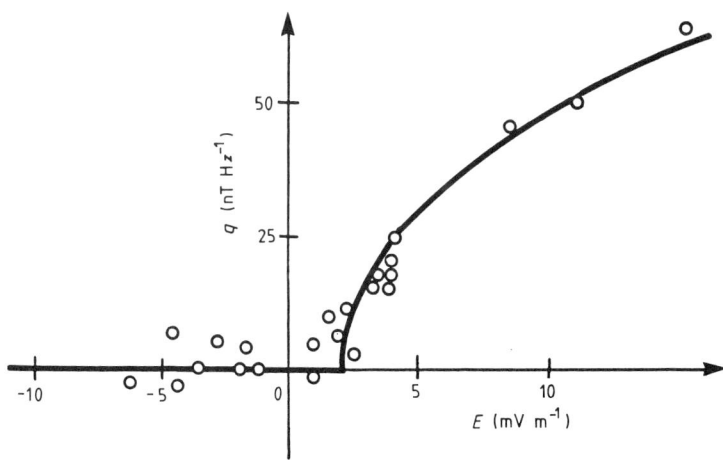

Figure 8.9. Bifurcation diagram in the modified RBM model. The experimental dots are taken from the paper by Burton *et al* (1975).

The Ginzburg–Landau potential is usually chosen on the basis of symmetry of the system. In the case at hand, we do not have such symmetry considerations to work with and at this point it is not clear just how we are to seek the 'actual' potential $W(q, E)$ unless we simply try various alternatives. At $E > E_0$ we must therefore set

$$q \propto (E - E_0)^x$$

and attempt to determine accurately the critical index x from experiments. Until we have done this, we can assert no more than the following: the RBM model gives us $x = 1$, which corresponds to the potential (8.19), while the Ginzburg–Landau potential (8.20) predicts $x = \frac{1}{2}$, which does not contradict observations.

We turn now to the sinks. The main question here is: how are we to understand and how are we to model the well-known empirical fact, that the heavier the storm, the shorter it is? Afanasieva and Kalinin (1960) assert that 'very heavy storms are very short'. Similar ideas expressed in other terms can be found in many texts, although opposing opinions have been offered on this subject.

The feature is modelled by substituting α = constant for $\alpha(D)$ in (8.12). It is convenient to expand $\alpha(D)$ in the vicinity of zero

$$\alpha(D) = \alpha_0 + \alpha_1 D + \dots \tag{8.22}$$

where $\alpha_1 \geq 0$, since otherwise the model will describe an unnatural self-amplification of D_{st}.

We shall associate the sense of the second term in (8.22) with the excitation of MHD waves due to the instability of the ring-current particles. The particles are scattered over the waves and escape the ring current the faster the higher the wave amplitude is, which, in its turn, is larger than the current intensity. Phenomena of this kind refer to the domain of critical phenomena. Hence two recommendations follow.

In the first place, it is reasonable to make a replacement $q \to qq_0/(q + q_0)$ in (8.12), where q_0 is one more phenomenological parameter of the system. The idea lies in the fact that D must reach the saturation $\sim q_0/\alpha$ at $q \to \infty$ in accordance with the concept of the 'stability limit' of the radiation belt. Such renormalization of the source may turn out to be more effective than the selection of a complicated dependence of the sinks on D.

Secondly, the theory of critical phenomena with the peculiarities of the magnetospheric structure taken into account allows us to reinterpret and possibly improve α_0. Similar to the RBM model when the source is time-varying, we may consider the sink also to be time-varying. But while the time dependence of the sink in (8.22) is implicit, the explicit dependence is introduced in the suggested variant: α_0 is the larger, the faster the plasmapause travels away from the Earth. Briefly the point is that the critical flux of energetic particles is larger from the external part of the plasmapause than from the internal part. Thus the outward motion of the plasmapause 'eats away' particles of the ring current and weakens the D_{st} variation. The ground control of the plasmapause position may be carried out using the methods presented in Chapter 9.

8.3.4 Multiplicative noise

The stochastic equivalent of the RBM model (8.18) gives us an opportunity to include information on geomagnetic pulsations in the model of a geomagnetic storm. In fact, the problem is reduced to an account of additive noises. But besides additive noises there are undoubtedly multiplicative noises present in the system. They appear, for example, in the form of fluctuations of the control parameter E in front of the magnetosphere. The ion cyclotron instability in the upstream region, which results in the excitation of Pc3 pulsations, is the cause of fluctuations.

For further discussion we shall need an explicit expression for $q(E)$. Let us choose this value in the form (8.21). However, we have to keep in mind that the actual form of the dependence q on E is not yet fixed. In other words let us assume the critical index x to be equal to $\frac{1}{2}$.

So, in the absence of interplanetary electric field fluctuations we have

$$q(E) = \sqrt{\eta} \int_{E_0}^{\infty} \delta(E' - E)\sqrt{E' - E_0}\, dE'. \qquad (8.23)$$

This is simply a new presentation of the function plotted in figure 8.9. From (8.23) it follows that if $E < E_0$ then $q = 0$ and if $E > E_0$ then $q = [\eta(E - E_0)]^{1/2}$ as was supposed before.

Actually E fluctuates. So in (8.23) we have to replace the δ-function by the proper distribution function

$$q(E) = \sqrt{\eta} \int_{E_0}^{\infty} f(E' - E)\sqrt{E' - E_0}\, dE'. \qquad (8.24)$$

If the interplanetary electric field fluctuations in front of the magnetosphere have a Gaussian distribution, the integral (8.24) may be expressed through the function of a parabolic cylinder $D_{-3/2}$

$$q(E) = \text{constant} \times \sqrt{\sigma}\, D_{-3/2}[\sqrt{2}\sigma^{-1}(E_0 - E)] \exp[-(E - E_0)^2/2\sigma^2] \quad (8.25)$$

where σ^2 is the intensity of interplanetary electric field fluctuations. The function (8.25) is plotted in figure 8.10. The dependence $q(E)$ for $\sigma = 0$ is given by the shaded line. For $E \to -\infty$ we have $q \to 0$, and for $E \to +\infty$ we have $q \propto \sqrt{E}$. It should be noted that the influence of multiplicative noises is strongly pronounced in the vicinity of the critical point E_0.

We have demonstrated once again the simplicity and contents of the phenomenological approach to the analysis of fluctuation and critical phenomena using the RBM model. The inclusion of characteristics of geomagnetic pulsations in the model leads to interesting generalizations and may improve the quality of the forecast of magnetic storms.

Phenomenological modelling sometimes contradicts the microscopic description or the search for empirical relations by means of regression analysis. We have already pointed out the difficulties faced by researchers engaged in a microscopic description. With regards to the regression method, we note that it is capable of solving applied problems, but since it is not oriented

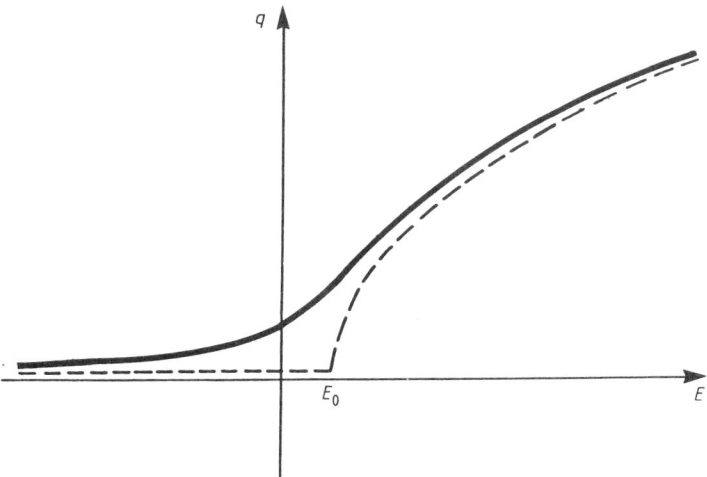

Figure 8.10. Dependence of the D_{st} source on the interplanetary electric field in the presence of multiplicative noises (solid line). The dashed line corresponds to the case of the absence of noise.

towards interpretation it suffers from a semantic vagueness and does not by itself enrich our understanding of geomagnetic phenomena. Phenomenology has the advantage of providing a general picture of magnetospheric processes, though at the cost of discarding details. Finally, phenomenological modelling enriches and deepens the traditional approach i.e. it sets new problems and outlines new ways for trying to find the sense of empirical data.

In conclusion it should be noted that when deriving (8.25) we used a hidden supposition that the fluctuations of the control parameter $E(t)$ are slow in the sense that the correlation interval τ_1 for these fluctuations is much longer than the time τ_2 of the source relaxation $q(E)$. An interested reader may rewrite the criterion $\tau_1 \gg \tau_2$ using the phenomenological parameters of the model in its explicit form. The second problem offered for discussion in connection with (8.25) is much more difficult, i.e. the case $\tau_1 \ll \tau_2$. Discussion of this problem brings us to the concept of the magnetic storm induced by multiplicative noises.

Bibliography

For section 8.1

The phenomenological approach to modelling of fluctuation and critical phenomena in geoelectromagnetism is developed in the series of papers by Guglielmi *et al* (see bibliography in the review Guglielmi 1989). The basic ideas were picked up from various sections of radiophysics. The pioneering paper by Gudzenko (1962), where the correlation method of investigating an uncontrolled auto-oscillatory system according to its signal was proposed, turned out to be most useful.

Kalisher (1975) investigated the fluctuations of PiC amplitude. She discovered that empirical distribution of fluctuations is approximated by a Rayleigh distribution better than by a Gaussian distribution. Kalisher and Polyakov (1984) investigated the fluctuations of the Pc1 group velocity. The Landau theory, on the basis of which the criterion (8.5) was obtained, is presented in the monograph by Landau and Lifshitz (1988a).

For section 8.2

The paper by Guglielmi *et al* (1983) where the ideas of Gudzenko (1962) are realized as applied to the Pc4 pulsations is presented here. The morphology of Pc4 was investigated by Samson and Rostoker (1972), Glaßmeier (1980), Sakurai *et al* (1981) and Hillebrand *et al* (1982).

Kiselev *et al* (1991) following Guglielmi *et al* (1983) constructed a phase diagram of Pc4 and arrived at a conclusion on the nonlinearity of oscillations. Kiselev *et al* (1991) also make a reference to Gudzenko, but they have not applied the proposed procedures. In particular the orbit stability of the phase picture was not investigated, so the conclusion on the nonlinearity of oscillations made by Kiselev *et al* (1991) cannot be considered to have been substantiated. We have to emphasize once again the obvious idea that it is impossible to derive any substantial information on the 'black box without an input' within the framework of a deterministic description.

Vörös *et al* (1994) applied a nonlinear time series analysis to geomagnetic pulsation data with the purpose of identifying deterministic chaos as manifested in the pulsation events.

For section 8.3

General information on magnetic storms may be found in the monograph by Akasofu and Chapman (1972). The phenomenological model (8.12) was suggested by Burton *et al* (1975) (see also Russell 1986). Specifications and stochastic generalizations of the model of Burton *et al* (1975) were suggested by Guglielmi (1988a, 1989), Potapov and Polyushkina (1992), Guglielmi and Pokhotelov (1994). Kalisher *et al* (1985) revealed the connection between the oscillation regime of the magnetosphere and the duration of the magnetic storm experimentally. General problems of phenomenological modelling of the critical phenomena are excellently presented in the monograph by Gilmore (1981). The reader will find information on the Ginzburg–Landau theory in it, a fragment of which was used to model the D_{st} variation.

Vörös (1994) and Vörös *et al* (1994) proposed a synergetic approach to the problem of geomagnetic storms and examined the chaotic attractors in the magnetosphere.

Chapter 9

Hydromagnetic diagnostics and geoelectric prospecting

9.1 Introductory comments

Hydromagnetic diagnostics represent a scientific method which furnishes the geophysicist or other interested user with qualitative and quantitative information for drawing conclusions about the state and possible evolution of the near-Earth medium on the basis of observations of geomagnetic pulsations, and also provides grounds, where necessary, to arrive at one or another decision about how to proceed. The data on polarization, phase and group wave velocities; on the magnetospheric oscillations spectrum; on amplification, damping and spreading of wave packets after covering a certain distance in the magnetosphere; on the spatial localization of a pulsation; on its amplitude and frequency modulation may all serve for diagnostic purposes. Sometimes the mere fact of the appearance of specific pulsations contains qualitative information on the dynamics of the magnetosphere.

Using the data on pulsations we may obtain information on the cold plasma density, localization and energy of resonant particles, the position of magnetospheric structural elements, the solar wind velocity and the magnitude of the interplanetary magnetic field. Also, the observation of pulsations allows us to investigate the processes of particle injection, bifurcations of the magnetosphere, the appearance of plasma instabilities and their consequences.

The laws of excitation and propagation of MHD waves in the near-Earth plasma serve as the physical background of hydromagnetic diagnostics. These laws are known partly from the theoretical analysis of simplified models and partly from experiment, and this knowledge is far from being complete. Let us explain how, under such uncertainty, the problem of hydromagnetic diagnostics should be understood. For that purpose we shall recall the overall classification of the problems that are solved in electrodynamics.

The first class is that of the direct problems or the problems of analysis. These are the familiar internal and external problems, where the sources are given and by them the wave field structure in the medium, the properties of which are considered to be known, is to be found. The second class of problems is that of synthesis problems. These are more specialized problems, where the sources which excite the given field in a medium are to be found. Finally, the third class is that of the so-called inverse problems. In these problems, one works from known sources and fields to find the structure of the medium.

It might seem that the problem of hydromagnetic diagnostics refers to the class of inverse problems. But above we have pointed out especially that we do not have reliable knowledge of either the sources or the fields. In most cases, we know something about the medium and something about the field and its sources. Therefore the problem of hydromagnetic diagnostics should be classified as a mixed problem. The idea of such problems and an approach to their solution was developed by Krasnushkin in a theory for the propagation of radio waves. A quote from Krasnushkin and Yablochkhin (1963) explains the essence of the matter 'In our case the properties of the medium are given incompletely, so it is necessary to solve a so-called mixed problem, in which one works from known and reliable, but incomplete data on the medium, and also from additional known data about the wave field to determine unknown data about the medium and the remaining field'. We believe that the theory of radiowave propagation gives us an example of the level we should aim at when developing hydromagnetic diagnostic methods.

Let us avail ourselves of another quotation from the monograph mentioned above, in order to express our definite scepticism about the use of various 'empirical laws' regarding the behaviour of geomagnetic pulsations for diagnostics purposes 'To make the theory useful for practical application, functional relationships between the field and the medium must be derived from the field equations and constitutive equations of the medium, not directly from the experimental data'.

The desired level has not, however, been reached at this point. In the meantime the diagnostic situation is of the nature of an art, and a fairly subtle one at that. There are no written recipes for it. Nevertheless, this art is a rational one, based on the methodical execution of several extremely important operations. More briefly, it has its own internal logic, which can be demonstrated best with specific examples. In selecting examples we were interested in simplicity as well as novelty.

Electromagnetic waves of natural origin are used not only for magnetospheric diagnostics but also in geoelectricity to study the Earth's crust by the method of magnetotelluric sounding (MTS). The essence of MTS lies in the estimation of the vertical distribution of the crust conductivity by the frequency dependence of the surface impedance. The surface impedance is found from observations of the geomagnetic pulsations.

In the past, these two directions developed independently. In diagnostics of the magnetosphere no use was made of information about the electric conductivity of the Earth's crust, and the work on magnetotelluric sounding made virtually no use of the wave structure of the inducing field.

In this chapter we shall describe the principal techniques of hydromagnetic diagnostics and magnetotelluric sounding and present the idea of a unity between these two methods. We will realize the idea of a unity with the help of the Leontovich impedance boundary condition.

9.2   Surface impedance of the Earth

The impedance ζ at the Earth's surface is introduced by the relation

$$E_t = \zeta b_t \times n \tag{9.1}$$

where E_t, b_t are the tangential components of the electric and magnetic fields, n is a unit normal to the surface directed earthwards[1]. The electromagnetic field is taken to be monochromatic. The impedance is a complex function of frequency. In Gaussian units the impedance is presented in dimensionless form. Within the range of geoelectromagnetic waves, $|\zeta|$ is small compared with unity. For this reason the Earth's surface in many cases may be considered to be an impedance surface proper, i.e. a surface with a fixed impedance. Naturally, such a concept is approximate, and, depending on the method of analysis, the surface impedance has some physical sense.

The simplest analysis consists in fixing the field structure over the surface. For example, we may imagine a field over the Earth which has the form of a superposition of vertical incident and vertical reflected plane waves[2]. It is clear that the surface should be considered to be planar in this case and the Earth's crust to be horizontally homogeneous. Assuming all this we arrive at the concept of the Cagniard–Tikhonov impedance, which is considered to be the basis of the classical version of magnetotelluric sounding.

It is more interesting to fix the field structure not over but under the Earth's surface. It is reasonable to suppose that the field penetrates into the depths of the Earth's crust in the form of a locally plane wave. Assuming this, we arrive at the concept of the Leontovich impedance. We shall confine ourselves to these simple approaches, and focus our attention on purely methodical matters.

9.2.1   Magnetotelluric sounding

In the classical model of MTS the relation (9.1) is rigorous. It represents the only possible linear relation between the polar vector E and the axial vector b. (The index t in (9.1) may be omitted, since all the vectors are parallel to the surface of the Earth within the framework of the given model.)

---

[1] In the literature on geoelectric prospecting, the magnetic field is usually designated as H, but here we retain the symbol b, as used in other sections of the book.

[2] In a more realistic model providing the same results, the source of the inductive field is a synphase current layer located over the Earth's surface, for example at the altitude of the ionosphere.

Based on general principles, we may express crude judgments about the complex coefficient ζ in (9.1). Thus the real part of the impedance is always positive: $\text{Re}\,\zeta > 0$. This is dictated by the requirement that the time-averaged Poynting vector must be directed into the Earth. Since E is real at real b then $\zeta(-\omega^*) = \zeta^*(\omega)$, where the asterisk denotes the complex conjugate value. If the conductivity of the rock is anisotropic, then ζ in (9.1) should be substituted for the two-dimensional tensor $\zeta_{\alpha\beta}$ and it may be inferred from the Onsager reciprocity relation that this tensor must be symmetrical: $\zeta_{\alpha\beta} = \zeta_{\beta\alpha}$.

The calculation of ζ requires, however, the solution of the respective electrodynamic problem. Let the z axis be directed downwards and the plane x, y coincide with the Earth's surface. At $z \geq 0$ the distribution of conductivity $\sigma(z)$ is given. It is convenient to introduce an auxiliary value $Z(\omega, z)$ which is defined by a relation similar to (9.1) and represents the impedance at the depth $z \geq 0$. The order of the initial equations will reduce if the equations for the field components are transformed into the equation for the impedance $Z(\omega, z)$

$$c\partial Z - 4\pi\sigma(z)Z^2 - i\omega = 0. \tag{9.2}$$

Here $\partial \equiv \partial/\partial z$. This is the Riccati equation. The boundary conditions are reduced to the condition of continuity Z at the points of function discontinuity $\sigma(z)$ and the condition $Z \to 0$ as $z \to \infty$. If the solution for this problem is found, then the surface impedance is also found $\zeta(\omega) = Z(\omega, 0)$ for the given distribution of conductivity $\sigma(z)$. The function $\sigma(z)$ is usually considered to be a piecewise function and the general solution is found using recurrent relations.

On the other hand, information on the function $\zeta(\omega)$ in a certain range of frequencies is obtained using (9.1) by evaluating the measurement data of the electromagnetic field component. Comparing this information with the calculation results, $\zeta(\omega)$ for various models $\sigma(z)$, we judge the conductivity distribution in the Earth's crust.

To illustrate this we shall consider the simplest case of the homogeneous half-space. Assuming $\sigma = $ constant, integrating (9.2) and using the condition $Z \to 0$ as $z \to \infty$ we find

$$\zeta = \left(\frac{\omega}{8\pi\sigma}\right)^{1/2}(1 - i). \tag{9.3}$$

A more complicated example is one in which a homogeneous layer with thickness h and conductivity σ_1 covers a homogeneous half-space with conductivity σ_2. We discuss the limiting cases of high and low frequencies without presenting a general formula for ζ.

High-frequency asymptotes for ζ have the form (9.3) if σ_1 is substituted for σ. This approximation is applicable if the depth of the magnetic field penetration into the layer

$$\delta_1 = \frac{c}{(2\pi\sigma_1\omega)^{1/2}} \tag{9.4}$$

is much shorter than the layer thickness ($h \gg \delta_1$).

With decreasing frequency δ_1 increases, the field begins to penetrate into the lower half-space and ζ experiences small oscillations originating from interference. When δ_1 grows noticeably higher than h, further behaviour of ζ will essentially depend on the value of the ratio σ_2/σ_1. If $\sigma_2 \ll \sigma_1$, i.e. the conductivity of the layer is higher than the conductivity of the lower medium, then at the beginning, with decreasing frequency

$$\zeta \simeq \frac{c}{4\pi \Sigma_1} \qquad (9.5)$$

where $\Sigma_1 = \sigma_1 h$ and then

$$\zeta \simeq \left(\frac{\omega}{8\pi \sigma_2}\right)^{1/2} (1 - i). \qquad (9.6)$$

The criteria of applicability of (9.5) and (9.6) have the form

$$\left[\frac{\sigma_2}{\sigma_1}\right]^{1/2} \ll \frac{h}{\delta_1} \ll 1$$

and

$$\frac{h}{\delta_1} \ll \left[\frac{\sigma_2}{\sigma_1}\right]^{1/2} \ll 1$$

respectively.

If $\sigma_2 \gg \sigma_1$, i.e. the conductivity of the layer is lower than the conductivity of the lower medium, then under the supplementary condition

$$\left[\frac{\sigma_1}{\sigma_2}\right]^{1/2} \ll \frac{h}{\delta_1} \ll 1$$

we have

$$\zeta \simeq -i\frac{\omega}{c}h \qquad (9.7)$$

and under the condition

$$\frac{h}{\delta_1} \ll \left[\frac{\sigma_1}{\sigma_2}\right]^{1/2} \ll 1$$

we have (9.6).

Usually the results of measurements are transformed into a so-called apparent specific resistivity of the rocks

$$\rho_a(T) = 2T |\zeta(T)|^2$$

where $T = 2\pi/\omega$. In the high-frequency limit ($\delta_1 \ll h$) we have $\rho_a = 1/\sigma_1$, i.e. using the value of the effective resistance we may judge the conductivity of the upper layer of the Earth's crust. At low frequencies ρ_a gives either the integral conductivity Σ_1 of the upper layer (the case (9.5)) or the thickness of the layer h (the case (9.7)).

9.2.2 The Leontovich boundary condition

In principle, the field may certainly be fixed outside a highly conducting body. For example, field sources may be chosen and fixed properly in the laboratory. But this procedure is impossible if we speak of the Earth and the fields as being of natural origin. Here it is more appropriate to imagine the field as if fixed inside the Earth. This is not only reasonable, but also useful from the heuristic point of view, since it motivates the search for modifications and new applications of the method.

The idea lies in the fact that the field inside well-conducting matter in the vicinity of its boundary should be considered as the field of a locally plane wave with wavefronts almost parallel to the boundary. Then, formally, relation (9.1) does not change but gains a different meaning. First of all, it now becomes approximate, since more or less arbitrary structure is admitted outside the body. Then, it becomes local, as ζ may smoothly change along the surface of the body. When thus understood, relation (9.1) is termed the approximate Leontovich boundary condition.

Let us use the Leontovich boundary condition on the surface of the Earth $(z = 0)$

$$E_x = \zeta b_y \qquad E_y = -\zeta b_x \qquad (9.8)$$

to derive the formula

$$b_z = i \frac{c}{\omega} \nabla \cdot (\zeta b_t) \qquad (9.9)$$

which may be effectively employed for the sounding of the Earth's crust and magnetospheric diagnostics. Here the Earth's crust is considered to be plane and the z axis is directed downwards, $\nabla \cdot$ denotes the surface divergence and $b_t = (b_x, b_y)$. From the induction equation

$$\nabla \times E = i \frac{\omega}{c} b$$

it follows that

$$b_z = \frac{c}{i\omega} \left(\frac{\partial E_y}{\partial x} - \frac{\partial E_x}{\partial y} \right). \qquad (9.10)$$

Substituting (9.8) into (9.10), we get (9.9).

Relation (9.1) is approximate and the problem, arising in this connection, consists of evaluating the applicability conditions (9.1). It is clear, that the depth δ of field penetration into the body must be small compared with the vacuum wavelength $\lambda = c/\omega$, the radii of curvature of the body surface, the characteristic change scales of the field and medium properties along the surface of the body. An exact answer to the question of applicability of the Leontovich boundary condition is difficult to present. In any case the answer will depend on the character of the problem being solved. There is always a risk that one or another applicability condition will be violated or, alternatively, that some condition will turn out to be too rigid.

Let us find the applicability condition using the 'machinery' of the parabolic equation. For simplicity we shall consider the Earth's crust as a homogeneous conducting half-space. The field components inside the body yield the Helmholtz equation, for example

$$\nabla^2 b_x + k^2 b_x = 0 \tag{9.11}$$

where $k = (1 + \mathrm{i})/\delta$. Let us direct the z axis vertically downwards, choose z as a direction variable and seek a solution of (9.11) in the form

$$b_x(x, z) = b_x^{(0)}(x, z)\mathrm{e}^{\mathrm{i}kz}. \tag{9.12}$$

Here it is assumed for simplicity that neither of the values depends on y. Suppose the dependence $b_x^{(0)}$ on z is considerably weaker than the dependence $\exp(\mathrm{i}kz)$. Substituting (9.12) into (9.11) and omitting the small term $\partial^2 b_x^{(0)}/\partial z^2$, we get a parabolic equation

$$2\mathrm{i}k\frac{\partial b_x^{(0)}}{\partial z} + \frac{\partial^2 b_x^{(0)}}{\partial x^2} = 0. \tag{9.13}$$

Let us consider a transverse-electric (TE) field. Its nonzero components E_y, b_x, b_z are connected through the equations of quasi-stationary electrodynamics

$$\frac{\partial b_x}{\partial z} - \frac{\partial b_z}{\partial x} = \frac{4\pi}{c}\sigma E_y$$

$$\frac{\partial E_y}{\partial x} = \mathrm{i}\frac{\omega}{c}b_z \qquad \frac{\partial E_y}{\partial z} = -\mathrm{i}\frac{\omega}{c}b_x.$$

Eliminating b_z from the first equation with the help of the second, we get

$$E_y + \frac{1}{k^2}\frac{\partial^2 E_y}{\partial x^2} = \frac{c}{4\pi\sigma}\frac{\partial b_x}{\partial z}. \tag{9.14}$$

The derivative $\partial b_x/\partial z$ we find from (9.12) using (9.13). After that (9.14) takes the form

$$E_y + \frac{1}{k^2}\frac{\partial^2 E_y}{\partial x^2} = -\zeta\left(b_x + \frac{1}{2k^2}\frac{\partial^2 b_x}{\partial x^2}\right) \tag{9.15}$$

where ζ is defined by (9.3). Considering the second terms on the left-hand and right-hand sides of (9.12) to be small, we shall finally write the boundary condition on the surface of the body

$$E_y = -\zeta\left(b_x + \frac{\mathrm{i}\delta^2}{4}\frac{\partial^2 b_x}{\partial x^2}\right). \tag{9.16}$$

Similarly we obtain an approximate boundary condition for the TM field

$$E_x = \zeta\left(b_y - \frac{\mathrm{i}\delta^2}{4}\frac{\partial^2 b_y}{\partial x^2}\right). \tag{9.17}$$

Hence it is seen that one of the conditions of the applicability of the impedance boundary condition (9.8) is the smoothness of change of the horizontal components of the magnetic field along the Earth's surface. For example, we should have

$$\frac{\delta^2}{4}\left|\frac{\partial^2 b_x}{\partial x^2}\right| \ll |b_x|. \tag{9.18}$$

9.2.3 The impedance equation

Let us focus our attention on (9.9). It is interesting because the range of its applicability is even wider than that of the Leontovich boundary condition. To demonstrate the point, we note that (9.9) does not change if (9.8) is replaced by

$$E_x = \zeta b_y + \frac{\partial f}{\partial x} \qquad E_y = -\zeta b_x + \frac{\partial f}{\partial y} \tag{9.19}$$

where f is some function of x, y and ω which is linear in the field and otherwise arbitrary. Substituting (9.19) into the z component of the induction equation, we get

$$b_z = i\lambda \nabla \cdot (\zeta b_t) \tag{9.20}$$

where $\lambda = c/\omega$.

The relations (9.19) on the Earth's surface $z = 0$ take place, for example, if the impedance surface proper, $z = h \geq 0$, with impedance ζ_0 is coated with a high-resistivity layer of variable thickness $h(x, y)$. Then instead of (9.8) we have (9.19) with $f = hE_z$ and $\zeta = \zeta_0 - i\omega h/c$, where E_z is the vertical component of the electric field directly below the $z = 0$ surface.

There are a diversity of applications of (9.20) in geoelectromagnetism. Let us assume that some of the quantities in (9.20) are known from experiment, while others are unknown and are to be determined. Then (9.20) will allow us either to calculate the unknown quantities immediately or impose certain restrictions on them. In other words, we shall employ various versions of 'reading' equation (9.20). Of course, in order to do this we shall have to introduce supplementary assumptions.

To begin with, we shall regard a scalar field of the surface impedance at some fixed frequency as unknown. Let us rewrite (9.20) in the form

$$A\frac{\partial \zeta}{\partial x} + B\frac{\partial \zeta}{\partial y} + C\zeta + D = 0. \tag{9.21}$$

We shall consider (9.21) to be a differential equation in order to seek the impedance $\zeta(\omega; x, y)$ under the condition that the coefficients

$$A = b_x \qquad B = b_y \qquad C = \nabla \cdot b_t \qquad D = \frac{i}{\lambda}b_z$$

are known from observations. By this the known method of geoelectric prospecting is generalized for the case of the horizontally inhomogeneous Earth, i.e. instead of the algebraic relation

$$\zeta = -\frac{D}{C} \tag{9.22}$$

the differential impedance equation (9.21) is introduced.

It is evident that the realization of this approach requires synoptical observations of the magnetic field oscillations over a fairly dense network of magnetometers. Here *a priori* information on the structure of the inducing field is not involved, since everything required for prospecting is obtained during the field observation by making use of this network.

9.2.4 Gradient of the surface impedance

The surface impedance gradient $\nabla\zeta$ is the simplest characteristic of the horizontal inhomogeneity of the Earth's crust. The trivial approach to the problem of measuring $\nabla\zeta$ is to carry out magnetotelluric sounding and to determining ζ at three or more points. In general of course, it is not possible to avoid the procedure of multipoint measurements of ζ. It is nevertheless interesting and useful to know that in certain special cases it is possible to measure $\nabla\zeta$ by observing and analysing the components of the electromagnetic field at only a single point.

For this purpose we shall use the relation (9.20) and rewrite it in the form

$$b_z = i\frac{c}{\omega}[(b_t \cdot \nabla\zeta) + \zeta(\nabla \cdot b_t)]. \tag{9.23}$$

Let us consider $\nabla\zeta$ to be unknown and b_z, b_t are assumed to be known values at a given point on the Earth's surface. To calculate $\nabla\zeta$ we lack information on $\nabla \cdot b_t$. The simplest solution is to discard the second term on the right-hand side of (9.23). Then the simplified formula

$$b_z = i\frac{c}{\omega}b_t \cdot \nabla\zeta \tag{9.24}$$

allows us to find $\nabla\zeta$.[3]

An ideal way of using (9.24) would be for the field, which is known *a priori* to be transverse, i.e. $\nabla \cdot b_t = 0$. The condition of transversity is satisfied for longitudinal and transverse resonances of the Earth–ionosphere cavity. This case has been discussed in Chapter 3 (formula (3.13) and exercise 3.2.2).

Let us return to the impedance equation (9.21). In the limits $l_\zeta \gg l_b$ and $l_\zeta \ll l_b$ we can replace (9.21) by (9.22) and (9.24), respectively (here l_ζ and

---

[3] To calculate both components of the complex vector $\nabla\zeta$ one needs two independent measurements of magnetic field components for different polarizations of b_t.

l_b are the length scale of the variations in the impedance and in the magnetic field). These two limiting cases, however, differ greatly in their importance to geosounding. Expression (9.24) provides nontrivial information about the structure of the Earth's crust on the basis of observations at one point, while (9.22) requires multipoint observations, which, broadly speaking, permit the more general approach based on (9.21).

9.2.5 On the search for an earthquake's precursors

The methods of induction electromagnetic sounding of the Earth's crust are used to study the structure of the Earth's crust as well as to solve applied problems associated with the search for minerals. It is known, for example, that oil and gas deposits are more often located in anticlines of underground layers of sedimentary rocks than in other places. These anticlines, in their turn, are usually situated over underlying crustal prominences. Knowing these features we may choose a definite strategy to apply induction sounding to search for areas likely to contain oil and gas. The methods of induction sounding as well as other geophysical methods are auxiliary but their application allows us to reduce the number of borings, and to increase the completeness and quality of reconnaissance.

There exists extensive literature dedicated especially to this topic and we shall not dwell upon it, but instead turn to another aspect of the problem. We shall discuss the attempts to apply induction sounding for monitoring the state of the Earth's crust in seismoactive regions, aimed at finding the variation of conductivity prior to an earthquake.

Observations testify to the fact that 2–3 months before an earthquake there begins a progressive increase of σ, and the general accretion $\Delta\sigma/\sigma$ reaches 15–20% before the moment of the earthquake.

This experimental fact finds reasonable explanation within the framework of the dilatancy–diffusion model of the onset of an earthquake. The essence of the model lies in the fact that with the increase of the deformation of the Earth's crust, proliferation of microcracks takes place long before fragile destruction begins. Filling of microcracks with water leads, on the one hand, to a decrease of the solidity of the rock and, on the other hand, to an increase of conductivity. The appearance of the porous fluid may be caused by its percolation from water-bearing horizons and by the dehydration of minerals. Laboratory tests with moist samples of rocks prove these notions[4].

A relative change of the conductivity of rocks over time by 15–20% is sufficient, generally speaking, to be caught by the methods of induction sounding. Attempts to discover the effect by measuring the surface impedance ζ according to the classical methods of magnetotelluric sounding have not yet met with success. Experiments of this kind should be continued since we have grounds

---

[4] The conductivity of dry samples of the rock does not increase, but decreases slightly with the growth of pressure on the sample, which is also in general agreement with the notion of dilatancy.

to expect positive results. Although monitoring of seismoactive regions by the methods of inductive sounding cannot cope with the problem of predicting earthquakes, it is advisable to use it as an auxiliary means together with other methods of geophysical control. In this connection we shall pay attention to an important advantage of the sounding method based on the formula (9.20). In contrast to magnetotelluric sounding, which provides a local value of ζ at the point of observation, this method allows us to measure $\nabla\zeta$ and thus it allows us to accomplish remote control of the state of the medium.

Exercises

Exercise 9.2.1.

Using the approximate Levi-Cività boundary condition find the impedance of a thin layer on the surface of a rock resting on a thick layer.

Solution 9.2.1.

Let us designate by σ_1 and h_1 the conductivity and thickness of the upper layer and similarly use σ_2 and h_2 for the lower layer. Using the formula (9.4) we shall calculate the skin length δ_1 for the upper layer and similarly δ_2 for the lower. The upper (lower) layer is considered thin (thick) implying that $\delta_1 \gg h_1$ ($\delta_2 \ll h_2$).

The thickness of the thin layer can be conveniently assumed to be infinitely small, but it should be borne in mind that when passing this infinitely thin film the horizontal components of the magnetic field experience discontinuity. The boundary condition at $z = 0$ will then have the form

$$b_x^+ - b_x^- = -(4\pi/c)\Sigma_1 E_y$$
$$b_y^+ - b_y^- = (4\pi/c)\Sigma_1 E_x. \tag{1}$$

Here $\Sigma_1 = \sigma_1 h_1$, the z axis is directed downwards and the signs $\pm$ correspond to the values over and under the film. The field E_t is continuous when passing through the thin film.

Since the lower layer is thick in the sense mentioned above, its thickness does not play any role and it may be considered infinitely large. Then under the film the following relations hold

$$E_x = \zeta_2 b_y^- \qquad E_y = -\zeta_2 b_x^- \tag{2}$$

where $\zeta_2 = 1/\sqrt{\varepsilon_2}$ and $\varepsilon_2 = 4\pi i\sigma_2/\omega$. Combining (1) and (2) we find the relationship between E_t and b_t^+ of the form (9.1), and

$$\zeta^{-1} = \frac{4\pi}{c}\Sigma_1 + \sqrt{\varepsilon_2}. \tag{3}$$

If σ_1 changes across the thin layer then the total conductivity in (3) should be defined as

$$\Sigma_1 = \int\limits_0^{h_1} \sigma_1(z)\,dz.$$

Exercise 9.2.2.

Let the body have spherical pores and the conductivity of the porous medium σ_1 differ from the conductivity of the environment σ_2. Show that the effective conductivity of the body σ_{ef} is smaller than the conductivity $\bar{\sigma}$, averaged over the volume of the body.

Solution 9.2.2.

Let m be the porosity, i.e. the volume concentration of pores. Averaged over the volume, the conductivity obviously equals

$$\bar{\sigma} = m\sigma_1 + (1 - m)\sigma_2. \tag{1}$$

The effective conductivity is equal to the proportionality coefficient between $\bar{j}$ and $\bar{E}$ by definition, where the bar signifies averaging over a volume V that contains a fairly large number of pores

$$\bar{j} = \sigma_{ef}\bar{E}. \tag{2}$$

According to the meaning of averaging, the following relation holds

$$\frac{1}{V}\int (j - \sigma_2 E)\,dV = \bar{j} - \sigma_2\bar{E}. \tag{3}$$

Here the left-hand side is proportional to porosity m, since the subintegral function differs from zero only inside the pores.

Let us assume that porosity is small, i.e. $m \ll 1$. Then we may approximately consider that the pores are located in the external electric field, which only slightly differs from $\bar{E}$. We shall ignore this distinction and avail ourselves of the known solution of the problem in the field E_1 inside the conducting sphere, placed in a homogeneous medium

$$E_1 = \frac{3\sigma_2}{\sigma_1 + 2\sigma_2}\bar{E}. \tag{4}$$

Substituting (2) and (4) into (3) we find

$$\sigma_{ef} = \sigma_2 + 3m\frac{(\sigma_1 - \sigma_2)\sigma_2}{\sigma_1 + 2\sigma_2}.$$

Comparing this with (1) we finally obtain[5]

$$\sigma_{\text{ef}} = \bar{\sigma} - m \frac{(\sigma_1 - \sigma_2)^2}{\sigma_1 + 2\sigma_2}.$$

Hence it is evident that $\sigma_{\text{ef}} < \bar{\sigma}$ independent of the sign of the difference $\sigma_1 - \sigma_2$.

9.3 Diagnostics of the magnetosphere

Magnetospheric plasma represents a multicomponent, unstable, inhomogeneous and extended medium.   Strong inhomogeneity in the distribution of the parameters of the medium and relatively rapid variability of these parameters hamper diagnostics.   There is no general method that would provide fairly complete information on the magnetosphere as a whole. The current state of the magnetosphere is evaluated by a complex of ground and satellite measurements of a wide set of physical parameters[6].

Hydromagnetic diagnostics is based on registering geoelectromagnetic waves. The methods of hydromagnetic diagnostics differ in frequency range, physical principles, means of realization, etc.   We shall confine ourselves to describing two methods of the ground control of plasma density in the magnetosphere.   For methods of diagnostics of other parameters, one may develop the ideas given in exercises 9.3.1–9.3.5.

9.3.1 The MHD locator

We shall describe the operation of a simple device that facilitates diagnostics of plasma density. The device contains a magnetometer and a set-up for registering the Earth's currents, i.e. an ordinary set of standard observatory equipment. In addition to that we consider a large volume of the Earth's crust in the vicinity of the observation point as an essential element of the device. The dimensions of the volume are of the order of the skin-length. The electrodynamic characteristics of the new element are taken into account when measurements are being carried out.

Previously the Earth's crustal properties were disregarded when accomplishing hydromagnetic diagnostics. To be more precise, only the following argument was taken into account.

Since the Earth is a relatively good conductor it is supposed to be ideally conducting for the aims of magnetospheric diagnostics, and the tangent projection of the electric field E_t and the normal projection of the magnetic field

---

[5] To avoid misunderstanding we shall point out that the result refers to the case where the pores are isolated, i.e. do not communicate. The rock in its natural arrangement contains a ramified system of interconnected pores and cracks side by side with isolated pores.

[6] In contrast to this the general state of interplanetary medium in front of the magnetosphere may be estimated by the satellite measurement data of a few parameters (solar wind density and velocity, magnitude and orientation of the interplanetary magnetic field).

b_n on the surface of the Earth are assumed to be zero. The tangent projection b_t is just a doubled field of external sources. This is a zero (in terms of the thickness of the skin-layer) approximation. In the first approximation we have (9.1), the field b_t is taken in the zero approximation as before.

Thus, b_t depends on the external sources and on the external medium, but is almost independent of the specific conductivity of the rocks in the vicinity of the observation point, whereas E_t and b_n vitally depend on both the external sources, the external medium and on the conductivity of the rocks. This understanding lies at the foundation of a widespread rule to use b_t for diagnostics of the upper half-space, and not E_t and b_n.

Our absolutely trivial idea is as follows. If we first study the conductivity of the rocks in the vicinity of the observatory, directed towards magnetospheric diagnostics, using the methods of geoelectrics, we may remove the indefiniteness in ζ and then the additional information which is contained in E_t and b_n will allow us to widen the range of diagnostic tools. We shall realize this idea in the framework of the problem on plasma density diagnostics by measuring resonant oscillations of the magnetosphere.

Geomagnetic parallels and meridians are plotted in figure 9.1. Here the x axis is directed northwards, the y axis points eastwards and the horizontal line is the projection of the magnetic shell over the Earth's surface. On the left-hand side the latitudinal profile of the amplitude of resonant oscillations of the chosen magnetic shell is shown schematically. In an elementary solution, a chain of magnetic observatories are located along the meridian, then the horizontal components of the magnetic oscillations are synchronously registered, the latitudinal amplitude profile is calculated using interpolation, and the latitude of the magnetic shell is determined by the amplitude maximum at the given frequency.

We present a method that allows us to solve this problem using observations at just one point (it is marked by a circle on figure 9.1). In other words, we suggest an algorithm for measurements and calculations that offers the distance x_R up to the projection over the Earth of the oscillating magnetic shell.

Let us locate the observational point ($x = y = 0$) in the centre of the region with a relatively homogeneous distribution of conductivity of the rocks in horizontal directions. We shall register the East–West component of the electric field E_y and the vertical component of the magnetic field b_z. Let us make a spectral analysis and designate $E = |E_y(\omega)|$, $Z = |b_z(\omega)|$ and $\lambda = \omega/c$. We find the distance from the observation point ($x = 0$) to the magnetic shell, resonating at the frequency ω, from the expression

$$x_R(\omega) = \lambda(E/Z)\sin\theta \tag{9.25}$$

where $\theta(\omega)$ is the phase difference between the spectral components $E_y(\omega)$ and $b_z(\omega)$.

Let us sketch a further step. We find a function, inverse to $x_R(\omega)$, and from it we find the plasma density $\rho(L)$ at the equator of the magnetic shell

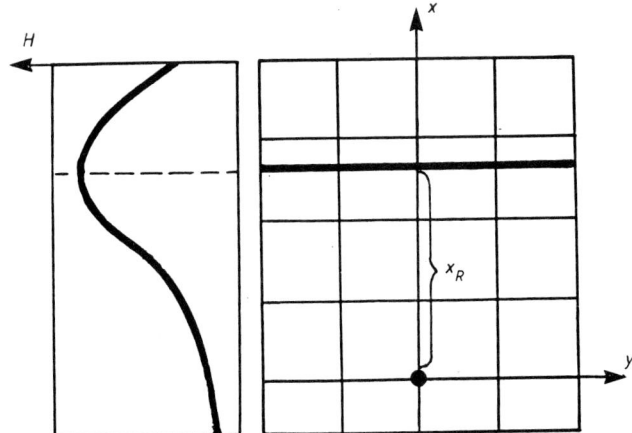

Figure 9.1. Latitudinal profile of the amplitude of resonant oscillations (left) and the location of the observatory relative to the projection of the resonating magnetic shell onto the Earth's surface (right).

with parameter L. (The parameter L is related in a known way to x_R.) We shall return to that problem but meanwhile we present a derivation of (9.25) from (9.9) with regard of the structure (4.69) of resonant oscillations.

Let Δ be the width of a resonance, then

$$b_x(x) = b_x(x_R)[1 - i(x - x_R)\Delta^{-1}]^{-1}. \tag{9.26}$$

Here we have taken into account the $\pi/2$ rotation of the polarization when oscillations are transmitting through the ionosphere. From (9.26) we find

$$\frac{\partial b_x}{\partial x} = \frac{b_x(0)}{x_R + i\Delta} \tag{9.27}$$

at $x = 0$. On the other hand

$$\frac{\partial b_x}{\partial x} = -i\frac{\omega}{c\zeta}b_z. \tag{9.28}$$

This follows from (9.9) in the case $\nabla\zeta = 0$ and remembering that in the vicinity of the resonant point we have $\nabla \cdot b_t \approx \partial b_x/\partial x$ within terms of the order of $(m\Delta)^2 \ln(m\Delta)$, where m is an azimuthal number and $m\Delta \ll 1$. Combining (9.8), (9.27) and (9.28) we get (9.25).

Similarly we find the width of the resonance

$$\Delta = \lambda(E/Z)\cos\theta. \tag{9.29}$$

Since Δ is an essentially positive value, then

$$-\pi/2 < \theta < \pi/2 \tag{9.30}$$

which provides a simple criterion for the selection of resonant oscillations.

Let us consider the restriction connected with omitting the term $\partial b_y/\partial y$. If ϕ is a geomagnetic latitude, then $\partial b_y/\partial y = imb_y(R_E \cos\phi)^{-1}$. Comparing this with (9.27), we find the corresponding criterion

$$\frac{|b_y|}{|b_x|} \ll \frac{R_E \cos\phi}{m(x_R^2 + \Delta^2)^{1/2}}. \tag{9.31}$$

Let us point out that (9.25) does not take into account the information on the geoelectric properties of the rocks in its explicit form (except the one concerning horizontal homogeneity of conductivity distribution). This is achieved by a specific compensation: as both components E_y and b_z are proportional to ζ, their ratio (E_y/b_z) does not depend on ζ. The idea of compensation is quite similar to the idea of MTS: since E_y and b_x are proportional to the intensity of external sources, their ratio (E_y/b_x) does not depend on this intensity and moreover it is proportional to ζ. (We recall that E_y is proportional to ζ, and b_x is almost independent of ζ.) In other words, while MTS is aimed at the compensation of the indefinite conditions in the upper half-space, our method is directed to the compensation of indefinite conditions in the lower half-space.

Let us consider possible generalizations of the method. The first evident generalization consists in the substitution of the scalar ζ for the tensor $\zeta_{\alpha\beta}$ in the initial formulae and the respective modification of the calculation formulae, i.e. taking account of the lower half-space anisotropy. The second generalization is associated with taking account of the horizontal inhomogeneity of the Earth's crust. We shall do this approximately, considering ζ to be a scalar and the inhomogeneity to be sufficiently weak. Then b_z should be simply substituted for $b_z - i\lambda(b_t \cdot \nabla\zeta)$ in (9.25). The realization of this version requires preliminary investigation of the geoelectric structure of the region in the vicinity of the observational point and calculation of $\nabla\zeta$ on this basis.

Then, it is reasonable to consider the possibility of completely abandoning the use of boundary conditions of the impedance type. For this we should study the conductivity distribution in the lower half-space near the point of observation by geoelectric methods, and then jointly solve the internal problem (for the Earth's crust) and the external (for the magnetosphere), matching the solutions on the boundary. In other words, the problem of hydromagnetic diagnostics should be considered not to be inverse, but to be a mixed problem in the sense stated in section 9.1. The emphasis here is on the preliminary study of electric conductivity of the lower half-space. Once this indefiniteness has been removed, it becomes possible to use additional relationships between the components of the electromagnetic field in order to improve the accuracy of magnetospheric diagnostics.

Finally, it is possible that the generalization of (9.25) is associated with more accurate modelling of resonant oscillations of the magnetic shells. This allows us to abandon the restriction $|x_R| < \Delta$, which is obligatory, generally speaking, when employing (9.25).

Let us compare the 'MHD location' method with the so-called 'gradient method'. Its essence is as follows: from the equality of spectral amplitudes $|b_x(\omega)|$ at two observatories, located close to each other along a meridian, we find the resonant frequency of the magnetic shell, passing strictly in between these observatories[7]. Compared with this the MHD locator method has two advantages: only one observatory is used instead of two; and it is not the frequency of one strictly fixed shell that is measured, but the frequency of any shell, crossing the Earth's surface in the distance range of the order $(-\Delta, +\Delta)$ in the vicinity of the observatory (the approximate value of the interval is 1000 km).

Comparison with the 'polarization method' is more interesting. It is based on the known property of the resonant oscillations: the rotation of the b_t vector changes by 180° when crossing the resonant line. Based on the polarization inversion, the position of the resonance may be evaluated. To realize the method, a number of observatories distributed along the meridian are used. Limited information may be obtained from one observatory. We refer to the case where shells in a certain latitude range resonate simultaneously, and the chosen observatory turns out to be in this range. Then, in the spectrum observed, there is a frequency above and under which the spectral components $b_t(\omega)$ are subject to reciprocally opposite rotation. Supposedly, this is a resonant frequency of the magnetic shell, crossing the Earth at the latitude of the observatory.

9.3.2 Diagnostics based on oscillation spectra

The spectrum of magnetospheric MHD oscillations depends on the spatial distribution of plasma and magnetic field. Since the magnetic field structure is relatively stable, the spectral data can be used to monitor the plasma density variations. Of interest in this regard are the Alfvén oscillations. Different parts of their spectrum are formed in different regions of the magnetosphere. It thus is possible to work from the known spectrum to reconstruct not only integral parameters of plasma distribution but also local parameters.

As always in the case of hydromagnetic diagnostics, here we should solve the respective inverse problem. However, for simplicity and due to the lack of necessary information on the wave field, a solution of the direct problem under the given hypothesis on the structure of the medium is usually sought. An appropriate (usually dipole) geomagnetic field approximation is chosen. The distribution of the plasma density is described by a certain set of test functions, and the calculation of the spectrum of Alfvén oscillations is carried out. The parameters of the model are selected so that the theoretical and experimental spectra should be as close as possible.

The oscillation frequency of the given magnetic shell comparatively weakly depends on the plasma distribution along the geomagnetic field lines, but strongly depends on the plasma density $\rho(L)$ at the equator of this shell. This allows us

---

[7] The gradient method is a simplified version of the method of interpolation of the amplitudes, measured over a chain of observatories distributed along a meridian.

to choose the plasma distribution along the field lines more or less arbitrarily, and then find $\rho(L)$ using the experimental data on the spectrum. In advance, each spectral peak $\omega_n(L)$ should be marked with the number n, corresponding to the harmonic number of the Alfvén oscillations. At small n (two–three first harmonics) it is not difficult to do this based on the *a priori* information on the magnetospheric spectrum. Difficulties may arise at large n.

The form of the connection between $\omega_n(L)$ and $\rho(L)$ was discussed in section 4.5. Inverting the corresponding formulae we find

$$N_{eq}(L) = \Lambda_n(L)[nT_n(L)/L^4]^2. \qquad (9.32)$$

Here $T_n = 2\pi/\omega_n$ and, for convenience, instead of ρ an equivalent plasma density $N_{eq} = \rho/m_p$ is introduced. In other words, N_{eq} is the plasma density if the mass is measured in terms of proton mass units. The function $\Lambda_n(L)$ is found from the solution of the corresponding eigenvalue problem. At fairly large L and n the function is indistinguishable from the constant $\Lambda \simeq 1.6 \times 10^3$ s^{-2} cm^{-3}.

It is reasonable to term the procedure described as hydromagnetic spectroscopy. It may be developed in still another direction, namely, we may try to extract information on the plasma distribution along the field lines by measuring nonequidistance of harmonics in the spectrum of magnetic shell oscillations. Besides, nonequidistance may be effectively used to identify the numbers of moderately high harmonics. (Very high harmonics are always equidistant.)

The major difficulty when accomplishing hydromagnetic spectroscopy lies in estimating L. A broadband external source excites the shells over a wide range of L. Owing to the spatial interference of the resonances, the observer will register a wide spectrum of oscillations. In this case the task is to 'shuffle' the spectral components of the registered signal about different L. If the source is narrow banded, then the task is to find L of the very shell that resonates at the frequency of the source. (It is clear that the position of the observer does not provide the necessary information due to the finite spatial resonance width.)

If the hydromagnetic spectroscopy is carried out using the MHD locator data, then x_R should be first substituted for L using the formula

$$L(\omega) = L_0 + \delta L(\omega) \qquad (9.33)$$

where

$$\delta L(\omega) = 2[x_R(\omega)/R_E] \sin \phi_0/\cos^3 \phi_0$$

in the case of the dipole field. Here ϕ_0 is a geomagnetic latitude of the observation point, $L_0 = \cos^{-2} \phi_0$, $\delta L \ll L$. The inversion of $L(\omega)$ gives L-dependence of the resonance frequency $\omega(L)$.

Let us present an example. The dependence $T = 2\pi/\omega$ on x_R is shown in figure 9.2. The calculations of x_R are made according to the data of magnetic pulsations observed in the Ukraine not far from the village Rakhny Sobovy

($\phi_0 = 43.9°$). All 20 events presented in figure 9.2, were registered on 30 August 1974 in the time interval 09.30–12.30 LT. Each event is a wave packet with an explicit carrier frequency and smooth amplitude envelope. There were 23 events of this kind found in this interval; but three of them were sorted out using the criterion (9.30). Besides the distance x_R, the calculation of impedance $\zeta = -E_y/b_x$ was made for each selected event.

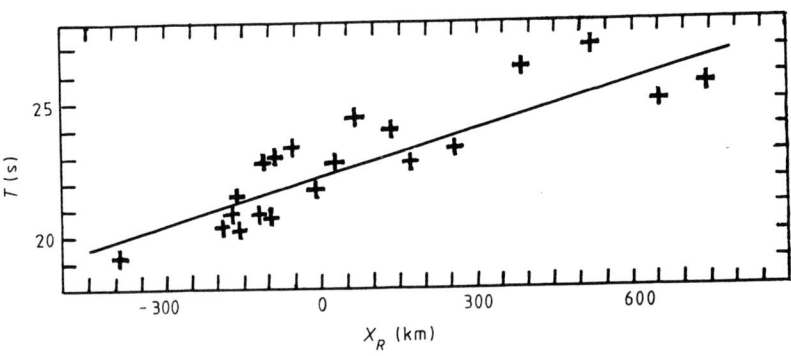

Figure 9.2. Result of the MHD location on 30 August 1974 at latitude $\phi_0 = 43.9°$. Vertical axis—the oscillation period, horizontal axis—the distance from the observation point to the resonating magnetic shell.

From figure 9.2 we see that the connection between x_R and T is quite close. The correlation coefficient $r = 0.87$ has the expected sign (the period increases with increasing of latitude). At the same time the connection between ζ and T has turned out to be rather weak. The correlation coefficient $r = 0.24$. This testifies to the fact that at least in the considered case the method of magnetospheric MHD location is more informative than the method of magnetotelluric sounding.

According to the data of figure 9.2 and using (9.32) and (9.33) we find the dependence $\rho(L)$ (see figure 9.3). The result agrees with the general ideas of the plasma distribution over the given interval L.

9.3.3 Diagnostics by the signal repetition period

Let us consider a Pc1 wave packet in one of the longitudinal waveguides in the magnetosphere. In the geometric–optic approximation the doubled time τ of the group delaying of the packet between the ends of the waveguide is given by (5.40). The value τ, as well as the wave-packet carrier frequency ω, is easily measured by processing a Pc1 sonogram. If in addition to that we measure the parameter L of the wave packet trajectory, then we may raise a question on diagnostics of plasma density in the magnetosphere.

Such a possibility arises owing to the fact that the value $\tau(\omega)/\tau(0)$ is a 'universal' function of the ratio ω/Ω_0, where Ω_0 is the gyrofrequency of

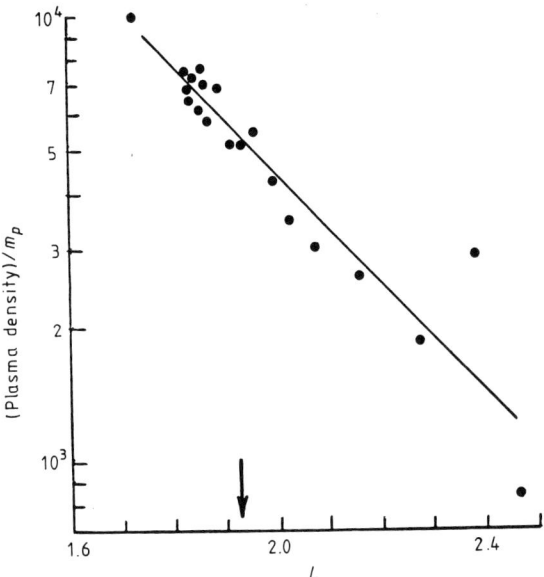

Figure 9.3. Dependence of the plasma density on the parameter of the magnetic shell. The arrow indicates the position of the MHD locator.

the protons at the top of the trajectory. The universality is understood in the following sense. The function depends in practice neither on L nor on the plasma density. Physically this means that the main contribution to the signal dispersion comes from the near-equatorial section of the trajectory where the difference $(\Omega - \omega)$ is minimum.

We introduce a dimensionless parameter, characterizing the signal dispersion

$$D = d \ln \tau / d \ln \omega.$$

It is quite clear that the function $D(\omega/\Omega_0)$ is universal in the same sense (figure 9.4).

The diagnostics procedure is as follows. Using the Pc1 sonogram we measure the carrier frequency ω, the repetition time τ at the carrier frequency ω and the dispersion parameter D. (The latter operation is complicated but sometimes successful.) With the help of the measured value of D we determine ω/Ω_0 and, knowing ω, find $L = 14.4\,\Omega_0^{-1/3}$. The plasma density at the top of the trajectory is found by the inversion of (5.40) with regard to the measured values of τ, ω and L.

If we fail to measure the dispersion D, then the parameter L should be found in a different way. We know that the Pc1 signals propagate horizontally inside the ionospheric waveguide for long distances from the end of the longitudinal

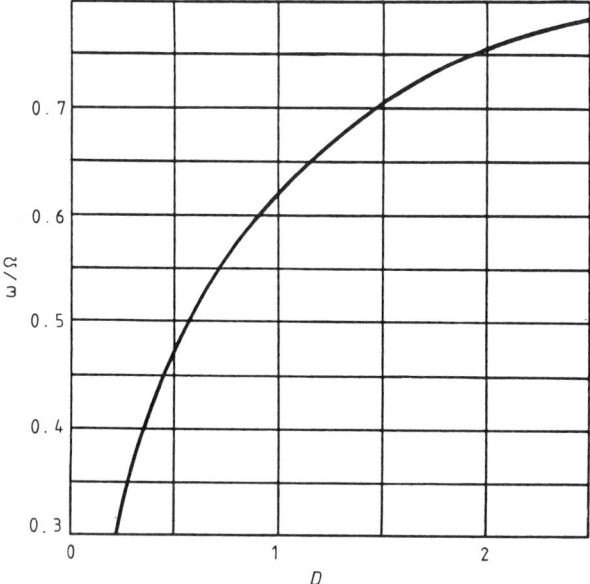

Figure 9.4. 'Universal' dependence of ω/Ω versus dispersion D for the ion cyclotron wave packet.

waveguide. Therefore the position of the observation point tells us nothing about L. We may try, for example, to find L by the indications of a couple of MHD direction-finders (see exercise 9.3.5).

Exercises

Exercise 9.3.1.

Find the spatial distribution of plasma density ρ using the latitudinal dependence of the period of toroidal oscillations, assuming the ambient magnetic field to be a dipole and the distribution ρ spherically symmetric (the symmetry centre coincides with the position of the magnetic dipole).

Solution 9.3.1.

We present the equation of toroidal oscillations (4.47) and the boundary condition in the spherical coordinate system (r, ϑ, φ)

$$-\frac{F_1}{4\pi}\frac{\mathrm{d}}{\mathrm{d}r}\left(\frac{F_2}{\rho}\right)\frac{\mathrm{d}\psi}{\mathrm{d}r} = \omega^2\psi \tag{1}$$

$$\mathrm{d}\psi/\mathrm{d}r_e|_{r=r_e} = 0 \tag{2}$$

where r_e is the radius of an ideally conducting sphere imitating the Earth. Other notations are as follows

$$F_1 = (r \sin \vartheta)^2 F \qquad\qquad F_2 = (r \sin \vartheta)^{-2} F$$

$$F = B[1 + (r d\vartheta/dr)^2]^{-1/2} \qquad\qquad \psi = b_\varphi r \sin \vartheta.$$

Here $\vartheta = \vartheta(r, r_0)$ is the equation of the magnetic field line, $B(r, r_0)$ is the value of the magnetic field along this line and r_0 is the field-line parameter, for example, the distance from the centre to its top. For the dipole field we have

$$\sin \vartheta = (r/r_0)^{1/2}$$

$$F = 2Mr^{-3}(1 - r/r_0)^{1/2}$$

where M is a dipole magnetic moment. According to the statement of the problem $\rho = \rho(r)$, i.e. all the coefficients in equation (1) depend only on r. This allows us to solve the problem avoiding any *a priori* assumptions on the form of the dependence $\rho(r)$.

Diagnostics are reduced to solving the inverse Sturm–Liuwille problem, i.e. ρ should be found with the help of the known spectrum of eigenvalues of ω. If we try an analytical solution we shall inevitably arrive at an approximate method. By replacing the variables

$$\psi = \rho^{1/4}(r \sin \vartheta)\phi \qquad\qquad \chi = 2\pi^{1/2} \int_{r_e}^{r} \frac{\sqrt{\rho}}{F} \, dr$$

we bring (1) to the normal form

$$\left[\eta - \frac{d^2}{d\chi^2} \right] \phi = \omega^2 \phi.$$

Here

$$\eta = 2(g'/g)^2 - g''/g$$

$$g = \rho^{1/4}(r \sin \vartheta).$$

The prime denotes differentiating over χ. The boundary condition will be written as

$$\phi'(0) = \phi'(\chi_0) = 0$$

where

$$\chi_0 = 4\pi^{1/2} \int_{r_e}^{r_0} \frac{\sqrt{\rho}}{F} \, dr.$$

The idea of bringing it to the normal form lies in the fact that the spectrum of the operator $-d^2/d\chi^2$ is known. So we may try to employ an ordinary perturbation theory procedure. Regarding the correction η to the unperturbed operator to be small, we obtain in the first approximation $\omega_n \chi_0 = \pi n$, $n = 1, 2, \ldots$ or

$$\frac{\pi^{1/2} n M}{2 r_0^{1/2} \omega_n(r_0)} = \int_{r_e}^{r_0} \frac{r^3 \sqrt{\rho(r)}}{\sqrt{r_0 - r}} dr. \tag{3}$$

We now consider the problem of how to restore the form of the function $\rho(r)$ using the dependence $\omega_n(r_0)$, which is assumed to be known. From the mathematical point of view the problem is brought to a solution of the integral equation (3), where $\rho(r)$ is regarded as an unknown function. This is the Abel equation whose solution has the form

$$\sqrt{\rho(r)} = \frac{Mn}{4\pi^{3/2} r^3} \int_{r_e}^{r} \left[\frac{dT_n}{dr_0} - \frac{T_n}{2r_0} \right] \frac{dr_0}{\sqrt{r - r_0}}$$

where $T_n = 2\pi/\omega_n$. To calculate ρ at the point r it is necessary to know the form of the function $T_n(r_0)$ over the interval from r_e to r.

This problem has only methodological importance, but if the magnetosphere of the Earth was not influenced by any external forces, except gravity, the plasma distribution would be spherically symmetric and the problem under consideration would have a geophysical meaning. (We should recall that in thermodynamic equilibrium the magnetic field does not influence the state of the substance.)

Exercise 9.3.2.

Provide a chemical analysis of the magnetospheric plasma using the data from a synchronous observation of Alfvén waves and whistlers.

Solution 9.3.2.

This problem, in contrast to the preceding one, may have practical importance for magnetospheric diagnostics. Of course, it cannot be solved in the whole volume and a series of conditions and restrictions should be imposed straight away.

First of all, a whistler must have a 'nose frequency', in order to use it for finding the parameter L of its trajectory in the magnetosphere. The second condition is obvious and we could almost go without mentioning it: the magnetic shell with parameter L must oscillate. But this is not enough. It is necessary to apply the observational means that allow us to choose the oscillation frequency of this shell out of the total spectrum of the magnetospheric oscillations. (This is just the situation in which to apply the MHD locator concept!) Finally, let us confine ourselves to an analysis within the framework of a simple hypothesis as regards the chemical composition of the plasma. Suppose it consists of electrons and ions of two types, namely, the protons H^+ of solar origin and single-charged ions of oxygen O^+ of ionospheric origin. The problem is to find the densities N_{H^+} and N_{O^+} of these ions.

Using the dispersion of the whistler we evaluate the electron number density N at the top of the trajectory. From the oscillation frequency of the magnetic shell we evaluate the plasma density ρ at the same place. Using the quasineutrality condition we find

$$N_{H^+} = (m_{O^+}N - \rho)/(m_{O^+} - m_{H^+})$$
$$N_{O^+} = (\rho - m_{H^+}N)/(m_{O^+} - m_{H^+}).$$

Exercise 9.3.3.

Define the energy and localization of resonant particles responsible for the excitation of Pc1 pulsations.

Solution 9.3.3.

According to the data on pulsations it is rather difficult to obtain complete information on the energetic particles in the magnetosphere. In fact, it is impossible to get any information on the energetic particles until they 'realize themselves' in the process of excitation of pulsations. But even in this favourable situation the hydromagnetic diagnostics provides data on the particles only in the limited range of variations of the parameter L and usually only at a narrow portion of the energetic spectrum, which corresponds to the observed frequency spectrum of pulsations.

However, the value of the information on the energetic particles in the magnetosphere is determined not only by the volume of the quantitative information. Sometimes the very fact of the appearance of specific pulsations, testifies to the injection of energetic particles into the magnetosphere and to the acceleration or abrupt change of the anisotropy of the particles available. Therefore, it is useful to develop the methods allowing us to estimate localization, average energy and the flow of particles responsible for the excitation of pulsations.

The solution of the problem is based on an analysis of the resonance condition (6.7). Let us take the generally accepted point of view, according to which Pc1 are excited as a result of the instability of energetic protons. The instability is due to the anisotropy of the proton distribution in the geomagnetic trap. The growth rate has a maximum at $\theta = 0$, so that with regard to the wave type we have $s = 1$. Then from (6.7) with regard to (4.5) it follows that

$$\mathcal{E}_\parallel/\mathcal{E}_m = (\Omega/\omega)^2(1 - \omega/\Omega)^3$$

where $\mathcal{E}_\parallel = mv_\parallel^2/2$ is the longitudinal energy of the resonant protons and $\mathcal{E}_m = B^2/8\pi N$ is the magnetic energy density per single particle. (For simplicity we consider plasma to consist of electrons and protons.)

To estimate $\mathcal{E}_\parallel$ we should know B and N in the generation region as well as the oscillation frequency ω. The position of the generation region (in fact, the parameter L of the magnetic shell) may then be found by the standard model of the geomagnetic field, proceeding from the assumption that the instability growth rate reaches a maximum at the equator of the Pc1 trajectory.

There are various other solutions to the problem. We present a method of estimating B using measurements of the jump of the carrier frequency Pc1 under sudden compressions and expansions of the magnetosphere. The frequency jump appears due to the fact that under compression, for example, the gyrofrequency Ω increases due to the enhancement of the geomagnetic field and radial drift of particles into the depths of the magnetosphere. In addition, the density of the background plasma and the energy of resonant particles changes.

The compressions of the magnetosphere under the action of the solar wind are accompanied by positive magnetic impulse SI^+ and SSC, and the expansions by negative impulses SI^- (sudden impulses and storm sudden commencements). We evaluate the frequency change $\Delta\omega$ at the given value of the magnetosphere deformation, characterized by the value of the magnetic impulse ΔH. (Here ΔH is the horizontal component of the geomagnetic field measured at the equatorial observatory.)

The unperturbed field is considered to be dipolar. We choose the simplest approximation of the field perturbation under the deformation of the magnetosphere. We shall consider the potential perturbation and retain in the Gaussian expansion of the perturbation only the first axial-symmetrical harmonic. Then the radial displacement of the particles close to the geomagnetic equator is

$$\Delta L \approx (\tfrac{1}{3}) L \Delta H / B.$$

The gyrofrequency of the shifted emitter is changed by the value

$$\Delta\Omega \simeq \tfrac{5}{3} \frac{e}{m_p c} \Delta H.$$

Similarly we estimate the change of the energy and the particle's pitch-angle (from the condition of conservation of the magnetic moment and the longitudinal invariant) and the change of plasma density (from the frozen-in condition). So we shall find the dependence $\Delta\omega$ on ΔH and ω/Ω, and by this dependence we shall find B and L as functions of ω, $\Delta\omega$ and ΔH. At $\omega \ll \Omega$ we get

$$B \simeq 2.17 \left(\frac{f}{\Delta f}\right) \Delta H$$

$$L \simeq 5.3 \left(\frac{\Delta f}{f} \frac{10^2}{\Delta H}\right)^{1/3}.$$

Here ΔH is expressed in nT and $f = \omega/2\pi$.

The values $\Delta f = f_2 - f_1$ and $f = (f_1 + f_2)/2$ are easily measured by analysing the Pc1 sonogram. Here f_1 is the carrier frequency before SI, and f_2

is the carrier frequency set after a time equal to two–three times the signal path across the magnetosphere after SI. The value ΔH is determined by a standard magnetogram. Figure 9.5 shows the dependence of Δf on ΔH constructed by the results of the analysis of the 12 Pc1 series during which $SI^{\pm}$ and SSC were observed. The dashed line designates the dependence $\Delta f = \chi \Delta H$ on the coefficient $\chi = 1.2 \times 10^{-2}$ Hz nT^{-1}, which is found by the method of least squares. The noticeable spread of the points in figure 9.5 is caused by the fact that χ depends on the value ω / Ω, which changes occasionally, rather than on the measurement errors. The theory gives $\chi \simeq 0.03(\omega / \Omega)$ at $\omega \ll \Omega$. Comparing this with the measured value we find that the ratio $\omega / \Omega \simeq 0.4$ is not small, at least in the events presented in figure 9.5. In these cases the presented formulae may be used for a rough estimate only and more accurate values of B and L should be calculated by the general formulae or by specially constructed diagrams.

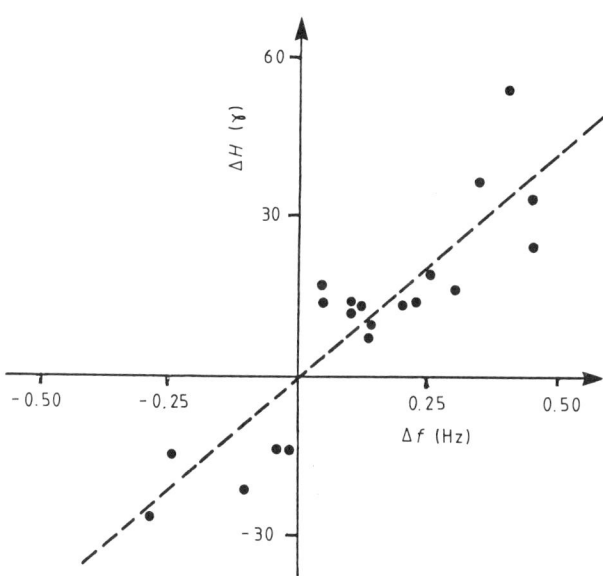

Figure 9.5. Dependence of the jump of the Pc1 carrier frequency on the SI magnetic impulse value.

Now we lack only N to estimate the energy. Knowing L, this value may be obtained by the period of repetition of signals $\tau(\omega)$. The typical result is $\mathcal{E} \simeq 20$ keV at $L \simeq 6$.

Exercise 9.3.4.

Estimate the azimuthal extent of the energetic particle injection region into the magnetosphere in the night sector at the beginning of a substorm according to the observational data on Pi1B and IPDP pulsations.

Solution 9.3.4.

The injection of energetic particles into the magnetosphere from the geomagnetic tail almost immediately excites the burst of noise Pi1B in the pre-midnight sector, and soon after excites pulsations of increasing frequency IPDP in the evening sector. Let $\Delta\varphi_0$ be the azimuthal extent of the injection region and $\Delta\varphi$ be the longitudinal distance between the maxima of intensity Pi1B and IPDP. The value $\Delta\varphi$ may be measured using the observation of pulsations at one observatory through averaging over many events. For a separate event, $\Delta\varphi$ may be measured using synoptic observations over the chain of observatories stretched along the geomagnetic parallel.

Let us show that if we take a number of more or less probable assumptions, then to an accuracy up to a coefficient of the order of unity $\Delta\varphi_0 = \Delta\varphi$.

The proof is based on the analysis of the formulae (6.11) and (6.74) for the instability growth rate, causing IPDP excitation. At $t > 0$ it gives a local growth rate of the increase of oscillations in the equatorial plane at the fixed magnetic shell. The increase occurs as a result of the Cherenkov instability of protons injected at the moment $t = 0$.

We shall choose the Gaussian distribution of particles over φ at $t = 0$

$$\Phi(\varphi) = (1/\sqrt{\pi}\Delta\varphi_0) \exp[-(\varphi/\Delta\varphi_0)^2].$$

Here the periodicity over φ is not taken into account. This is possible at small dimensions of the injection region ($\Delta\varphi_0 \ll 2\pi$) over time intervals shorter than the azimuthal drift period of the main bulk of injected particles ($t \ll 2\pi/\dot\varphi_\perp$).

The formula (6.74) with respect to $\Phi(\varphi)$ contains an integral

$$I = \int_0^\infty \xi J_1^2(a\sqrt{\xi})e^{-f(\xi)}\,d\xi$$

where

$$f = \xi + (\tau/\Delta\varphi_0)^2\left[(\varphi/\tau) - \zeta - \xi\right]^2.$$

Here we introduced the dimensionless variables $\xi = v_\perp^2/w^2$, $\zeta = v_z^2/w^2$, $\tau = t\dot\varphi_\perp$ and $a = k_\perp w/\Omega$. The form of the subintegral expression prompts the application of the steepest descent method. From the condition $f'(\xi_0) = 0$ we find the pass point

$$\xi_0 = (\varphi/\tau) - \zeta - \Delta\varphi_0^2/2\tau^2$$

and the values

$$f_0 = \xi_0 + \Delta\varphi_0^2/4\tau^2 \qquad f_0'' = 2\tau^2/\Delta\varphi_0^2.$$

The approximate value of the integral is

$$I \approx (2\pi/f_0'')^{1/2}\xi_0 J_1^2(a\sqrt{\xi_0})e^{-f_0}.$$

Substituting this into (6.11) and (6.74) we find the growth rate which depends on ω and k_z. Let us find the maximum γ by k_z and ω

$$\gamma(\tau,\varphi) \simeq \Omega\frac{N'w}{Nc_A}\left(\frac{\varphi}{\tau} - \frac{\Delta\varphi_0^2}{2\tau^2}\right)\frac{1}{\tau}\exp\left[-\left(\frac{\varphi}{\tau} - \frac{\Delta\varphi_0^2}{4\tau^2}\right)\right].$$

The absolute maximum $\gamma_m \simeq (N'w/Nc_A)(\Omega/\Delta\varphi_0)$ is reached at $\varphi_m \simeq \sqrt{2}\Delta\varphi_0$ and $\tau_m \simeq \Delta\varphi_0/\sqrt{2}$. Thus we have

$$\Delta\varphi_0 \simeq \varphi_m/\sqrt{2}.$$

Now the formal analysis is over. We want to supply the result with meaningful content in terms of the morphology of geomagnetic pulsations, having identified φ_m and $\Delta\varphi$. For this we have to introduce a number of assumptions. We suppose that the azimuths of the intensity maxima Pi1B and IPDP are equal to the azimuth of the injection region centre ($\varphi = 0$) and the azimuth φ_m, by which the maximum γ is reached. The first of these assumptions has a reliable empirical basis, but the second one still requires substantiation in the framework of a more complete model that takes account of nonlocality of the excitation, nonlinear effects, etc. Finally, assuming $\Delta\varphi_0 \simeq \varphi_m/\sqrt{2}$ we implicitly introduce the assumption that the dimension of the injection region does not correlate with the position of its centre.

The most probable times for the appearance of IPDP and Pi1B pulsations are at 21 LT and 23 LT. Hence it follows that $\Delta\varphi_0 \simeq 20°$.

Exercise 9.3.5.

Find the arrival direction of the Pc1 signals, propagating in the ionospheric waveguide in relation to the observation point.

Solution 9.3.5.

First of all we should measure the surface impedance of the Earth ζ and its gradient $\nabla\zeta$ at the observation point within the Pc1 range using the methods of geoelectrics, and register b_z and b_t within the same range. Then we avail ourselves of the formula (9.23). According to the condition of the problem, the signals arrive at the observation point propagating in the ionospheric waveguide. Locally the waveguide may be considered to be horizontally stratified. This allows us to substitute the operator ∇ in (9.23) for ik_t, where $k_t = (k_x, k_y)$ is the local wave vector of horizontal propagation. We suppose k_x, k_y to be unknown and solve the equation

$$Ak_x + Bk_y + C = 0$$

with the complex coefficients

$$A = \zeta b_x \qquad B = \zeta b_y \qquad C = \frac{\omega}{c} b_z - ib_t \cdot \nabla \zeta$$

which are known from the experiment. The bearing, as the angle between the magnetic meridian (the x axis) and the direction of propagation, is $\vartheta = \tan^{-1}(k_y/k_x)$.

The MHD direction-finder enables us to find the phase velocity ω/k_t of the horizontal propagation of Pc1. Then, using the fluctuations of the phase velocity and the angle of arrival ϑ we may judge the ionospheric inhomogeneities along the trace of propagation. Finally, two MHD direction-finders, distributed along the longitude, allow us to determine the coordinates of the end of the longitudinal magnetospheric waveguide and, consequently, the parameter L of the trajectory of Pc1 in the magnetosphere. In fact we shall have to introduce a correction for the lateral refraction of rays in the ionospheric waveguide.

We have discussed the last (fourth) example where (9.9) is used to solve the problem of hydromagnetic diagnostics and geoelectric reconnaissance. Depending on what values in (9.9) are considered unknown, this formula either reduces to the impedance equation, or serves as the basis for the arrangement of the MHD locator, direction-finder or gradientometer.

9.4 Interplanetary medium diagnostics

9.4.1 Interplanetary magnetic field

In the early 1970s the discovery of the extramagnetospheric origin of Pc3 stimulated the search for methods of remote control of the state of an interplanetary medium in front of the magnetosphere. At present such methods may be of auxiliary importance only, since the control is much more reliable when carried out by means of satellites. The application of ground methods allows us, for example, to fill in the blanks in the known King catalogue, determine the variations of parameters for 10–15 years before carrying out regular satellite observations, etc.

In the light of the information presented in section 6.5, it is natural to try to develop a method for diagnostics of the interplanetary magnetic field B using the data on the oscillation frequency f. If we simply use the formula $B = f/g$, then the results will not be accurate. The accuracy of the diagnostics increases if we use f in combination with the K_p-index. According to these two values a so-called B-index is formed

$$B_* = af + bK_p + c. \qquad (9.34)$$

The coefficients in (9.34) are calculated using the data of simultaneous observations of Pc3 on the ground and the interplanetary magnetic field on board the satellite. Generally speaking, they depend on the geomagnetic latitude of the observation point, the period of averaging, local time, and it may also depend on the season and on the phase of the 11 year cycle of solar activity. When averaged over an hour in the morning and the midday hours at latitude $53°$ we have $a = 0.15$, $b = 0.16$, $c = 0.7$, if B_* is measured in nT (or gamma) and f in millihertz. The correlation coefficients between f, B and K_p are

$$r(f, B) = 0.76 \pm 0.02$$
$$r(B, K_p) = 0.41 \pm 0.04$$
$$r(f, K_p) = 0.48 \pm 0.04.$$

The evaluation of the magnitude of the interplanetary magnetic field by the B-index gives not bad results. As an example let us consider the characteristics of B_* variation when the Earth crosses the boundaries between the interplanetary magnetic field sectors. Figure 9.6 shows the variations of the B-index (left scale) and the frequency of Pc3 (right scale) when passing from the sector with negative polarity of the interplanetary field into the sector with positive polarity. At the sectors' boundary (vertical line) the B-index is minimum and 2–3 days later reaches a maximum. This picture agrees with the data of direct measurements.

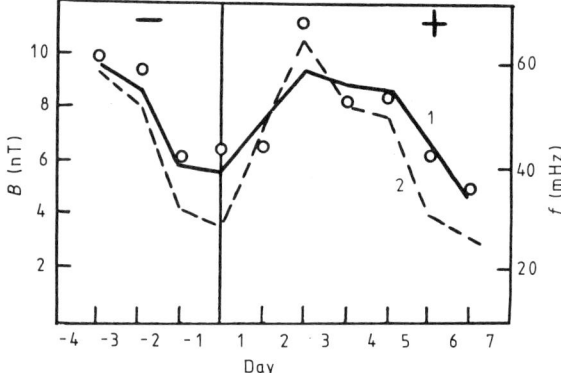

Figure 9.6. Variation of the B-index (1) and variation of Pc3 frequency (2) in the time interval November 30–December 9 1967. Circles represent the results of direct measurements. Days are counted from the moment at which the boundary between the sectors of the interplanetary magnetic field is passed.

The method of diagnostics of B using the frequency Pc3 does not possess high accuracy, but it may turn out to be useful when analysing the interplanetary situation, when for some reason or other direct measurements are absent.

The theory of Pc3 generation indicates the dependence of the oscillation amplitude on the orientation of the interplanetary magnetic field. We fail to determine this dependence in theory, since the amplitude depends on many factors and they hardly come under analysis. Nevertheless, it seems probable that the spectrum of the Pc3 amplitude modulation contains information on small-scale inhomogeneities of the interplanetary magnetic field.

The analysis of the ground observations testifies to the fact that in the frequency range 10^{-3}–10^{-4} Hz the spectrum of the amplitude envelope of Pc3 is a power spectrum with the index $n = 1.5$–2. The spectrum of magnetic inhomogeneities in the interplanetary medium has a similar form (according to *in situ* measurements on board satellites). This gives some hope that it will be possible to obatin information on the inhomogeneities of interplanetary magnetic field from the data on the Pc3 amplitude modulation spectrum.

A rough estimation of magnetic inhomogeneities is based on the analysis of deep Pc3 amplitude fadings. Average values of the duration of fadings τ_1 and the time interval τ_2 between the fadings allow us to evaluate the typical dimension of inhomogeneities $l_1 = V\tau_1$, the distance between the inhomogeneities $l_2 = V\tau_2$ and the parameter $\xi = \tau_2/\tau_1$, if the wind velocity V is known. (For the evaluation of V from the Pc3 data see below.) The characteristic values $\tau_1 = 1.5 \times 10^3$ s, $\tau_2 = 4 \times 10^3$ s at $V = 4 \times 10^7$ cm s^{-1} give $l_1 = 6 \times 10^{10}$ cm, $l_2 = 1.6 \times 10^{11}$ cm and $\xi = 2.67$.

9.4.2 Solar wind velocity

Experiment shows that the amplitude of the daytime permanent pulsations of the magnetosphere in the Pc3-4 frequency range as a rule increases with increase of the solar wind velocity. This feature of pulsations stimulates the search for remote monitoring of the solar wind by means of observation of oscillation amplitude variations on the ground in the Pc3-4 frequency range. As in the case of diagnostics of the interplanetary magnetic field, the ground methods cannot compete with direct measurements of the solar wind velocity on board the satellites. They should be considered as a useful and comparatively inexpensive auxiliary means of control of the state of the near-Earth environment.

Whereas when diagnosing the magnetic field B from the data on the oscillation frequency f the theory points out the character of the relation between f and B, there are no such indications for the given case. We fail to establish the dependence of the oscillation amplitude A on the solar wind velocity V theoretically. The relation between A and V is established empirically and on this foundation one or another possibility of diagnosing V by A is sought.

So, first we must find the dependence of A on V. Usually this dependence is sought in the class of linear models of the type

$$A = a + bV \tag{9.35}$$

or in the class of exponential models

$$\ln A = \alpha + \beta V. \tag{9.36}$$

Here A is the amplitude of pulsations, averaged over one or another interval of time, or the spectral density, or some other appropriate characteristics of oscillations intensity. The parameters a, b or α, β are sought by means of retrospective analysis of ground and satellite measurements.

Then one has to study the 'behaviour' of the model in terms of the local time, the seasonal effects, etc. Based on this experiment we choose the most informative interval of the local time, the spectral range and the period of averaging, and also determine the accuracy of measuring V, on which we may rely. Sometimes, besides A, supplementary diagnostic features are employed, for example, the K_p-index. In this case methods of multifactor dispersion analysis, the equations of multiple regression, etc, are applied when posing a diagnosis. Figure 9.7 illustrates the situation.

We shall not dwell on these details but refer to the relations (9.35) and (9.36). At a first glance they seem quite appropriate to describe the empirical connection of A with V. However, it is known from experiment that $a < 0$ in (9.35). But this is certainly inadmissible, as it leads to a negative value of A at $V = 0$. The model (9.36) is also inadequate (see below).

The pulsation amplitude must certainly be zero when the solar wind velocity equals zero. It is quite easy to construct a model, satisfying this requirement

$$\frac{A}{A_0} = \frac{V}{V_0} \exp\left(-\frac{V_0}{V}\right). \tag{9.37}$$

Here A_0 and V_0 are two phenomenological parameters, analogous to a and b in (9.35) and α and β in (9.36). Let us present arguments in favour of (9.37).

First we suppose $a = -A_0$ and $b = A_0/V_0$ and rewrite the linear relation (9.35) in the form

$$\frac{A}{A_0} = \frac{V}{V_0}\left(1 - \frac{V_0}{V}\right). \tag{9.38}$$

We shall try to generalize (9.38) with regard to the natural requirement $A = 0$ at $V = 0$. Let us present a generalized model in the form

$$\frac{A}{A_0} = \frac{V}{V_0} \Phi\left(\frac{V_0}{V}\right)$$

where the function $\Phi(\xi)$ has the following features: (1) $\Phi(\xi) \geq 0$ for all ξ; (2) $F(\xi) = 1 - \xi + \ldots$ for $\xi \to 0$ in accordance with (9.38); (3) $\Phi(\xi)$ is finite when $\xi \to \infty$. The simple function that satisfies all these conditions, is $\Phi = \exp(-\xi)$. This leads to (9.37).

Compare (9.37) with the exponential model (9.36). We can put $\alpha = \ln A_0$ and $\beta = 1/V_0$ and rewrite (9.36) in the form

$$\frac{A}{A_0} = \exp\left(\frac{V}{V_0}\right).$$

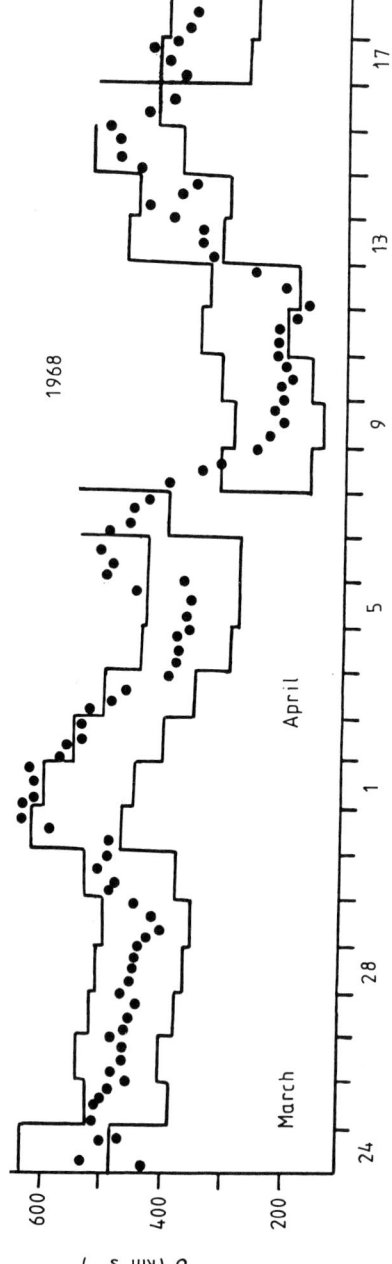

Figure 9.7. Broken line—interval estimation of the solar wind velocity on the basis of data on geomagnetic pulsations; points—results of direct observation. After Potapov and Polyushkina (1982).

One can see that the exponential model (9.26) is totally incompatible with (9.37).

Now we shall try to find physical arguments in favour of the model (9.37). In order to specify the subject, we should refer to Pc4 with respect to which the concept of the generation at the magnetopause due to Kelvin–Helmholtz instability does not seem improbable. Let us assume a simple hypothesis that the amplitude A_m of the surface waves at the magnetopause is proportional to the solar wind velocity, i.e.

$$A_m \propto V. \tag{9.39}$$

The attenuation of pulsations in the magnetosphere is proportional to the factor

$$\exp[-k(R_m - L)] \tag{9.40}$$

where R_m is the distance from the Earth's centre to the magnetopause, L is the parameter of the magnetic shell and k is the wave number of the surface waves on the magnetopause. It is known that

$$k \simeq \omega/V \tag{9.41}$$

for the surface waves. Combining (9.39)–(9.41) and knowing that $\omega \simeq c_A/L$ for the oscillations of the magnetic shell we get (9.37) and

$$V_0 \simeq c_A(R_m - L)/L \tag{9.42}$$

where c_A is the effective Alfvén velocity in the magnetosphere.

These qualitative assumptions support the idea that the model (9.37) is a fairly reasonable imitation of A–V dependence. The approximate values of the model parameters for the observational conditions during the daytime at mid-latitudes are $A_0 \simeq 2.5 \, \gamma$ and $V_0 \simeq 380 \text{ km s}^{-1}$.

The MHD locator, described in the previous section, may be used to improve the method. Let the observational point be situated to the south of the magnetic shell (for the northern hemisphere). Then the observed amplitude $|b_x|$ will be weakened $q \equiv [1 + (x_R/\Delta)^2]^{1/2}$ times as compared with the amplitude in the resonance (see (9.26)). Therefore, it is reasonable to substitute the modified amplitude $A = q|b_x|$ into (9.37).

9.4.3 On the applications of hydromagnetic diagnostics

Hydromagnetic diagnostics have an original and fairly interesting philosophy. But why is the diagnosis needed?

Some papers on hydromagnetic diagnostics give the impression that the diagnosis is considered to be an end in itself: diagnosis for the sake of diagnosis! From the point of view of pure science this sounds tempting. Indeed, science implies that we think first of all, and the diagnostics pose sharply nonstandard, interesting and difficult questions. But if we proceed from our definition of hydromagnetic diagnostics, finding the diagnosis is not purely research work.

Out of the context of the user's 'real world', diagnostics turns into a meaningless exercise. In order to be of benefit it must be relevant to the everyday needs of the scientist's stimulating work. If interest in the development of diagnostics had been poorly motivated in the past this science would not have existed for so long.

Thus, the purposes of hydromagnetic diagnostics are set from outside. First, by the problems of geophysics itself, and second, by the requirements of the applied sciences. Let us consider these two problems separately.

In geomagnetism diagnostics are of undoubted benefit. First of all they are of benefit to the physics of geomagnetic pulsations. This is clear since the attempt to extract original information on the magnetosphere compels the researcher to account for his or her ideas on the origin of wave phenomena critically. In the monograph by Guglielmi and Troitskaya (1973) we read 'In the course of diagnostics ... critical verification of the established notions on the origin of various kinds of pulsations are carried out. Essentially, the hydromagnetic diagnostics may be considered to be the most active means for the verification of specific models of pulsations'.

The years since then have proven the correctness of these words. The method of tests and criteria to settle the alternatives, which have been brought into obligatory common use when posing the diagnosis and when studying the nature of geomagnetic pulsations, provided interesting results. During this development, for example, the extramagnetospheric origin of Pc3 has been discovered, the probability of MHD wave penetration from the interplanetary medium into the magnetosphere has been experimentally proved, convective and injection varieties of IPDP have been found and their association with the large-scale processes in the magnetosphere during magnetic storms have been proved, an approach to the understanding of auto-wave phenomena in the magnetosphere has also been contemplated as well as a great deal more.

Hydromagnetic diagnostics can be successfully applied to other branches of geomagnetism too. For example, Pi2 are used for an exact countdown to the beginning of a substorm. Moreover, observations of geomagnetic pulsations are employed when studying modulation phenomena in the radiation belts and in short-term forecasting of magnetic storms, etc. The list of examples of this kind can be extended.

The case is quite otherwise when we proceed to purely applied aspects of hydromagnetic diagnostics. True, in geoelectrics data on the horizontal structure of the wave field of pulsations are used to improve the methods of induction sounding of the Earth's crust with a view to search for minerals, but this is, perhaps, the only example of a successful practical application of diagnostics, and in this case too the research is mainly of a methodical character and has not been introduced into the practice of geological prospecting.

Therefore, hydromagnetic diagnostics have practically no other users apart from cosmophysicists. Those who may may need magnetospheric diagnosis, can hardly image the potentialities of hydromagnetic diagnostics within the context of their business. A potential user should bear in mind that the method of hydromagnetic diagnostics is not universal: it is capable of giving some data and incapable of giving others.

It will be imprudent to require from the hydromagnetic diagnostics what it cannot give. But in reality the method is of no interest to the user. The result is the only important thing for the user, i.e. the specific diagnosis, and if the user can attain this object in some other way that seems more obvious, for example, by launching a satellite into orbit, then that user will prefer that method. However, in practice, situations that demand taking urgent measures may occur. In these cases it is extremely important to be able to evaluate, at least approximately, the state and evolution tendency of the near-Earth medium correctly, in a proper time and with reasonable expenditure, and this is what may be achieved by watching geomagnetic pulsations. In this connection we shall mention a typical example, described in the literature on heliobiology more than once: for the supervisor of the medical institution who believes in unfavorable influence of space factors on his or her patients it is important to know the state of the near-Earth medium in order not to miss signs of impending danger[8].

To conclude this chapter on hydromagnetic diagnostics we present a table which lists the parameters of the near-Earth medium that may be estimated from the data on geomagnetic pulsations to some degree of reliability. Analogous tables have been compiled before, but in comparison with them this table contains more parameters and more detailed indications of the methods of diagnosis. Nevertheless, it should be admitted that no major changes in the tendencies and methods of hydromagnetic diagnostics, which were established in the mid-1970s, have occurred since then. In our opinion a kind of saturation has been reached in this domain and although the search for new methods and means should not be stopped, we think it would be advisable to make a major effort to perfect reliable diagnostics procedures and introduce them into the practice of geophysical observations.

Table 9.1. Diagnostics of the magnetosphere and interplanetary medium

Parameter or process	Type of pulsations	Way of diagnosing
Magnitude of interplanetary magnetic field	Pc3	By the dependence of carrier frequency on the magnitude of interplanetary magnetic field
Solar wind velocity	Pc2-4	By the dependence of amplitude on the solar wind velocity

[8] We have not had the opportunity here to discuss the problem in detail (e.g. Vladimirsky *et al* 1994). We only note that we do not doubt the trustworthiness of the reports on the influence of geomagnetic perturbations on various systems of organisms, for example changes in blood system, changes in cutaneous potentials, changes of reflex levels, etc.

Table 9.1. (*continued*)

Parameter or process	Type of pulsations	Way of diagnosing
Inhomogeneities of the solar wind	Pc2-4	By the analysis of amplitude and frequency modulation
Plasma density at the top of the field line and the energy of resonance protons	Pc1	By dispersion analysis of the wave packets
The plasma density at the tops of the field lines and the distribution of plasma along the field lines	Pc4-5	By the latitudinal dependence of frequency and by the nonequidistance of the spectrum
Chemical composition of plasma	Pc1-2	By the thin structure of dynamic spectra
Position of the dayside cusp	Ipcl	By the latitudinal dependence of the amplitude
Beginning of the substorm and the position of structural elements of its current system	Pi1B Pi2	By the forefront of the impulse and by the spatial distribution of the phase and polarization
Position of plasma-pause, layers and fibres of plasma	Pc1	By polarization and phase bearing of the waveguide end
	Pc3-4	By latitudinal profile of the amplitude
Small-scale inhomogeneities of plasma	Pc1	By fluctuations of the signal repetition period

Table 9.1. (*continued*)

Parameter or process	Type of pulsations	Way of diagnosing
Electric field of convection	Pc1	By the nonstationarity of the spectrum
Azimuthal extent of the injection region	Pi1B	By the azimuthal dependence of the amplitude
Parameters of the cloud of injected protons	IPDP	By the velocity of the 'westward drift' frequency
Intensity of magneto-spheric convection in the morning sector	PilC	By the velocity of displacement of structural elements of dynamic spectrum eastwards
Duration of the magne-tic storm	Pc1-4 Pi2	By the change of the oscillation regime

Bibliography

For section 9.1

The research work on hydromagnetic diagnostics at first appeared to satisfy the needs of geomagnetism. Although the methods and techniques of approaching such problems were known before, systematic wide application of hydromagnetic diagnostics did not start until the 1960s. The monographs by Guglielmi and Troitskaya (1973), Nishida (1978) and the reviews by Troitskaya and Guglielmi (1967, 1970), Aubry (1970) and Guglielmi (1974) contain extensive bibliographies relevant to the given subject. The idea of commonality of the problems of hydromagnetic diagnostics and geoelectric prospecting was suggested in the review by Guglielmi (1989).

For section 9.2

The method of magnetotelluric sounding of the Earth's crust was invented by L Cagniard and A Tikhonov independently in 1950. The modern state of the problem may be studied in the monographs by Rokityansky (1982), Wait (1982) and Berdichevsky and Zhdanov (1984). Chetaev (1970) and co-workers developed the original approach to the problem of MTS (see also Chetaev *et al* (1975), Chetaev (1975) and Savin *et al* (1984)).

The Leontovich boundary condition (1948) was known to the specialists from the propagation of radio waves in the 1930s, but it was published only after World War II. Rytov (1940), with reference to Leontovich, derived the approximate boundary conditions on the surface of a good conductor, using the asymptotic theory of perturbations. Until recently Rytov's theory was unknown in geoelectromagnetism. Guglielmi (1984, 1989) took notice of it and stressed that the results of Rytov's theory (1940) may be usefully employed in geoelectric prospecting, carried out by means of the method of magnetotelluric sounding.

The gradient method of inductive sounding of the Earth's crust and the upper mantle, based on formula (9.22) was developed by Berdichevsky *et al* (1969), Schmucker (1970), Kuckes (1973), Lilley and Sloan (1976), Jones (1980, 1981), Rokityansky (1982) and Pajunpää (1988). Instead Guglielmi (1989a) suggests the equation (9.21). The observational materials gained during field-work may be reinterpreted based on equation (9.21) without carrying out supplementary measurements.

The Guglielmi *et al* (1989c) method of determining $\nabla \zeta$ by the observational data of the Shumann resonances is similar to the methods known as Parkinson plane, Wilhelm ellipsoid, Wiese vector and others. An excellent review and comparative analysis of these and other known methods were presented by Gregori and Lanzerotti (1980). A distinctive peculiarity of the method of defining $\nabla \zeta$ is that it is based on the asymptotic theory of the skin effect. In this sense it, first, allows generalizations and, secondly, does not take us out of the limits of phenomenological electrodynamics.

Other methods operate with supplementary geometric objects supposedly reflecting the internal structure of the Earth, i.e. they vaguely imply certain interpretations of the measurements. The use of the method of defining $\nabla \zeta$ leads to an interpretation problem, but this problem can be dealt with as a problem independent of the measurements. The measurements, on the other hand, are carried out completely and in a unified way in terms of surface impedance.

The beginning of research into inductive electromagnetic phenomena in thin conducting films goes back to the paper of Levi-Città, published in 1902 (see the monograph by Bateman (1955)). Film modelling was introduced into geoelectromagnetism by Sheinman (1947) and Price (1949). The application of the so-called Price equation that discusses the horizontal inhomogeneity of the film turned out to be particularly successful. Many publications are dedicated to the generalizations and, especially, to applications of the film model. One of the aspects of the problem is associated with the necessity to take into account the influence of thin conducting upper layers (seas, oceans, low-Ohm sections of the sedimentary cover) when putting the programme of deep inductive sounding of the Earth into practice (Fainberg and Zinger 1981).

The experimental and theoretical associations between the conductivity and deformation of rocks are discussed by Parkhomenko (1967), Barsukov and Sorokin (1973), Bolt (1978), Kasahara (1981), Chelidze (1984) and Mogi (1985).

For section 9.3

The idea of the MHD locator was proposed by Guglielmi (1989, 1992b). It appeared during the polemics with Baransky *et al* (1985), who suggested a gradient method of diagnostics. The data of field observations to calculate x_R and construct figure 9.2 were generously granted by Professor D Chetaev.

The inversion of polarization of oscillations when crossing the resonance line was observed by Lanzerotti (1976). This property is the basis of the polarization method of diagnostics.

Obajashi (1958) was the first to pay attention to the possibility of determining plasma density in the magnetosphere by latitudinal dependence of the period of geomagnetic pulsations. The oscillation spectrum was calculated in the WKB approximation and the radial profile of plasma density was sought in the class of functions

$$\rho(r) = \alpha \exp(\beta/r)$$

where α and β are the parameters to be determined (see also Kitamura 1965). Meanwhile when the problem is thus posed the profile $\rho(r)$ may be found without any *a priori* suppositions on its form (Troitskaya and Guglielmi 1967). With respect to methodology, it is an interesting and almost unique example of the problem of hydromagnetic diagnostics being solved exactly like an inverse problem.

However, the assumption of spherical symmetry, i.e. of the fact that ρ depends only on r, is too strong an idealization. It has had to be rejected. At the same time the axially symmetric model, although not quite reproducing reality, is considered admissible and is usually employed when diagnosing (Jacobs 1970).

Extensive literature is dedicated to the results of experimental research on a latitudinal dependence of the spectrum of magnetospheric oscillations. We present a reference to the recent research carried out by Takahashi *et al* (1990b) with the help of the *AMPTE* satellite. It turned out that at daytime hours there are no sharp jumps of resonant frequency of Pc3-4 within the interval $2 < L < 6$. From this observation the authors drew a conclusion about the absence of sharp differences in the plasma density when crossing the plasmapause during daylight hours. This gives us an example of the application of hydromagnetic spectroscopy for the qualitative study of the plasma in the magnetosphere. We mention one more important result obtained by Takahashi *et al* (1990b) which refers to the problem of hydromagnetic spectroscopy. It turned out that the oscillations of the magnetic field have quasi-toroidal structure implying that they are polarized mainly in the azimuthal direction (during daylight hours). This property is traced from high magnetic shells up to $L < 3$, which gives us grounds to count on the effectiveness of the operation of the MHD locator over a wide interval of geomagnetic latitudes. Lanzerotti *et al* (1974) demonstrated the efficiency of the meridian chain of magnetometers when following displacements of the plasmapause at rapid changes of the geomagnetic situation.

The method of diagnosing plasma concentration by means of dispersion analysis of Pc1 was proposed by Watanabe (1965), Dowden and Emery (1965) and developed in the works by Liemohn *et al* (1967), Kenney *et al* (1968), Feygin *et al* (1970a), Troitskaya and Guglielmi (1970) and Gendrin *et al* (1971).

The history of low-frequency wave diagnostics of the magnetosphere begins with the classical work by Storey (1953), where a correct evaluation of the concentration of electrons at high altitudes using the data on whistlers was made for the first time. The discovery of 'nose' whistlers, made by Helliwell *et al* (1956), expanded the potential of the dispersion method, and it allowed Carpenter (1963) to discover plasmapause (the so-called Carpenter's 'knee').

When composing exercise 9.3.2 we availed ourselves of the idea of the concentration of electrons by the dispersion of the whistlers. The diagnostic potentialities of simultaneous observation of geomagnetic pulsations and whistlers was demonstrated by Webb *et al* (1977). However, the authors did not attempt to carry out chemical analysis.

Let us point out a number of works of our choice, where other tendencies of diagnosing have been developed.

The forefront of the burst of pulsations Pi1B-Pi2 warns of the beginning of the magnetospheric substorm (Samson and Harrold 1985). Bol'shakova and Troitskaya (1977) proposed to register the daily variation of long-period pulsations of Ipcl (interval of continuous pulsations of long period) over the meridional chain of high-latitudinal observatories to carry out the monitoring of the day polar cusp.

When tested, the method proved its high efficiency. Fraser (1968) and Offen (1972) suggested that a large-scale electric field in the magnetosphere be estimated by nonstationarity of the carrier frequency of Pc1 (see also Guglielmi 1974). Finally, Kalisher and Polyakov (1984) discussed the possibility of evaluating small-scale inhomogeneities of the magnetosphere by means of measuring fluctuations of the repetition period of Pc1.

For section 9.4

Guglielmi *et al* (1973) applied the method of multifactor dispersion analysis to study the correlation of geomagnetic pulsations with the parameters of the interplanetary medium in the vicinity of the Earth's orbit. The analysis allows us to estimate the degree of influence of the factors and their combinations on the 'response' of the system and allows us to clear up the authenticity of the influence if it has been discovered. It was established that the carrier frequency f of oscillations in the Pc3 frequency range depends most of all on the value B of the interplanetary magnetic field in the front of the magnetosphere. Taking all this into account Guglielmi and Bol'shakova (1973) suggested using the information on the frequency f to diagnose B. Russell and Fleming (1976) supported and developed the idea despite some criticism. Further development is associated with the solution of the problem of a reliable selection of pulsations of extramagnetospheric origin. Plyasova-Bakounina and Münch (1991) proposed a method of monitoring B notable for the fact that the observation of Pc3 was carried out simultaneously at two observatories that were situated at one latitude but were $\geq 20°$ apart in longitude. This allows one to filter out interference of magnetospheric origin, since it has a comparatively small radius of coherence.

The method of evaluating parameters of magnetic inhomogeneities in the solar wind by means of amplitude modulations of Pc3 was suggested by Troitskaya and Guglielmi (1969). It was timely in connection with the problem of diffusion of solar cosmic rays in the interplanetary magnetic field and in connection with the dimensions of the solar system region that forms the modulation of galactic cosmic rays. At present there is no longer any need for the ground data on the interplanetary medium to solve these problems. The stress is laid on the search and physical explanation of the association between the steepness of the spectrum of geomagnetic pulsations and the steepness of the spectrum of magnetic fluctuations in the interplanetary medium (De Lauretis *et al* 1991).

Snyder *et al* (1963) suggested that the average daily solar wind velocity V be estimated by the data on $\sum K_p$ using the empirical formula of the form

$$V \text{ (km s}^{-1}) = 8.44 \sum K_p + 330.$$

On the other hand, at the dawn of research on geomagnetic pulsations a positive correlation between the intensity of pulsations A and the K_p-index was established. It encouraged the search for the relationship between A and V and the attempt to construct on its basis a ground control system of the solar wind. Saito (1964), Vinogradov and Parkhomov (1975), Singer *et al* (1977), Greenstadt *et al* (1979), Wolfe (1980), Wolfe *et al* (1980, 1989), Takahashi *et al* (1981), Verö *et al* (1985), Miyake *et al* (1987), Slawinski *et al* (1988), Lanzerotti *et al* (1989), Kalisher and Rusakova (1990), Olson *et al* (1991) and Morrison (1991) all investigated the A–V relationship in the Pc3-4 range. This relationship is studied by observations not only at mid-latitudes (Vinogradov and Parkhomov, 1975), but also at the latitudes of northern and southern polar cusps (Wolfe *et al* 1989, Olson *et al* 1991), as well as in the magnetosphere over the geostationary orbit (Takahashi *et al* 1981). The coefficient of correlation between A and V is rather large. Wolfe (1980) reports the coefficient of correlation $r = 0.74$ at six-hour averaging of the data within the range of the period of pulsations $30.7-60$ s. When calculating r the model (9.35) was employed. Within the framework of the same model Olson *et al* (1991) found $r = 0.83$ under the condition that the time intervals where $V > 400$ km s^{-1} are chosen. This is a reasonable restriction, since at $V \rightarrow 0$ the model (9.35) provides an unacceptable result. Kalisher and Rusakova (1990) discovered the dependence of r on the local time within the framework of the model (9.35); at mid-latitudes the correlation is maximum within the intervals 09–11 LT and 13–15 LT. The authors recommend inferring V in these intervals of the local time precisely. They also propose an alternative model

$$V = \alpha + \beta\sqrt{A}$$

where $\alpha = 274$ km s^{-1} and $\beta = 153$ km s^{-1} $\gamma^{-1/2}$ at latitude 53°. The model (9.37) was proposed by Guglielmi and Potapov (1994).

There are lively discussions about the 'A–V problem' in the literature (see, for example, the discussion between Lanzerotti and Wolfe 1988 and Yumoto and Miyake 1988).

When investigating the relationship of the properties of pulsations with the parameters of the interplanetary medium the catalogue by King (1989) was widely used.

Chapter 10

Epilogue: geoelectromagnetic waves and man

From our, inevitably subjective, point of view geoelectromagnetic waves are fascinating primarily as a source of diverse and difficult problems. Somebody else may be attracted more by the beauty and singularity of wave structures or the benefit that may be derived from magnetotelluric sounding. In any case scarcely anyone remains indifferent to this phenomenon of Nature after getting to know it better.

While analysing our own sensations and comparing notes with our colleagues, we asked ourselves a question: can the fascination of geoelectromagnetic waves be connected with their particular property, namely, with the specific structure of spatio-temporal scales of electromagnetic field? Invisible and inaudible, can they affect our subconsciousness, like the rustling of foliage in a dense forest or ocean waves or the glimmering of stars. Every living thing on Earth has been subject to the constant, though small, influence of geoelectromagnetic oscillations for millions of years. Isn't this why there is a surprisingly close accordance between the frequencies of geoelectromagnetic oscillations and some of the frequencies inherent in biological organisms? Such questions arise from time to time. Everybody may have their own opinion as regards this but nobody has a definitive answer yet ...

The relations between humans and geoelectromagnetic waves have also more prosaic but not less important aspects. Lanzerotti (1979, 1983) collected and analysed the information on the influence of geomagnetic perturbations on technological systems, especially on extended communications. In particular he reported a wonderful observation, made by telegraphists back in the middle of the last century. During heavy magnetic storms the telegraph communication was either put out of action or started operating even when switched off from the feed source. Moreover, spontaneous modulation of the regime of the telegraph correlated with pulsating forms of aurora within the range 3–30 mHz.

Variable geoelectromagnetic field, affecting long conductors such as rails, air pipes, gas pipe-lines, oil pipe-lines, braided long cables, air wires, etc, may cause even more menacing consequences. It is probable that under extreme conditions electromagnetic induction or skidding of high potential are able to cause Joule heating or even spark-formation at points of poor contact, for example in flanges, at loose bolt junctions or in corroded sections.

All these effects can be successfully fought by applying special methods of technical protection. We would have liked to end this book on an upbeat note, but unfortunately the theme takes a turn that reveals a rather gloomy perspective. We are referring to the appalling pressure that Nature itself is subjected to from the industrial and military activities of humans. Against the background of general pollution of the environment, the, at present, almost unnoticeable modification of the spectrum of geoelectromagnetic waves may not seem worth anyone's attention. But do we fully realize the role of natural electromagnetic phenomena in the life of people and other beings? Many people speak of electromagnetic ecology, but does anybody know for sure the mechanism of influence of the external electromagnetic field on organisms? In due course knowledge will come, but that might be too late ...

Addenda

A.1 Geophysical values

Here we present the necessary information on the Sun, the Earth and the near-Earth space. It is extracted from many sources but primarily from the book by Allen (1955). Details of the system of physical units are also given here.

Like Allen we give an average, round, 'typical' value sometimes without indicating the variability interval, error, etc, and without explaining the procedure of obtaining such meaning.

The Sun and the Earth as a whole

Average distance from the Earth to the Sun (astronomical unit)	$A{=}AU{=}1.496 \times 10^{13}$ cm
Radius of the Sun	$R_\odot = 6.96 \times 10^{10}$ cm
Mass of the Sun	$M_\odot = 1.99 \times 10^{33}$ g
Angular velocity of the Sun's rotation at the equator	2.9×10^{-6} rad s^{-1}
Radius of the Earth	$R_E = 6371$ km
Mass of the Earth	$M_E = 5.977 \times 10^{27}$ g
Angular velocity of the Earth's rotation around the axis	7.292×10^{-5} rad s^{-1}
Average velocity of the Earth's movement along the orbit	29.77 km s^{-1}
Acceleration due to gravity near the Earth's surface (standard value)	980.665 cm s^{-2}
Centrifugal acceleration at the equator	3.39 cm s^{-2}
Magnetic moment of the Earth	8.06×10^{25} abs. u.

The Earth's crust

Effective thickness	35 km
Continental crust	35–75 km

Ocean crust	5–10 km
Elastic wave velocity	3–7.5 km s^{-1}
Conductivity	10^3–10^{10} s^{-1}

The crust forms the upper layer of the lithosphere, which is the silicate shell of the Earth, and is approximately 100 km thick. The lithosphere consists of plates or blocks. They are also called crustal blocks or rigid blocks. The boundaries of the lithospheric plates are the zones of seismic activity.

The rocks from which the Earth's crust is made are mineral aggregates. We know of more than 7000 kinds of minerals, but most of them are seldom found. The major rock-forming minerals are silicates (800 minerals), oxides and hydroxides (200 minerals), carbonates (80 minerals) and also sulphates, sulphides and haloids.

By origin the rocks are divided into three classes: magmatic, sedimentary and metamorphic. Magmatic and metamorphic rocks predominate. The mass of the sedimentary rocks makes up only 5% of the mass of the Earth's crust, however, these rocks cover 75% of the Earth's surface.

Ocean

Average depth	3770 m
Maximum depth	11022 m
Average conductivity	4×10^{10} s^{-1}.

Conductivity is approximately in linear dependence on salinity. At temperatures of several °C the empirical formula $\sigma = 10^9 S$ applies, where S is the concentration of salts in g l^{-1}. In ocean waters $S \simeq 36$ g l^{-1}.

Arctic Ocean	$S = 30$ g l^{-1}	$\sigma = 3 \times 10^{10}$ s^{-1}
Black Sea	$S = 10$ g l^{-1}	$\sigma = 10^{10}$ s^{-1}
Finnish Bay	$S = 3.5$ g l^{-1}	$\sigma = 3.5 \times 10^9$ s^{-1}.

When sea water freezes, its conductivity remains fairly high, since the ice is pierced by cells filled with strong brine. Conductivity of a thick layer of ice is anisotropic: in the vertical direction σ is lower by an order than in the horizontal direction.

Atmosphere

Near the surface of the Earth

Standard temperature	$T = 0$ °C
Standard pressure	$p = 1.013$ bar $= 760$ mm
Air density	$\rho = 1.29 \times 10^{-3}$ g cm^{-3}

Molecular weight	$\mu = 28.97$
Sound velocity	$c_s = 331$ m s^{-1}
Concentration of molecules	$N = 2.69 \times 10^{19}$ cm^{-3}
Free path length	7×10^{-6} cm
Height of the homogeneous atmosphere at 0° C	$H = 8$ km
Humidity at the tropics	3%
Humidity at the Antarctic	2×10^{-5}%
Electric field in an area of good weather (directed downwards)	$E = 130$ V m^{-1}
Density of the negative surface charge of the Earth (in the same place)	3.5×10^{-4} abs. u.
Charge of the Earth	$Q = -3 \times 10^5$ C
Difference of potential between the Earth and the ionosphere	200–250 kV
Average conductivity	$\sigma = 2.5 \times 10^{-4}$ s^{-1}
Ratio of conductivities	$\sigma_+/\sigma_- = 1.25$
Concentration of light ions	$N_+ = 600$ cm^{-3}, $N_- = 500$ cm^{-3}
Concentrations of heavy ions over the sea	200 cm^{-3}
Mobility of light ions	10^{-4} m^2 s^{-1} V^{-1}
Mobility of average ions	10^{-6} m^2 s^{-1} V^{-1}
Mobility of heavy ions	$< 10^{-7}$ m^2 s^{-1} V^{-1}
Average lifetime of light ions	70 s
Density of vertical current conductivity (towards the Earth)	$j = 2.5 \times 10^{-12}$ A m^{-2}
Total current over the whole surface of the Earth	$J = 1800$ A
Resistance between the Earth and the ionosphere	200 Ω

The variables p and ρ decrease roughly exponentially with height. The electric field E over the sea also decreases approximately exponentially and does not exceed several V m^{-1} at an altitude of 10 km. Over the land E slightly increases at the beginning (up to an altitude of several hundred of metres), and then decreases approximately exponentially. The conductivity σ increases with height approximately exponentially.

Average length of a lightning discharge	5 km
Extreme length	100 km
Average height of a thunderstorm discharge	5 km
Average peak intensity of current	20 kA
Extreme intensity of current	500 kA
Energy of a lightning flash	2×10^{17} erg

In the mountains and in the sea at $E \geq 10^3$ V m^{-1} corona discharges called St. Elmo's fire appear. The current of the corona $J \geq 10$ mA.

Ionosphere

Arrangement of the layers according to their height

D	70–90 km
E	90–130 km
F	> 130 km

Characteristic values of the electron concentration N:

during daylight (the height is indicated in brackets after the letter)

D (70 km)	100–200 cm^{-3}
E (110 km)	$(1.5–3) \times 10^5$ cm^{-3}
F_1 (180 km)	$(3–5) \times 10^5$ cm^{-3}
F_2 (280 km)	$(2–50) \times 10^5$ cm^{-3}

at night

D(70 km)	10 cm^{-3}
E(110 km)	3×10^3 cm^{-3}
F(320 km)	$(1–3) \times 10^5$ cm^{-3}

The value N strongly depends on the season, the phase of the solar activity cycle and the geomagnetic activity.

Ion composition

D	NO^+, O^+_2, $(H_2O)_n H^+$, O_2^-, NO_3^-, HCO_3^-
E	NO^+, O_2^+
F	O^+

At altitudes of 90–100 km thin layers of Mg^+, Fe^+, Si^+, Na^+, Ca^+ of meteorite origin appear sporadically. They are designated by E_s. The typical concentration of metallic ions is 2×10^4 cm^{-3}. An extreme value is 10^6 cm^{-3}.

Above 500–1000 km H^+ ions predominate. Here a conditional boundary between the ionosphere and magnetospheric plasma is usually drawn.

Integral Pedersen conductivity	$\Sigma_\perp = 2–5 \ \Omega^{-1}$
Integral Hall conductivity	$\Sigma_H = 5–10 \ \Omega^{-1}$

Magnetosphere

The main magnetic field is approximated by the dipole field $M = 8.06 \times 10^{25}$ G cm^3, displaced 450 km from the centre of the Earth towards the Pacific ocean. The dipole axis is deviated 11.5° from the axis of rotation of the Earth.

Horizontal component of the field H at the magnetic equator	0.31 G
Vertical component of the field Z:	
at the North Pole	0.6 G
at the South Pole	0.7 G

Whether the dipole approximation is applied depends on the distance (3–5)R_E from the centre of the Earth. At greater distances magnetic fields of external sources are taken into account one way or another.

The distance from the centre of the Earth to the magnetopause at the day side	8–12 R_E
Length of the geomagnetic tail	$\simeq 10^3 R_E$
Distance from the subsolar point over the magnetopause to the bow shock wave	$\simeq 2$–3 R_E
Magnetic field B over the magnetopause at the subsolar point	40–60 nT
Magnetopause thickness	100–200 km
Width of the daytime polar cusp:	
over the magnetopause	$\simeq 10^3$ km
near the Earth's surface	$\simeq 10$ km

The background plasma of ionospheric origin (at shorter distances) and solar origin (at greater distances) spreads relatively freely along the geomagnetic field lines and at the same time it is subject to complex convective drifts across the field lines under the influence of large-scale electric fields. As a result a so-called plasmasphere, limited by the plasmapause, is formed.

McIllwain parameter L for the plasmapause	3–6
Thickness of the plasmapause	500–1000 km
Concentration of plasma N on the inside of the plasmapause	10^2–10^3 cm^{-3}
Concentration of plasma N on the outside of the plasmapause	5–10 cm^{-3}
Concentration of plasma at the periphery of the magnetosphere	1–10 cm^{-3}

Besides a relatively cold background plasma, the magnetosphere contains energetic electrons and ions (radiation belts, particles of ring current, auroral particles, etc). Their simple description and brief numerical characteristics are difficult (see, for example, Roederer 1970, Akasofu and Chapman 1972).

Interplanetary medium

Specific values of the parameters along the orbit of the Earth

Solar wind velocity	$V = 400$ km s^{-1}
Plasma concentration	$N = 6$ cm^{-3}
Relative content of He^{++}	5 %
Electron temperature	$T_e = 1.5 \times 10^5$ K
Ion temperature	$T_i = 5 \times 10^4$ K
Magnetic field	$B = 5$ nT
Alfvén velocity	$c_A = 45$ km s^{-1}
Parameter β	1.2

All the parameters fluctuate, deviating considerably from the average values. So sometimes V reaches 800 km s^{-1} and more, B may reach 40 nT and N can be many dozens of particles in 1 cm^3. The relative concentration of α-particles in the flare flow of the solar plasma may reach 25%.

Indices of activity

For different purposes we use different scales, by means of which the intensity of the earthquakes is defined. Usually the 12 force scale is used. At force 4–5 tableware begins to tinkle and chandeliers rock. At force 6–7 cracks appear and destruction starts. From force 8 onwards, there is disastrous destruction.

According to Richter (1958) the magnitude of an earthquake is

$$M = \log a + 1.656 \log \Delta + 1.82.$$

Here a is the maximum amplitude of the surface wave in μm, Δ is the angular distance to the epicentre in degrees. For example, the strong earthquake in Alaska in 1964 had $M = 8.6$. The destructive earthquake of 1966 in Tashkent had $M = 5.3$.

The seismic energy E (erg) and the length of discontinuity L (km) in the earthquake centre are estimated by the value M

$$\log E = 11.8 + 1.5M \qquad M = 6.03 + 0.76 \log L.$$

The visual estimate of the sea roughness is given by the force 6 scale as shown in the following table.

Number on scale	Height of waves (m)	State of the water surface
0	0	Smooth
1	0–0.25	Ripples, small crests
2	0.25–0.75	Crests begin overturning, spume is glassy but not white
3	0.75–1.25	Crests overturn, here and there 'white-caps' are seen
4	1.25–2.0	'White-caps' are seen everywhere
5	2.0–3.5	Wind tears the spume away from the crests of waves

Wind strength is given by the Beaufort scale.

Number on scale	Description or velocity (m s^{-1})
0	calm
1	light air
2	breeze
3	gentle breeze
4	temperate wind
5	fresh wind
6	strong wind
7	high wind
8	fresh gale
9	gale
10	strong gale
11	storm
12	hurricane
13	39
14	44
15	49
16	54
17	59

Kinetic energy of dangerous atmospheric phenomena

Waterspout	4×10^7 J
Tornado	4×10^{10} J
Squall	4×10^{12} J
Hurricane	4×10^{17} J

Estimate of tornado by the Fujita scale

Number on scale	0	1	2	3	4
Max wind velocity (m s^{-1})	18	33	50	70	93
Length of the track (km)	0.5	1.6	5.1	16.1	50.1

Geomagnetic activity is characterized by means of global terms K_p, AE and D_{st}.

The planetary K_p index is calculated by averaging local K indices, measured at 12 observatories. The values $0, 1 \ldots 9$ are attributed to the index K so that $K = 9$ corresponds to variations of the magnetic field of largest amplitude, and $K = 0$ to those of smallest amplitude at the given observatory within the limits of every three hours during the day. The index K_p has 28 gradations: from 0_0, 0_+, 1_-, 1_0, 1_+, 2_- and so on to 8_+, 9_-, 9_0. The total of eight three-hour intervals within the given 24 hours is designated ΣK_p.

AE characterizes the activity of the auroral electrojet. It is calculated by H-components of the magnetic field variations at auroral and subauroral observatories and is expressed in nT. Usually the AE-index is adduced for hour-long intervals.

D_{st} was introduced by Sugiura (1972) for characteristics of the ring current in the magnetosphere. It is calculated by H-components of the magnetic field at the observatories at Honolulu, San Juan, Kakioka and Hermanus. It is expressed in nT and is published regularly in the form of a continuous series of hour-long values. Under quiet conditions the index approaches zero. During the magnetic storms D_{st} reaches large negative values (dozens and even hundreds of nT).

Solar activity is characterized by the Wolf number

$$W = 10g + s$$

where g is the number of clusters of spots and s is the total number of separate spots over the visible hemisphere of the Sun. In years of minimum activity $W \simeq 5$, in years of maximum activity $W \simeq 100$. The average duration of a spot-forming cycle is 11 years.

Chromospheric flares are estimated in numbers in a scale. The highest number 4 is attributed to flares of extraordinary intensity. They happen once or twice within the 11 year cycle. The energy of such a flare reaches 10^{32} erg.

Units

In the CGS system, which is also termed the Gaussian system or absolute system, the basic values are cm, g, s. The electric charge is expressed through the basic values using Coulomb's law where the coefficient of proportionality is chosen to be dimensionless and equal to unity. The dimensions of charge are $[e] = g^{1/2} \, cm^{3/2} \, s^{-1}$. The unit of magnetic field, Gauss, follows from the previous units. One Gauss is the field which acts on 1 cm of a current-carrying conductor in one absolute unit with a force of $(1/c)$ dyn, if the conductor is situated perpendicular to the magnetic field lines. Here c is the velocity of light. The dimensions of the magnetic field are $[H] = g^{1/2} \, cm^{-1/2} \, s^{-1}$. Dielectric and magnetic permeabilities of the vacuum are dimensionless and both equal unity.

Besides the CGS system the SI system is sometimes used in the litereature as well as various out-of-system units. Translation of one unit into another can cause difficulties. The table below should help.

Length	$1 \, \mu m = 10^{-4} \, cm$
Energy	$1 \, J = 10^7 \, erg$
	$1 \, eV = 1.602 \times 10^{-12} \, erg$
Power	$1 \, W = 10^7 \, erg \, s^{-1}$
Pressure	$1 \, bar = 10^6 \, dyn \, cm^{-2}$
	$1 \, Pa = 10 \, dyn \, cm^{-2}$
Temperature	$0 \, °C = 237 \, K$
	$1 \, eV = 11606 \, K$
Electric charge	$1 \, C = 3 \times 10^9 \, abs. \, u.$
Electric current	$1 \, A = 3 \times 10^9 \, abs. \, u.$
Potential	$1 \, V = 1/300 \, abs. \, u.$
Electric field	$1 \, V \, m^{-1} = \frac{1}{3} \times 10^{-4} \, abs. \, u.$
Resistance	$1 \, \Omega = \frac{1}{9} \times 10^{-11} \, abs. \, u.$
Specific resistance	$1 \, \Omega \, m = \frac{1}{9} \times 10^{-9} \, s$
Specific conductivity	$1 \, S \, m^{-1} = 9 \times 10^9 \, s^{-1}$
Magnetic field	$1 \, T = 10^4 \, G$
	$1 \, \gamma = 10^{-5} \, G$

A.2 Classification

One of the most important tasks is to describe and classify waves and the conditions under which they arise. The classification does not introduce additional information as compared with that contained in the whole complex of the initial data, but it is convenient, as it allows us to express the diversity of waves in a limited number of regulated and well identified types.

None of the classifications is perfect. (Let us recollect, for instance, the ancient classification of temperaments.) In our case the first stage of classification, namely the ascertainment of a large group (we shall term them classes), is overcome quite easily. The waves may be distributed, for example, according to the state of a substance in which the sources are arranged. Solid, liquid and gaseous states correspond to the ground, water and air respectively, and plasma state to the ionosphere, magnetosphere and interplanetary medium. Here everyone may not agree. For example, waves with their sources located in condensed media, i.e. in the Earth's crust and sea water, may be placed in the first class; waves with their sources in the neutral and weakly ionized gas (atmosphere and ionosphere) may be placed in the second class and so on.

Let us discuss further stages of classification. It is necessary to construct a hierarchy of categories; but what indication shall we employ to divide the given data into kinds, varieties, etc? Shall we avail ourselves again of the same criterion? For example, objects of the given class may be arranged by groups (we shall term them kinds) depending on whether the sources are found, for example, inside or outside the plasmasphere, or in the magnetopause or in the geomagnetic tail. In the case of magnetospheric waves this has a certain sense, but for water waves we would have to speak of 'Pacific Ocean Waves' or 'Bay of Biscay Waves', etc. Some other approach is necessary here. For the classification we should take not one, but several indications. It is reasonable to take the following supplementary indications.

(i) the mechanism of wave generation;
(ii) mode structure of the wave field;
(iii) morphological indications.

For example, for waves in the Earth's crust we know induction, electrokinetic, piezomagnetic and other mechanisms of electromagnetic field generation. Therefore, we may divide the class into kinds according to (i) above. It is advisable to employ (ii) for further division into varieties (surface waves, body waves and so on).

Point (iii) may seem unnecessary, since it is determined by (i) and (ii), but the generation mechanism and (or) the mode structure of the field are sometimes unknown. Then morphological properties are used for classification (continuous or impulse oscillation regime, narrow or broad frequency band, presence or absence of the frequency modulation, etc).

We should dwell especially upon the classification of geomagnetic pulsations. The system of classification, the foundation of which was laid in 1963 at the XIII General Assembly of IUGG is widely used in the literature (Troitskaya and Guglielmi 1967, Jacobs 1970). Seven kinds of pulsations, well studied by that time, were given special abbreviated names (see below). The kinds are grouped into two classes, termed Pc (continuous pulsations) and Pi (irregular pulsations). In other words, an attempt to apply the nomenclature principle of Carl Linné is made here. However, this attempt cannot be considered satisfactory, since the properties of continuity and irregularity do not form binary opposites.

Classification of geomagnetic pulsations

Type	Range of periods (s)
Pc1	0.2–5
Pc2	5–10
Pc3	10–45
Pc4	45–150
Pc5	150–600
Pi1	1–40
Pi2	40–150

The inconvenience of the system, which hampers accumulation and transmission of the information, lies in the fact that it turned out to be poorly adapted to the changes of classification and to its modification by new kinds and varieties of wave regimes. Soon after publication, the system began accumulating abbreviations like PiB, PiC, IPDP, Ipcl, iprp, Ps6, Pip, SE, DS, etc, and the terms 'gyroharmonics', 'pearls', 'hydromagnetic hiss', 'hydromagnetic howling', 'rib-emission' (sometimes simply 'ribs'). Nishida (1978) and Guglielmi (1979) count 20–30 kinds and varieties of pulsations. In reality there are apparently even more.

The accuracy of the language of science is of extraordinary importance and we express concern over the state of systematization of pulsations particularly in connection with the problems of hydromagnetic diagnostics, and forecast that an accurate description of waves is necessary that will be intelligible not only to humans but also to computers.

We shall not dwell on this problem any longer since its solution must be decided by the special international commission and be approved by the authoritative International Congress.

A.3 Methods of research[1]

Magnetic field oscillations are measured by magnetometers of various constructions. An induction magnetometer contains a multiloop coil as a sensor. Usually the coil is supplied with a ferromagnetic core to concentrate the magnetic flux. Using three mutually orthogonal coils all three components of the magnetic field are measured at a given point in space. Quantum magnetometers with optical pumping, possessing heightened sensibility, are also used. However, these devices measure the modulus of the magnetic field vector only. They are termed helium, caesium, rubidium, etc, depending on the type of working substance. Record sensibility of the order of 10^{-16} T is intrinsic to

---

[1] See also a special issue of *Journal of Geophysics* (1984, volume **55**, number 2) which is dedicated to the experimental techniques.

superconducting magnetometers, the action of which is based on the Josephson effect.

Horizontal components of the electric field of oscillations over the Earth's surface are measured using nonpolarizing electrodes, which are buried in the ground. The vertical component in the upper section of the range of geoelectromagnetic waves is measured by means of an aerial. Measuring the vertical component in the lower section of the range will entail great difficulties. Electrodes, for example, may be put down into a specially bored well. But at the ground–air boundary the vertical component of the electric field experiences a considerable jump. It is difficult to judge the electric field over the Earth's surface since the distribution of conductivity in the interior of the Earth is rather indefinite.

To overcome this difficulty, the electrodes are sunk into the sea, but sea roughness prevents us from measuring there. Vinogradov (1956) carried out measurements in winter, from ice, at Lake Baikal, which has a depth of 1741 m.

Generally speaking, the problem emerges only when we intend to single out signals that originate in space. The total electric field E_z (mainly of atmospheric origin) is registered steadily, for example, by a collector in the form of a radioactive probe, raised at a low altitude in an open glade, or in the form of an electrostatic fluxmeter the operation of which is based on measuring the induced charge of a conductor.

Atmospheric noises disguise cosmic signals of E_z whose presence was guessed fairly early (Bauer 1924). The solution of the problem is sought in heightening the sensitivity of devices, improving the methods of observation and the processing of the signals.

The standard equipment of a geoelectromagnetic observatory is a magnetometer and a two- or three-component device to register the electric field. Usually an observatory is also equipped with other geophysical equipment: ion probe, riometer, seismometer, etc. If the observatory is directed towards investigating atmospheric electricity, then meteorological devices are an obligatory component of the complete set.

In time many of the properties of geoelectromagnetic waves were determined according to the data obtained by separate observatories. The most important property of this kind within the range of geomagnetic pulsations lies in the fact that the permanent oscillation regime predominates in the day time and the impulse oscillation regime predominates at night. This property is undoubtedly associated with the asymmetry of the magnetosphere relative to the plane of the morning–evening meridian.

However, the potentialities of one observatory are rather limited. Modern methods of research are characterized by the organization of purposeful experiments in order to solve specific problems. Ground magnetometers are arranged in a particular way: for example, in lattice junctions, covering the given area, satellites and radars are used, etc. Synoptical observations even at two scattered points increase the effectiveness of research.

For example, here is the method applied by Ochadlick (1990) to measure the correlation length of oscillations of magnetospheric origin. He used two aeroplanes, flying at an altitude of about 7 km over the ocean and gradually moving away from each other. Recording of the oscillations within the range 2–25 s was carried out by means of airborne helium magnetometers. It was found out from the results that the correlation length equals approximately 100–150 km.

What were the measurements over the ocean carried out for? They were carried out in order to avoid the interfering influence of geological inhomogeneities of the Earth's crust. But ocean waves also cause magnetic interference, lowering the correlation radius. This can be disregarded at an altitude of 7 km, as the magnetic effect of the ocean waves has a rapid exponential decrease with altitude. We give this example to emphasize a simple idea: it is useful to know the whole complex of geoelectromagnetic processes when planning certain experiments in this domain.

Let us consider another example of effective application of a pair of magnetometers, but this time they were placed in opposite hemispheres in magnetic conjugate points[2]. One point was at the village Sogra in the Arkhangelsk region, the other was at Kerguelen Island in the India Ocean (Gendrin and Troitskaya 1965). Finally it was ascertained that the Pc1 signals emerge at conjugate points by turns. This suggested that we were dealing with a wave packet that oscillates along the geomagnetic field lines reflecting periodically from the ionosphere in the conjugate regions.

The magnetometers placed on the satellites increased the informativeness of the observation of geoelectromagnetic waves. But even a brief description of satellite methods would have led us away from the subject[3]. To illustrate this we shall mention two–three examples. They were analysed in detail in the preceding chapters.

Figure 6.5 shows a spectrum of magnetic oscillations, which are probably never observed over the Earth's surface. They were registered by the magnetometers on board the *OGO-3* satellite when it was crossing the equatorial vicinity of the plasmapause (Russell *et al* 1970) by attention to the typical toothed structure of the spectrum within the range 100 Hz. These teeth appear due to amplification of waves over the harmonics of the gyrofrequency of ions (Guglielmi 1979). The waves are localized in the toroidal waveguide under the arch of plasmasphere and that is why they do not reach the Earth.

Figure 7.13 gives an idea about the variation of the horizontal components of the electric vector along the trajectory of the satellite *IC-Bulgaria-1300* when it was crossing the auroral zone at an altitude of 830 km (Chmyrev *et al* 1988).

---

[2] Magnetic conjugate or simply conjugate points are located at one and the same geomagnetic field line.

[3] A special issue of the journal *Geoscience and Remote Sensing* (1985 volume **GE-23**, number 3) makes it possible to frame an idea on various aspects of organizing and accomplishing satellite research using the experiment of AMPTE as an example.

The waves have vortex structure (Petviashvili and Pokhotelov 1985). Such waves do not penetrate to the Earth due to the low order of magnitude of the horizontal scale of vortex structure.

Observations of this kind testify to the existence of a rich spectrum of waves, localized in the magnetosphere and not reaching the Earth. Many other important inferences, which changed the concept of geoelectromagnetic waves, were made by means of satellites. We shall mention one more that is particularly interesting in the context of this book.

Day by day Pc3 are registered over the sunlit side of the Earth. For years they have been considered to be waves of magnetospheric origin (see, for example, the monograph by Jacobs 1970). But there appeared reliable information on the interplanetary medium from satellites and it was found that Pc3 are of extramagnetospheric origin (see the reviews by Guglielmi 1974 and Yumoto 1985).

At the early stages of investigation the information was processed mainly by sight. The general characteristics of the oscillation regime were investigated by looking through the oscillograms. The carrier frequency and amplitude of oscillations was then measured. Later on various kinds of spectral analysis, methods of statistical processing of signals, etc, were put into practice. When studying frequency-modulated signals a so-called spectral-temporal analysis, which provides the dynamic spectrum of oscillations, proved to be most efficient. Figures 4.8, 5.4, 6.7 and 6.9 show examples of dynamic spectra. They were obtained using analogue methods by means of sonograph and they are called sonograms. Frequency is given along the vertical axis of the sonogram and time along the horizontal axis. The spectral oscillation density is proportional to the blackening of special paper. Numerical methods to obtain dynamic spectra are also employed. R R Heacock compiled a rich atlas of dynamic spectra which can be found in the Final Report, Grant No. GA–4059, Geophysical Institute of the University of Alaska, National Science Foundation, Washington, DC 20550, August 1970.

Sometimes even simple methods of analysing the information led to interesting results. For example, it was enough to depict the distribution of the perturbation vector of the electric field along the satellite trajectory to see the vortex structure in the auroral ionosphere (figure 7.13) and it was enough to draw a phase picture of Pc4 to see the saw-like character of oscillations (figure 8.4).

A.4 Man-made waves

Although this book is dedicated to waves of natural origin, we present some extracts from the literature on waves engendered by man-made activity, to make the picture complete.

Nuclear explosions, particularly in the upper atmosphere, perturb the geomagnetic field. A special issue of *Journal of Geophysical Research* (1963, volume **68**, number 3) is dedicated to the effects caused by the nuclear explosion over Johnstone Island. A chemical explosion of considerable strength is also capable of provoking a noticeable reaction in the ionosphere. Liperovsky *et al* (1992) reported the electromagnetic effect of a superimposed explosion of a strength of 250 t TNT carried out in Middle Asia in the vicinity of Alma-Ata.

Explosions of lesser strength are also capable of provoking considerable modification to the environment and deform the spectrum of natural electromagnetic radiations if they follow each other for a long period of time. Artificial perturbations of such kind distort geoelectromagnetic background not to mention the damage that they cause.

Returning to the effects of nuclear and powerful chemical explosions, we cannot but mention that probably despite our expectations they provided surprisingly little new information concerning geoelectromagnetic waves. Of course the analysis of such experiments is interesting for the solution of applied problems, in particular for the development of a system to discern nuclear explosions in order to distinguish them from natural calamities, such as the explosion of a large meteorite. It is difficult even to imagine what could have happened if the Tunguska meteorite, whose energy of explosion exceeded 10^{23} erg, had entered the atmosphere not above the uninhabited taiga but above a densely populated area, and not in 1908 but in the years of nuclear confrontation, and if the system of detection and discernment of nuclear explosion had not existed. But, we repeat, reading the literature on the given subject leaves us with the impression that the experimental results have proved *a priori* ideas at best, without providing anything essentially new. It is clear that we speak here only of geoelectromagnetic waves.

On the other hand, 'low-power', but well-organized experiments sometimes lead to fascinating discoveries. The experiments carried out by Ivanov (1939, 1940) on exciting seismoelectric signals may serve as a classic example. The source of seismic waves was created by a chemical explosion of low power, or still more easily, by dropping a heavy metal pig on the ground. As a result a so-called seismoelectric effect of the second kind, which served as a stimulus to Frenkel (1944) to create the known theory that enriched geoelectrics and seismology, was created.

At present experiments of such kind are accomplished by applying a more effective source, namely a seismo-vibrational device. The operation of the oscillatory regime (scanning by frequency, switching of power) enables us to investigate delicate nonlinear effects such as the generation of electric signals of a combination of frequencies, hysteresis phenomena, etc (Guglielmi *et al* 1989a).

Magnetic signals of combination frequencies within the range that is of interest to us are excited by periodic influence of the radiation from a ground radiotransmitter upon the ionosphere. Experiments within the ranges of kilohertz, hertz and millihertz are discussed in the articles by Getmantsev *et al* (1974), Guglielmi *et al* (1985) and Stubbe and Kopka (1981) respectively.

Fraser-Smith (1981), who did a great deal to advance this domain, gave a brief but informative review of the earlier works. One will find many witty ideas there such as the idea of exciting waves in the ionosphere and the magnetosphere, passing strong alternating current through a conducting loop. It is clear that the loop must cover a fairly large area. So, the idea of the 'peninsula method' lies in using sea water at the shore of a peninsula isthmus as a conductor and placing the feeding electrodes at both sides of this isthmus. The experiments were carried out at North Neck, Massachusetts and at Rybachiy Peninsula near Murmansk.

We give another example. Fraser-Smith (1981) analysed the perturbations of geoelectromagnetic field as an after-effect of the functioning of technical devices which were not meant especially for this purpose (radiotransmitters, electric power lines, DC-power rapid transient systems, etc) and arrived at the following conclusion 'Mankind may already be influencing natural ULF [ultra low frequency] geometric activity and pulsations over wide areas of the world. This threat can only grow worse in the future, unless efforts are made to protect the ULF band of the electromagnetic spectrum from interference'. This is an alarming idea with its opponents and adherents; but what is it based on?

The major argument is the so-called 'weekend effect'. It turns out that on Saturdays and Sundays, when the industrial activity calms down everywhere, total geomagnetic activity increases as compared to other days of the week. This gives the impression that industrial activity suppresses the activity of the magnetosphere or rather suppresses the kinds of activities that form the weekend effect.

Influenced by these ideas, Guglielmi *et al* (1978) as well as Zotov and Kalisher (1979) undertook the search for man-made effects, suppressing, at the same time, involuntarily stimulation of the wave activity of the magnetosphere. They proceeded from the fact that the beginning of every hour and other chosen moments of time serve as peculiar signals that synchronize the operation regime of the world network of radiotransmitters, in particular those that are used in radiosounding of the ionosphere. The effects of 'hour marks' and 'world days' on Pc1 activity were singled out at the statistical level. A discussion on the reliability of these effects is given in the articles by Menk (1985) and Guglielmi and Zotov (1986).

Our excursion into the domain of man-made waves is almost over, but we must mention the AMPTE (Active Magnetospheric Particle Tracer Explorers) experiment. The purposes and results of this project are grandiose in their conception and excellent in their realization and are presented in the works by Haerendel (1986), Krimigis and Dassoulas (1984) as well as in a series of articles under the general title 'An artificial comet', published in *Nature* (1986, volume **320**, number 6064, pp 700–23). In the course of the experiment a wide spectrum of electromagnetic waves that were excited by the injection of barium and lithium clouds from the satellite into the magnetosphere and the interplanetary plasma was observed.

References

Afanasieva V I and Kalinin Yu D 1960 Large and very large magnetic storms *Geomagnetic Perturbations* (Moscow: AN SSSR) p 5

Akasofu S-I 1960 On the ionospheric heating by hydromagnetic waves connected with geomagnetic micropulsations *J. Atmos. Terr. Phys.* **18** 160

Akasofu S-I and Chapman S 1972 *Solar–Terrestrial Physics* (Oxford: Oxford University Press)

Akhiezer A I, Akhiezer I A, Polovin R V, Sitenko A G and Stepanov K N 1974 *Plasma Electrodynamics* (Moscow: Nauka)

Alfvén H and Arrhenius G 1975 *Structure and Evolutionary History of the Solar System* (Dordrecht: Reidel)

Alfvén H and Fälthammar C-G 1963 *Cosmical Electrodynamics* (London: Oxford University Press)

Allan W 1992 Ponderomotive mass transport in the magnetosphere *J. Geophys. Res.* **97** 8483

Allan W, Manuel J R and Poulter E M 1991 Magnetospheric cavity modes: some nonlinear effects *J. Geophys. Res.* **96** 11461

Allen C W 1955 *Astrophysical Quantities* (London: University of London/ The Athlone Press)

Altman C and Fijalkow E 1969 The transmission of electromagnetic waves through the ionosphere at micropulsation frequencies *Alta Freq.* **38** 183

Anderson B J, Erlandson R E and Zanetti L J 1992 A statistical study of Pc1-2 magnetic pulsations in the equatorial magnetosphere, 1 equatorial occurrence distribution; 2 wave properties *J. Geophys. Res.* **97** 3075

Arnoldy R L, Lewis P B Jr and Cahill L J Jr 1979 Polarization of Pc1 and Ipdp pulsations correlated with particle precipitation *J. Geophys. Res.* **84** 7091

Asbridge J R, Bame S J and Strong I B 1968 Outward flow of protons from the earth's bow shock *J. Geophys. Res.* **73** 5777

Aubry M P 1970 Diagnostics of the magnetosphere from ground based measurement of electromagnetic waves *Ann. Geophys.* **26** 341

Balser M and Wagner C A 1960 Observations of the Earth–ionosphere cavity resonances *Nature* **188** 638

Banks P M and Holzer T E 1969 High-latitude plasma transport: the polar wind *J. Geophys. Res.* **74** 6317

Baransky L N, Borovkov Yu E, Gokhberg M B and Krylov S M 1985 High resolution method of direct measurement of the magnetic field line resonances in the magnetosphere *Planet. Space Sci.* **33** 1369

Barrington R E and Fejer J A 1965 Ionospheric noise and geomagnetic pulsations *Physics of the Earth's Upper Atmosphere* ed C A Hines *et al* (Englewood Cliffs, NJ: Prentice-Hall) p 176

Barrington R E, Belrose J S and Mather W F 1966 A helium whistler observed in the Canadian satellite Alouette II *Nature* **210** 80

Barsukov O M and Sorokin O N 1973 The changes of rock resistivity in the Garm seismoactive region *Izv. AN SSSR, Ser. Fiz. Zhemli* 100

Bateman H 1955 *The Mathematical Analysis of Electrical and Optical Wave-Motion on the Basis of Maxwell's Equations* (Cambridge: Trinity College; New York: Dover)

Bauer L A 1924 Correlation between solar activity and atmospheric electricity *Terr. Magn. Atmos. Electr.* **29** 23

Baumjohann W 1986 Merits and limitations of the use of geomagnetic indices *Solar Wind–Magnetosphere Coupling* ed Y Kamide and J A Slavin (Tokyo: TERRAPUB) p 3

Baumjohann W and K-Glaßmeier 1984 The transient response mechanism and Pi2 pulsations at substorm onset—review and outlook *Planet. Space Sci.* **32** 1361

Beal H T, and Weaver J T 1970 Calculations of magnetic variations induced by internal ocean waves *J. Geophys. Res.* **75** 6847

Belcher J W 1971 Alfvénic wave pressure and the solar wind *Astrophys. J.* **168** 509

Belov S V, Migunov N I and Sobolev G A 1974 Magnetic effects which accompany strong earthquakes at Kamchatka *Geomagn. Aeron.* **14** 380

Belyaev P P and Polyakov S V 1980 Boundary conditions for MHD waves on the ionosphere *Geomagn. Aeron.* **20** 637

Belyaev P P, Polyakov S V, Rapoport V O and Trakhtengertz Yu V 1987 Discovery of the resonant structure of the spectrum of atmospheric electromagnetic noise background in the range of short period geomagnetic pulsations *Dokl. AN SSSR* **297** 840

Benioff H 1960 Observations of geomagnetic fluctuations in the period range 03 to 120 seconds *J. Geophys. Res.* **65** 1413

Berdichevsky M N, Vanyan L L and Fainberg E B 1969 Magnetic variation sounding using the space derivatives of the field *Geomagn. Aeron.* **9** 229

Berdichevsky M N and Zhdanov M S 1984 *Advanced Theory of Deep Geomagnetic Sounding* (Amsterdam: Elsevier)

Berthold W K, Harris A K and Hope H J 1960 Worldwide effects of hydromagnetic waves due to Argus *J. Geophys. Res.* **65** 2233

Bhatnagar P L 1979 *Nonlinear Waves in One-Dimensional Dispersive Systems* (Oxford: Clarendon)

Blanc E 1985 Observations in the upper atmosphere of infrasonic waves from natural or artificial sources: a summary *Ann. Geophys.* **3** 673

Bliokh P V, Galyuk Yu P, Hunninen E M, Nikolaenko A P and Rabinovich L M 1977 On resonance phenomena in the Earth–ionosphere cavity *Izv. Vissch. Uchebn. Zaved. Ser. Radiofiz.* **20** 501

Bliokh P V, Nikolaenko A P and Filippov Yu F 1980 *Schumann Resonances of the Earth–Ionosphere Cavity (Institute of Electrical Engineering Waves Series)* vol 9 (London: Peter Peregrinus)

Bockris J O M and Reddy A K N 1970 *Modern Electrochemistry* vol 2 (New York: Plenum)

Bol'shakova O V and Troitskaya V A 1968 The connection of the magnetic field direction with the regime of the continuous oscillations *Dokl. AN SSSR* **180** 334

Bol'shakova O V and Troitskaya V A 1977 Dynamics of the dayside cusp based on the observations of long period geomagnetic pulsations *Geomagn. Aeron.* **17** 1076

Bolt B A 1978 *Earthquakes* (San Francisco: Freeman)

Bomke H A, Ramm W J, Goldblatt S and Smith H W 1960 Global hydromagnetic wave ducts in the exosphere *Nature* **185** 299

Bondarenko N M and Guglielmi A V 1976 Discrete electromagnetic signals in the range 01–1 Hz in the polar cap *Geomagn. Aeron.* **16** 316

Bonifazi C, Moreno G, Lazarus A J and Sullivan J D 1980 Deceleration of the solar wind in the Earth's foreshock region: ISEE-2 and IMP-8 observations *J. Geophys. Res.* A **85** 6031

Booker H G 1962 Guidance of radio and hydromagnetic waves in the magnetosphere *J. Geophys. Res.* **67** 4135

Booker H G and Dyce R B 1965 Dispersion of waves in a cold magnetoplasma from hydromagnetic to whistler frequencies *Radio Sci.* **69A** 463

Bösinger T, Alanko K, Kangas J, Opgenoorth H and Baumjohann W 1981 Correlation between PiB type magnetic micropulsations, auroras and equivalent current structures during two isolated substorms *J. Atmos. Terr. Phys.* **43** 933

Bossen M, McPherron R L and Russell C T 1976 Simultaneous Pc1 observations by the synchronous satellite ATS-1 and ground stations; Implications concerning IPDP generation mechanisms *J. Atmos. Terr. Phys.* **38** 1157

Brekke A and Egeland A 1983 *The Northern Light* (New York: Springer)

Brinton H C, Pickett R A and Taylor H A Jr 1968 Thermal ion structure of the plasmasphere *Planet. Space Sci.* **16** 899

Buchert S, Haerendel G and Baumjohann W 1990 A model for the electric fields and currents during a strong Ps6 pulsation event *J. Geophys. Res.* A **95** 3733

Bud'ko N I, Karpman V I and Pokhotelov O A 1972 Nonlinear theory of circularly polarized VLF and ULF waves in the magnetosphere *Cosmic Electrodyn.* **3** 165

Buldyrev V S, Kovtun A A and Van F-Chi 1973 Magnetosonic resonator of Pc2-3 pulsations in the Earth's magnetosphere *Geomagn. Aeron.* **13** 136

Bullough R K and Caudrey P J (ed) 1980 *Solitons* (Berlin: Springer)

Burridge R and Weinberg H 1977 Horizontal rays and vertical modes *Wave Propagation and Underwater Acoustics* ed J B Keller and J S Papadakis (Berlin: Springer)

Burton R K, McPherron R L and Russell C T 1975 An empirical relationship between interplanetary conditions and D_{st} *J. Geophys. Res.* **80** 4204

Campbell W H and Stiltner E C 1965 Some characteristics of geomagnetic pulsations of frequencies near 1 c/s *Radio Sci.* D **69** 1117

Campbell W H 1967 Geomagnetic pulsations *Physics of Geomagnetic Phenomena* vol 2, ed S Matsushita and W H Campbell (New York: Academic) p 821

Carovillano R L and McClay J F 1965 Hydromagnetic eigenmodes in multipole fields *Phys. Fluids* **8** 2006

Carovillano R L, Radoski H R and McClay J F 1966 Poloidal hydromagnetic resonances *Phys. Fluids* **9** 1860

Carovillano R L and Radoski H R 1967 Latitude-dependent plasmasphere oscillations *Phys. Fluids* **10** 225

Carpenter D L 1963 Whistler evidence of 'knee' in the magnetospheric ionization density profiles *J. Geophys. Res.* **68** 1675

Carpenter D L 1966 Whistler studies of the plasmapause in the magnetosphere. 1. Temporal variations in the position of the knee and some evidence on plasma motion near the knee *J. Geophys. Res.* **71** 693

Chang D B and Pearlstein L D 1965 On the effect of resonant magnetic-moment violation on trapped particles *J. Geophys. Res.* **70** 3075

Chapman S and Bartles J 1940 *Geomagnetism* (Oxford: Oxford University Press)

Chave A D 1984 On the electromagnetic fields induced by oceanic internal waves *J. Geophys. Res.* C **89** 10519

Chelidze T L 1984 Anomalously high tensosensitivity of the electric conductivity of inhomogeneous media *Sov. Phys.–JETP* **87** 635

Chen A J 1970 Penetration of low-energy protons deep into the magnetosphere *J. Geophys. Res.* **75** 2458

Chen L and Hasegawa A 1974 A theory of long-period magnetic pulsations. 1. Steady state excitation of field line resonance *J. Geophys. Res.* **79** 1024

Chen L and Cowley S C 1989 On field line resonances of hydromagnetic Alfvén waves in dipole magnetosphere *Geophys. Res. Lett.* **16** 895

Chetaev D N 1970 On the field structure of short-period geomagnetic variation and magnetotelluric sounding *Izv. AN SSSR Ser. Fiz. Zhemli* **2** 52

Chetaev D N 1978 On the local structure of magnetotelluric field *Izv. AN SSSR Ser. Fiz. Zhemli* **14** 751

Chetaev D N, Fedorov E N, Krylov S M, Lependin V P, Morgunov V A, Troityskaya V A and Zybin K Yu 1975 On the vertical electric component of the geomagnetic pulsation field *Planet. Space Sci.* **23** 311

Chew G F, Goldberger M L and Low F E 1956 The Boltzmann equation and the one-fluid hydromagnetic equations in the absence of the particle collisions *Proc. R. Soc.* A **236** 112

Chibisov G V 1976 Astrophysical upper limits on the photon mass *Usp. Fiz. Nauk* **119** 551

Chmyrev V M, Bilichenko S V, Pokhotelov O A, Marchenko V A and Stenflo L 1988 Alfvén vortices and related phenomena in the ionosphere and magnetosphere *Phys. Scr.* **38** 841

Chmyrev V M, Pokhotelov O A, Marchenko V A, Stenflo L, Streltsov A V and Steen Å 1991 Vortex structures in the ionosphere and the magnetosphere of the Earth *Planet. Space Sci.* **39** 1025

Chmyrev V M, Marchenko V A, Pokhotelov O A, Shukla P K, Stenflo L and Streltsov A V 1992 The development of discrete active auroral forms *IEEE Trans. Plasma Sci.* **PS-20** 764

Chrzanowski P, Green G, Lemmon K T and Young J M 1961 Traveling pressure waves associated with geomagnetic activity *J. Geophys. Res.* **66** 3727

Churilov S M and Shukhman I G 1983 On the waveguide propagation of nonlinear magnetosonic waves *Phys. Plasmy* **9** 827

Cole K D, Morris R T, Matveeva E T, Troitskaya V A and Pokhotelov O A 1982 The relationship of the boundary layer of the magnetosphere to IPRP events *Planet. Space Sci.* **30** 129

Cornwall J M 1965 Cyclotron instabilities and electromagnetic emission in the ultra low frequency and very low frequency ranges *J. Geophys. Res.* **70** 61

Cornwall J M 1966 Micropulsations and the outer radiation zone *J. Geophys. Res.* **71** 2185

Cornwall J M, Coronoti F V and Thorne R M 1970 Turbulent loss of ring current protons *J. Geophys. Res.* **75** 4699

Couzens D A and King J H 1986 *Interplanetary Medium Data Book, NSSDS, WDC-A-R&S* (Greenbelt, MD: Goddard Space Flight Center)

Crews A and Futterman J 1962 Geomagnetic micropulsations due to the motion of ocean waves *J. Geophys. Res.* **67** 299

Criswell D R 1969 Pc1 micropulsation activity and magnetospheric amplification of 0.2 to 5.0 Hz hydromagnetic waves *J. Geophys. Res.* **74** 205

Cummings W D, O'Sullivan R J and Coleman P J Jr 1969 Standing Alfvén waves in the magnetosphere *J. Geophys. Res.* **74** 778

D'Angelo N 1975 Have Pc2-4 micropulsations an extramagnetospheric origin? *Geomagn. Aeron.* **15** 1062

Davis L and Chang D B 1962 On the effect of geomagnetic fluctuations on trapped particles *J. Geophys. Res.* **67** 2169

Delahay P 1965 *Double Layer and Electrode Kinetics* (New York: Interscience)

De Lauretis M, Villante U, Vellante M and Wolfe A 1991 An analysis of power spectral indices in the micropulsation frequency range at different ground stations *Planet. Space Sci.* **39** 975

Dessler A J 1959a Upper atmosphere density variations due to hydromagnetic heating *Nature* **184** 261

Dessler A J 1959b Ionospheric heating by hydromagnetic waves *J. Geophys. Res.* **64** 397

Dirac P A M 1975 *General Theory of Relativity* (New York: Wiley)

Dmitrienko I S and Mazur V A 1985 On waveguide propagation of Alfvén waves at the plasmapause *Planet. Space Sci.* **33** 471

Dowden R L and Emery M W 1965 The use of micropulsation 'whistler' in the study of the outer magnetosphere *Planet. Space Sci.* **13** 773

Dragt A J 1961 Effect of hydromagnetic waves on the lifetime of Van Allen radiation protons *J. Geophys. Res.* **66** 1641

Duffus H J, Nasmyth P W, Shaud J A and Wright Ch 1958 Subaudible geomagnetic fluctuations *Nature* **181** 1258

Duncan R R 1961 Some studies of geomagnetic micropulsations *J. Geophys. Res.* **66** 2087

Dungey J W 1954 Electrodynamics of the outer atmosphere *Report 69* Ionospheric Research Laboratory, Pennsylvania State University, University Park, PA

Dungey J W 1958 *Cosmic Electrodynamics* (Cambridge: Cambridge University Press)

Dungey J W 1963 Hydromagnetic waves and the ionosphere *Proc. Int. Conf. Ionosphere* (London: Institute of Physics) p 230

Dungey J W and Southwood D J 1970 Ultra low frequency waves in the magnetosphere *Space Sci. Rev.* **10** 672

Dungey J W and Southwood D J 1975 Ultra low frequency waves in the magnetosphere *Phil. Trans. R. Soc.* A **280** 131

Dyson F 1971 *Neutron Stars and Pulsars (Fermi Lectures 1970)* (Rome: Accademia Nazionale del Lincei)

Eleman F 1965 The response of magnetic instruments to earthquake waves *J. Geomagn. Geoelectr.* **18** 43

Engebretson M J and Cahill L J Jr 1981 Pc5 pulsations observed during the June 1972 geomagnetic storm *J. Geophys. Res.* A **86** 5619

Engebretson M J, Anderson B J, Cahill L J Jr, Arnoldy R L, Rosenberg T J, Carpenter D L, Gail W B and Eather R H 1990 Ionospheric signatures of cusp latitude Pc3 pulsations *J. Geophys. Res.* A **95** 2447

Ershkovich A I and Nusinov A A 1972 Geomagnetic tail oscillations *Cosmic Electrodyn.* **2** 471

Ershkovich A I, Nusinov A A and Chernikov A A 1972 Nonlinear waves in geomagnetic wake *J. Geophys. Res.* **77** 6907

Fainberg, E B and Zinger B Sh 1981 Electromagnetic induction in the spherical model of the Earth with a real distribution of near-surface conductivity *Phys. Earth Planet. Interiors* **25** 52

Fairfield D H 1969 Bow shock associated waves observed in the far upstream interplanetary medium *J. Geophys. Res.* **74** 3541

Fermi E 1960 *Notes on Quantum Mechanics* (Chicago, IL: University of Chicago Press)

Feygin F Z and Yakimenko V L 1969 Mechanism of pearl generation and development during cyclotron instability of the outer proton zone *Geomagn. Aeron.* **9** 700

Feygin F Z and Yakimenko V L 1970 On the fine structure of micropulsations of Pc1 type *Geomagn. Aeron.* **10** 558

Feygin F Z, Gokhberg M B, Troitskaya V A, Yakimenko V L, Gendrin R, Lacourly S and Roux A 1970a The determination of some magnetosphere parameters based on data from ground observations of Pc1 type micropulsations *Ann. Geophys.* **26** 383

Feygin F Z, Gokhberg M B and Matveeva E T 1970b Comparison of satellite data with the occurrence of Pc1 pulsations *Ann. Geophys.* **26** 903

Feygin F Z and Yakimenko V L 1971 Appearance and development of geomagnetic Pc1 type micropulsations ('pearls') due to cyclotron instability of proton belt *Ann. Geophys.* **27** 49

Feygin F Z and Kurchashov Yu P 1975 A quasilinear dynamics of Pc1 geomagnetic pulsations (pearls) *J. Geomagn. Geoelectr.* **26** 539

Field E C and Greifinger C 1965 Transmission of geomagnetic micropulsations through the ionosphere and lower exosphere *J. Geophys. Res.* **70** 4885

Francis W E and Karplus R 1960 Hydromagnetic waves in the ionosphere *J. Geophys. Res.* **65** 3593

Fraser B J 1968 Temporal variations in Pc1 geomagnetic micropulsations *Planet. Space Sci.* **16** 111

Fraser B J and McPherron R L 1982 Pc1-2 magnetic pulsation spectra and heavy ion effects at synchronous orbit: ATS-6 results *J. Geophys. Res.* **87** 4560

Fraser B J and Sentman D 1991 Ion-cyclotron and lower ELF wave phenomena *ANARE Research Notes 80* Antarctic Division, Australia p 333

Frazer-Smith A C 1970 Some statistics on Pc1 geomagnetic pulsation occurrence at middle latitudes: inverse relation with sunspot cycle and semiannual period *J. Geophys. Res.* **75** 4735

Fraser-Smith A C 1981 Effects of man on geomagnetic activity and pulsations *Adv. Space Res.* **1** 455

Frazer-Smith A C 1982 ULF/Lower-ELF electromagnetic field measurements in the polar caps *Rev. Geophys. Space Phys.* **20** 497

Frenkel J 1944 On the theory of seismic and seismoelectric phenomena in moist soil *Izv. AN SSSR* No 4 133; *J. Phys.* **8** 230

Fujita S 1988 Duct propagation of hydromagnetic waves in the upper ionosphere, 2 Dispersion characteristics and loss mechanism *J. Geophys. Res.* A **93** 14674

Fujita S and Tamao T 1988 Duct propagation of hydromagnetic waves in the upper ionosphere, 1 Electromagnetic field disturbances in high latitudes associated with localized incidence of a shear Alfvén wave *J. Geophys. Res.* A **93** 14665

Fukunishi H 1969 Occurrence of sweepers in the evening sector following the onset of magnetospheric substorm *Rep. Ionos. Space Res. Japan* **23** 21

Fukunishi H 1973 Occurrence of IPDP events accompanied by cosmic noise absorption in the course of proton aurora substorms *J. Geophys. Res.* **78** 3981

Fukunishi H and Toya T 1981 Morning IPDP events observed at high latitudes *J. Geophys. Res.* A **86** 5701

Fukunishi H, Toya T, Koike K, Kuwashima M and Kawamura M 1981 Classification of hydromagnetic emissions based on frequency–time spectra *J. Geophys. Res.* A **86** 9029

Galejs J 1964 Terrestrial extremely-low-frequency propagation *Natural Electromagnetic Phenomena Below 30 kc/s* ed D F Bleil (New York: Plenum) p 205

Galejs J 1972 *Terrestrial Propagation of Long Electromagnetic Waves* (New York: Pergamon)

Gendrin R and Troitskaya V A 1965 Preliminary results of a micropulsation experiment at conjugate points *Radio Sci.* D **69** 1107

Gendrin R, Lacourly S, Troitskaya V A, Gokhberg M B and Shepetnov R V 1967 Characteristiques des pulsations irrégulieres de periods descroissante (IPDP) et leurs relations avec les variations du flux des particules piègees dans la magnetosphere *Planet. Space Sci.* **15** 1239

Gendrin R 1970 Substorm aspects of magnetic pulsations *Space Sci. Rev.* **11** 54

Gendrin R, Lacourly S, Roux A, J Solomon, Feygin F Z, Gokhberg M B, Troitskaya V A and Yakimenko V L 1971 Wave packet propagation in an amplifying medium and its application to the dispersion characteristics and to the generation mechanisms of Pc1 events *Planet. Space Sci.* **19** 165

Gendrin R 1975 Waves and wave–particle interactions in the magnetosphere: a review *Space Sci. Rev.* **18** 145

Gendrin R 1981 General relationships between wave amplification and particle diffusion in a magnetoplasma *Rev. Geophys. Space Phys.* **19** 171

Gershman B N, Ginzburg V L and Denisov N G 1957 Propagation of electromagnetic waves in a plasma (ionosphere) *Usp. Fiz. Nauk* **61** 561

Getmantsev G G, Zuikov N A, Kotik D C, Mironenko L F, Mityakov N A, Rapoport V O, Sazonov Yu A, Trakhtengertz V Yu and Eidman V Ya 1974 Observation of combination frequencies during interaction of powerful short-wave radiation with ionospheric plasma *Pis'ma JETP* **20** 229

Getmantsev G G, Guglielmi A V, Klaine B I, Kotik D S, Krylov S M, Mityakov N A, Rapoport V O, Trakhtengertz V Yu and Troitskaya V A 1977 Excitation of magnetic pulsations during the action on the ionosphere by the radiation of powerful short-wave transmitter *Izv. Vyssh. Uchebn. Zaved. Ser. Radifiz.* **20** 1017

Gil'denburg V B 1964 On the plasma resonances in the inhomogeneous media *J. Tech. Phys.* **34** 372

Gilmore R 1981 *Catastrophe Theory for Scientists and Engineers* (New York: Wiley-Interscience)

Gintzburg M A 1961a Electromagnetic radiation from solar corpuscular streams *Phys. Rev. Lett.* **7** 399

Gintzburg M A 1961b On one new mechanism of micropulsation excitation of the Earth's magnetic field *Izv. AN SSSR, Ser. Geophys.* No 11 1679

Gintzburg M A 1963 Low frequency waves in multi-component plasma *Geomagn. Aeron.* **3** 610

Gintzburg M A 1964 Structure of the equations of cosmic electrodynamics *Sov. Astron. J.* **7** 536

Gintzburg M A 1974 Longitudinal photons in a plasma *Sov. Astron. J.* **51** 218

Ginzburg V L 1971 *Propagation of Electromagnetic Waves in a Plasma* (New York: Pergamon)

Ginzburg V L 1973 On the lows of energy and pulse conservation at the radiation of electromagnetic waves (photons) in medium and on the energy-pulse tensor in macroscopic electrodynamics *Usp. Fiz. Nauk* **110** 309

Ginzburg V L 1975 *Theoretical Physics and Astrophysics* (Moscow: Nauka)

Glaßmeier K-H 1980 Magnetometer array observations of giant pulsation event *J. Geophys. Res.* **48** 127

Gogatishvili Ya M 1984 Geomagnetic precursors of strong earthquakes in the spectrum of geomagnetic pulsations with frequencies 0.02–1 Hz *Geomagn. Aeron.* **24** 697

Gokhberg M B, Karpman V I and Pokhotelov O A 1972 To the nonlinear theory of the evolution of pearls *Dokl. AN SSSR* **204** 848

Gokhberg M B, Pilipenko V A, Pokhotelov O A and Troitskaya V A 1981 On the problems of the interaction between Pc1/Pi1 and Pc4,5 hydromygnetic waves *J. Geophys. Res.* **86** 833

Gokhberg M B, Krylov S M and Levshenko V T 1989 Electromagnetic field of the earthquake center *Dokl. AN SSSR* **308** 62

Gold T 1959 Motions of the magnetosphere of the Earth *J. Geophys. Res.* **64** 1219

Golikov Yu V, Plyasova-Bakounina T A, Troitskaya V A, Chernikov A A, Pustovalov V V and Hedgecock P C 1980 Where do solar wind-controlled micropulsations originate? *Planet. Space Sci.* **28** 535

Gorbachev L P and Surkov V V 1987 Perturbations of external magnetic field by the surface Rayleigh wave *Magn. Hydrodyn.* No 2, 3

Gorelic G S 1959 *Oscillations and Waves* (Moscow: GIFML)

Gosling J T, Asbridge J R, Bame S J, Pashman G and Scopke N 1978 Observations of two distinct populations of bow shock ions in the upstream solar wind *Geophys. Res. Lett.* **5** 957

Grant I F, McDiarmid D R and McNamara A G 1992 A class of high-m pulsations and its auroral radar signature *J. Geophys. Res.* A **97** 8439

Green C A, Odera T J and Stuart W F 1983 The relationship between the strength of the IMF and the frequency of magnetic pulsations on the ground and in the solar wind *Planet. Space Sci.* **31** 559

Greenstadt E W, Green I M, Inouye T, Hundhausen A J, Bame S J and Strong I B 1968 Correlated magnetic field and plasma observations of the Earth's bow shock *J. Geophys. Res.* **73** 51

Greenstadt E W and Olsen J V 1979 Geomagnetic pulsation signals and hourly distributions of IMF orientations *J. Geophys. Res.* A **84** 1493

Greenstadt E W, Olsen J V, Loewen P D, Singer H J and Russell C T 1979 Correlation of Pc3, 4 and 5 activity with solar wind speed *J. Geophys. Res.* A **84** 6694

Greenstadt E W 1981 Upstream hydromagnetic waves and their association with backstreaming ion populations: ISEE 1 and 2 observations *J. Geophys. Res.* A **86** 4471

Gregori G P and Lanzerotti L J 1980 Geomagnetic depth sounding by induction arrow representation: a review *Rev. Geophys. Space Phys.* **18** 203

Greifinger C and Greifinger P 1965 Transmission of micropulsations through the lower ionosphere *J. Geophys. Res.* **70** 2217

Greifinger C and Greifinger P S 1968 Theory of hydromagnetic propagation in the ionospheric wave guide *J. Geophys. Res.* **73** 7473

Greifinger C and Greifinger P S 1973 Waveguide propagation of micropulsations out of the plane of the geomagnetic meridian *J. Geophys. Res.* **78** 4611

Gudzenko L I 1962 Statistical method of determination of auto-oscillation system characteristics *Izv. Vyssh. Uchebn. Zaved. Ser. Radiofiz.* **5** 572

Guglielmi A V 1963 On the group velocity of slow waves in a moving magnetoactive plasma *Geomagn. Aeron.* **3** 754

Guglielmi A V 1967a The peculiarity of proton whistlers dispersion *Geomagn. Aeron.* **7** 344

Guglielmi A V 1967b On the nature of hydromagnetic whistlers *Dokl. AN SSSR* **174** 1076

Guglielmi A V 1968 Propagation of slow waves in a moving plasma *Ann. Geophys.* **24** 761

Guglielmi A V 1970a Annular trap for low-frequency waves in the Earth's magnetosphere *Sov. Phys.–JETP Lett.* **12** 25

Guglielmi A V 1970b Spectrum of Alfvén oscillations of the magnetosphere *Geomagn. Aeron.* **10** 234

Guglielmi A V 1970c Polarization splitting of Alfvén spectrum of the magnetosphere *Geomagn. Aeron.* **10** 524

Guglielmi A V 1971 Cyclotron instability of the outer radiation belt of the Earth under conditions of self-modulation of growing waves *Sov. Phys.–JETP Lett.* **13** 57

Guglielmi A V 1973 Super-Alfvénic travel of hydromagnetic impulses in the Earth's radiation belt *Geomagn. Aeron.* **13** 128

Guglielmi A V and Bol'shakova O V 1973 Diagnostics of interplanetary magnetic field by ground data on micropulsations of Pc2-4 type *Geomagn. Aeron.* **13** 535

Guglielmi A V and Troitskaya V A 1973 *Geomagnetic Pulsations and Diagnostics of the Magnetosphere* (Moscow: Nauka)

Guglielmi A V, Plyasova-Bakounina T A and Schepetnov R V 1973 On the relation between periods of Pc3-4 geomagnetic pulsations and parameters of interplanetary medium on the Earth's orbit *Geomagn. Aeron.* **13** 382

Guglielmi A V 1974 Diagnostics of the magnetosphere and interplanetary medium by means of pulsations *Space Sci. Rev.* **16** 331

Guglielmi A V and Dovbnya B V 1974 Hydromagnetic emission of the interplanetary plasma *Astrophys. Planet. Space Sci.* **31** 21

Guglielmi A V and Zolotukhina N A 1975 Quasi-periodic drift around the Earth of the energetic proton clouds on the Ipdp data *Dokl. AN SSSR* **221** 1086

Guglielmi A V, Klaine B I and Potapov A S 1975 Excitation of magnetosonic waves with discrete spectrum in the equatorial vicinity of the plasmapause *Planet. Space Sci.* **23** 279

Guglielmi A V 1976 On extramagnetospheric origin of Pc3 geomagnetic pulsations *Geomagn. Aeron.* **16** 744

Guglielmi A V and Repin V N 1978 MHD solitons in the geomagnetic tail *Geomagn. Aeron.* **18** 1089

Guglielmi A V and Zolotukhina N A 1978 Excitation of MHD waves of increasing frequency in the Earth's magnetosphere *Geomagn. Aeron.* **18** 307

Guglielmi A V, Dovbnya B V, Klaine B I and Parkhomov V A 1978 Induced excitation of Alfvén waves in near Earth plasma by impulse emission *Geomagn. Aeron.* **18** 179

Guglielmi A V 1979 *MHD Waves in the Plasma Environment of the Earth* (Moscow: Nauka)

Guglielmi A V 1980 Modulation instability of the magnetosonic waves in the radiation belt *Geomagn. Aeron.* **20** 968

Guglielmi A V and Repin V N 1981 Alfvén solitons in a nonequilibrium plasma *Geomagn. Aeron.* **21** 214

Guglielmi A V 1982 Interpolation formula for the whistling atmospherics *Geomagn. Aeron.* **22** 354

Guglielmi A V and Polyakov A R 1983 On the discreteness of spectrum of Alfvén oscillations *Geomagn. Aeron.* **23** 281

Guglielmi A V and Zolotukhina N A 1983 Estimation of the dimension of proton injection region based on Ipdp data *Geomagn. Aeron.* **23** 98

Guglielmi A V, Klaine B I and Polyakov A R 1983 Dynamic parameters of auto-oscillation model of geomagnetic pulsations *Geomagn. Aeron.* **23** 510

Guglielmi A V 1984 On the pre-history of magnetotelluric sounding method *Izv. AN Solid Earth Phys.* **20** 245

Guglielmi A V 1985a The origin of discrete MHD signals in the polar caps *Geomagn. Aeron.* **25** 337

Guglielmi A V 1985b The ray theory of MHD waves propagation *Geomagn. Aeron.* **25** 356

Guglielmi A V, Zotov O D, Klaine B I, Rusakov N N, Belyaev P P, Kotik D S, Polyakov S V and Rapoport V O 1985 Excitation of geomagnetic pulsations at periodic heating of the ionosphere by powerful SW radioemission *Geomagn. Aeron.* **25** 102

Guglielmi A V 1986 Excitation of electromagnetic field oscillations by elastic waves in the conductive body *Geomagn. Aeron.* **26** 383

Guglielmi A V and Zotov O D 1986 On the 'world days' geomagnetic effect *Geomagn. Aeron.* **26** 870

Guglielmi A V 1988a The problems of phenomenological modeling of D_{st} variation *Geomagn. Aeron.* **28** 272

Guglielmi A V 1988b The connection coefficient of Pc3 frequency with the value of IMF *Geomagn. Aeron.* **28** 465

Guglielmi A V 1989 Hydromagnetic diagnostics and geoelectric sounding *Sov. Phys. Usp.* **32** 678

Guglielmi A V, Kamshilin A N, Volkova E N and Chirkov E B 1989a Seismovibration excitation of geoelectric signals of combined frequencies *Dokl. AN SSSR* **309** 575

Guglielmi A V, Kalisher A L and Rusakova T B 1989b Reaction of magnetospheric oscillations in the range Pc2 on a change of IMF magnitude *Geomagn. Aeron.* **29** 33

Guglielmi A V, Rusakov N N and Ruban V F 1989c Electromagnetic sounding of the Earth at resonant frequencies of the Earth-ionosphere endovibrator *Izv. AN SSSR Ser. Fiz. Zhemli* **25** 116

Guglielmi A V and Ruban V F 1990 To the theory of induction seismomagnetic effect *Izv. AN SSSR Ser. Fiz. Zhemli* **26** 395

Guglielmi A V 1991 Magnetic structure of an elastic wave front in a conducting medium *Izv. AN SSSR Ser. Fiz. Zhemli* **27** 301

Guglielmi A V 1992a Elastomagnetic waves in a porous medium *Phys. Scr.* **46** 433

Guglielmi A V 1992b Hydromagnetic diagnostics of the near-Earth medium *Izv. RAN Ser. Fiz. Zhemli* **28** 393

Guglielmi A V 1992c Ponderomotive forces in the crust and magnetosphere of the Earth *Izv. RAN, Ser. Fiz. Zhemli* **28** 577

Guglielmi A V 1992d Tolman–Stewart effect in the Earth crust *Izv. RAN Ser. Fiz. Zhemli* **28** 907

Guglielmi A V and Levshenko V T 1993 Inertial mechanism generation of seismomagnetic signals *Dokl. AN RAN* **329** 432

Guglielmi A V and Pokhotelov O A 1993a Estimation of the photon mass based on the data on global resonances of Earth-ionosphere endovibrator *Dokl. RAN* **311** 615

Guglielmi A V and Pokhotelov O A 1993b Geophysical methods of photon mass estimation *Izv. RAN Ser. Physica Zhemli* **29** 1116

Guglielmi, A V, Pokhotelov O A, Stenflo L and Shukla P K 1993 Modifications of the magnetospheric plasma due to ponderomotive forces *Astrophys. Space Sci.* **200** 91

Guglielmi A V and Levshenko V T 1994 Electromagnetic signals caused by the earthquakes *Izv. RAN Ser. Fiz. Zhemli* **30** 365

Guglielmi A V and Pokhotelov O A 1994 Nonlinear problems of physics of the geomagnetic pulsations *Space Sci. Rev.* **65** 5

Guglielmi A V and Potapov A S 1994 Note on the dependence of Pc3-4 activity on the solar wind velocity *Ann. Geophys.* **12** 1192

Guillemin V and Sternberg S 1977 *Geometric Asympotics* (Providence, RI: American Mathematical Society)

Gurnett D A, Shawhan S D, Brice N M and Smith R L 1965 Ion cyclotron whistlers *J. Geophys. Res.* **70** 1665

Gurnett D A and Shawhan S D 1966 Determination of hydrogen ion concentration, electron density and proton gyrofrequency from the dispersion of proton whistlers *J. Geophys. Res.* **71** 741

Gurnett D A and Rodrigues P 1970 Observations of 8-amu/unit charge ion cyclotron whistlers *J. Geophys. Res.* **75** 1342

Gurnett D A 1976 Plasma wave interaction with energetic ions near the magnetic equator *J. Geophys. Res.* **81** 2765

Haerendel G 1986 *Active plasma experiments (AMPTE Sci. Publ. ASP-49)* (Garching: Max-Planck-Institut für Extraterrestrial Physik)

Harang L 1936 Oscillations and vibrations in magnetic records at high-latitude stations *J. Geophys. Res.* **41** 329

Hasegawa A 1969 Drift mirror instability in the magnetosphere *Phys. Fluids* **12** 2642

Hasegawa A 1970 Stimulated modulational instabilities of plasma waves *Phys. Rev.* A **1** 1746

Hasegawa A and Chen L 1974 Theory of magnetic pulsations *Space Sci. Rev.* **16** 347

Hasegawa A 1975 *Plasma Instabilities and Nonlinear Effects* (Berlin: Springer)

Hasegawa A and Mima K 1976 Exact solitary Alfvén wave *Phys. Rev. Lett.* **37** 690

Hasegawa A and Lanzerotti L J 1978 On the orientation of hydromagnetic waves in the magnetosphere *Rev. Geophys. Space Phys.* **16** 263

Hasegawa A and Mima K 1978 Pseudo-three-dimensional turbulence in magnetized nonuniform plasma *Phys. Fluids* **21** 87

Hasegawa A, Tsui K H and Assis A S 1983 A theory of long period magnetic pulsations. 3. Local field line oscillations *Geophys. Res. Lett.* **10** 765

Hasegawa A and Maclennan C G 1990 Field-aligned electric field accompanied by drift Alfvén waves in an inhomogeneous plasma *Geophys. Res. Lett.* **17** 1605

Hayashi K, Kokubun S, Oguti T, K Tsuruda, S Machida, Heacock R R and Hessler V P 1962 Pearl-type telluric current micropulsations at College *J. Geophys. Res.* **67** 3985

Hayashi K, Kokubun S, Oguti T, Tsuruda K, Machida S, Kitamura T, Saka O and Watanabe T 1981 The extent of Pc1 source region in high latitudes *Can. J. Phys.* **59** 1097

Heacock R R 1963 Notes on pearl-type micropulsations *J. Geophys. Res.* **68** 589

Heacock R R and Hessler V P 1965 Pearl-type micropulsations associated with magnetic storm sudden commencements *J. Geophys. Res.* **70** 1103

Heacock R R 1967 Evening micropulsation event with a rising midfrequency characteristics *J. Geophys. Res.* **72** 399

Heacock R R 1970 An atlas of micropulsation spectra *Final Report* Grant No GA-4059, Geophys. Inst. University Alaska, August 1970

Heacock R R 1971 The relation of the Pc1 micropulsation source region to the plasmapause *J. Geophys. Res.* **76** 100

Heacock R R, Henderson D J, Reid J S and Kivinen M 1976 Type IPDP pulsation events in the late evening–midnight sector *J. Geophys. Res.* **81** 273

Helliwell R A, Crary H, Pope H and Smith R L 1956 The 'nose' whistlers—a new high latitude phenomena *J. Geophys. Res.* **61** 139

Helliwell R A 1965 *Whistlers and Related Ionospheric Phenomena* (Stanford, CA: Stanford University Press)

Hillebrand O, Münch J and McPherron R L 1982 Ground-satellite correlative study of a giant pulsation event *J. Geophys.* **51** 129

Hollweg J V and Völk H J 1970 New plasma instabilities in the solar wind *J. Geophys. Res.* **75** 5297

Hoppe M and Russell C T 1980 Whistler mode wave packets in the Earth's foreshock region *Nature* **287** 417

Hoppe M M, Russell C T, Frank L A, Eastman T E and Greenstadt E W 1981 Upstream hydromagnetic waves and their association with backstreaming ion populations: ISEE-1 and 2 observations *J. Geophys. Res.* A **86** 4471

Hoppe M M and Russell C T 1982 Particle acceleration at planetary bow shock waves *Nature* **295** 41

Hoppe M M and Russell C T 1983 Plasma rest frame frequencies and polarizations of the low-frequency upstream waves: ISEE-1 and 2 observations *J. Geophys. Res.* A **88** 2021

Horita R E, Barfield J N, Heacock R R and Kangas J 1978 Satellite observations of protons involved in the generation of IPDP and Pc *Space Res.* **18** 301

Horita R E, Barfield J N, Heacock R R and Kangas J 1979 IPDP source region and resonant proton energies *J. Atmos. Terr. Phys.* **41** 293

Hruška A 1968 The magnetohydrodynamic toroidal waves *Planet. Space Sci.* **16** 1305

Hu Y D, Fraser B J and Olson J V 1991 Amplification of Pc1-2 waves along a geomagnetic field line *ANARE Res. Notes* **80** 55

Hughes W J 1974 The effect of atmosphere and ionosphere on long period magnetospheric micropulsations *Planet. Space Sci.* **22** 1157

Hughes W J and Southwood D J 1976a The screening of micropulsation signals by the atmosphere and ionosphere *J. Geophys. Res.* **81** 3234

Hughes W J and Southwood D J 1976b An illustration of modification of geomagnetic pulsation structure by the ionosphere *J. Geophys. Res.* **81** 3241

Hughes W J, McPherron R L and Russell C T 1977 Multiple satellite observations of pulsation resonance structure in the magnetosphere *J. Geophys. Res.* **82** 492

Hughes W J, McPherron R L and Barfield J N 1978 Geomagnetic pulsations observed simultaneously on three geostationary satellites *J. Geophys. Res.* **83** 1109

Hughes W J 1983 Hydromagnetic waves in the magnetosphere *Rev. Geophys. Space Phys.* **21** 508

Hultqvist B, Lundin R, Stasiewicz K, Block L, Lindqvist P-A, Gustafsson G, Koskinen H, Bahnsen A, Potemra T A and Zanetti L J 1988 Simultaneous observation of upward moving field-aligned energetic electrons and ions on auroral zone field-lines *J. Geophys. Res.* **93** 9765

Infeld E 1969 Solutions of the linearized equations of magnetohydrodynamics in nonhomogeneous magnetic fields *Phys. Fluids* **12** 1845

Inoue Y 1973 Wave polarizations of geomagnetic pulsations observed in high latitudes on the Earth's surface *J. Geophys. Res.* **75** 2959

Ishizu M, Saka O, Kitamura T-I, Fukunishi H, Sato N and Fujii R 1981 Polarization study of Pc1 and Pc1-2 band pulsations at conjugate stations *Mem. Nat. Inst. Polar Res.* Spec. No 18, 118

Istomin Ya N and Pokhotelov O A 1979 Nonlinear theory of low frequency geomagnetic pulsations *Planet. Space Sci.* **27** 249

Itonaga M, Saka O and Kitamura T-I 1981 Effects of the sunrise on polarization characteristics of low-latitude Pc3-4 band micropulsations *Mem. Nat. Inst. Polar. Res.* Spec. No 18, 152

Ivanov A G 1939 Effect of electrization of Earth layers by elastic waves passing through them *Dokl. AN SSSR* **24** 42

Ivanov A G 1940 Seismo-electrical effect of the second kind *Izv. AN SSSR Ser. Geogr. Geophys.* No 5, 699

Ivanov V N and Pokhotelov O A 1987 Flute instability in the plasma sheath of the Earth's magnetosphere *Sov. J. Plasma Phys.* **13** 833

Ivanov V N and Pokhotelov O A 1988 Vortex tubes in a dipole magnetic field *Sov. J. Plasma Phys.* **14** 694

Ivanov V N, Pokhotelov O A, Feygin F Z, Roux A, Perrout S and LeQuea D 1992 Ballooning instability in the Earth's magnetosphere at variable pressure and finite β *Geomagn. Aeron.* **32** 68

Jacobs J A and Jolley E J 1962 Geomagnetic micropulsations with periods 03-3 sec ('pearls') *Nature* **194** 641

Jacobs J A and Watanabe T 1962 Propagation of hydromagnetic waves in the lower exosphere and the origin of short period geomagnetic pulsations *J. Atmos. Terr. Phys.* **24** 413

Jacobs J A and Watanabe T 1964 Micropulsation whistlers *J. Atmos. Terr. Phys.* **26** 825

Jacobs J A 1970 *Geomagnetic Micropulsations* (New York: Springer)

Jones D L and Kemp D T 1971 The nature and average magnitude of the sources of transient excitation of Shumann resonances *J. Atmos. Terr. Phys.* **33** 557

Jones A G 1980 Geomagnetic-induction studies in Scandinavia—I Determination of the inductive response function from the magnetometer array data *J. Geophys.* **48** 181

Jones A G 1981 Geomagnetic induction studies in Scandinavia—II Geomagnetic depth sounding, induction vectors and coast-effect *J. Geophys.* **50** 23

Junginger H, Geiger G, Haerendel G and Melzner F 1984 A statistical study of dayside magnetospheric electric field fluctuations with periods between 150 and 600 s *J. Geophys. Res.* A **89** 5495

Kadomtsev B B 1966 Hydromagnetic plasma stability *Reviews of Plasma Physics* vol 2, ed M A Leontovich (New York: Consultants Bureau)

Kadomtsev B B and Pogutse O P 1974 Nonlinear helical perturbations of tokamak plasmas *Sov. Phys.–JETP* **38** 283

Kadomtsev B B 1976 *Collective Phenomena in Plasma* (Moscow: Nauka)

Kahalas S L 1969 On toroidal mode eigenfrequencies *Planet. Space Sci.* **71** 1281

Kaladze T D, Mikhailovskii A B, Potapov A S and Pokhotelov O A 1976 Role of longitudinal magnetic field inhomogeneity in cyclotron instability of the plasmasphere *Sov. J. Plasma Phys.* **2** 370

Kaladze T D, Petviashvili V I and Pokhotelov O A 1986 Condensation of Alfvén waves into vortices in an inhomogeneous plasma *Sov. Phys.–JETP* **64** 62

Kaladze T D, Marchenko V A, Pokhotelov O A and Petviashvili V I 1987 Negative energy vortices in an inhomogeneous plasma *Plasma Phys. Controlled Fusion* **29** 580

Kalisher A L 1975 Distribution function of the amplitude of Pi1C geomagnetic pulsations *Geomagn. Aeron.* **15** 952

Kalisher A L, Kurchashov Yu P, Troitskaya V A and Feygin F Z 1982 Heavy ions O$^+$ in the Earth's radiation belt as a possible source of Pc1-2 geomagnetic pulsations *Geomagn. Aeron.* **22** 879

Kalisher A L and Polyakov A R 1984 Fluctuations of group delay of the Pc1 geomagnetic pulsations *Geomagn. Aeron.* **24** 772

Kalisher A L, Sizova L Z, Troitskaya V A and Shevnin A D 1985 The connection of geomagnetic pulsations wih the ring current *Geomagn. Aeron.* **25** 97

Kalisher A L and Rusakova T B 1990 On the connection of geomagnetic pulsations amplitude with the solar wind velocity *Geomagn. Aeron.* **30** 412

Kangas J, Lukkari L and Heacock R R 1974 On the westward expansion of substorm-correlated particle phenomena *J. Geophys. Res.* **79** 3207

Kangas J, Lukkari L and Heacock R R 1976 Observation of evening magnetic pulsations in the auroral zone during the substorm *J. Atmos. Terr. Phys.* **38** 1177

Kangas J 1982 On the sources of short-period magnetic pulsations *Rep. Dept. Geophys. University Oulu* No 5, 57

Kangas J, Bösinger T and Pikkaraiinen T 1984 Short-period magnetic pulsations during the substorm *Proc. Conf. Achievements of the IMS, (Graz, Austria, 1984)* ESSP A-217, 599

Kangas J, Aikio A and Pikarainen T 1988 Radar electric field measurement during an IPDP plasma wave event *Planet. Space Sci.* **36** 1103

Karplus R, Francis W E and Dragt A J 1962 The attenuation of hydromagnetic waves in the ionosphere *Planet. Space Sci.* **9** 771

Kasahara K 1981 *Earthquake Mechanics* (Cambridge: Cambridge University Press)

Kavanagh L D Jr, Freeman J W Jr and Chen A J 1968 Plasma flow in the magnetosphere *J. Geophys. Res.* **73** 5511

Kawamura M, Kuwashima M and Toya T 1981 Comparative study Pc1: features low latitude Pc1 *Mem. Nat. Inst. Polar Res.* Spec. No 18, 83

Keilis-Borok V I and Monin A S 1959 Magnetoelastic waves at the boundary of the Earth's core *Izv. AN SSSR Ser. Geophys.* No 11, 1529

Kennel C F and Petschek H E 1966 Limit on stability trapped particle fluxes *J. Geophys. Res.* **71** 1

Kennel C F and Sagdeev R Z 1967 Collisionless shock waves in high β plasmas, 1 *J. Geophys. Res.* **72** 3303

Kennel C F and Scarf F L 1968 Thermal anisotropies and electromagnetic instabilities in the solar wind *J. Geophys. Res.* **73** 6149

Kenney J F and Knaflich H B 1967 A systematic study of structured micropulsations *J. Geophys. Res.* **72** 2857

Kenney J F, Knaflich H B and Liemohn H B 1968 Magnetosphere parameters determined from structured micropulsations *J. Geophys. Res.* **73** 6737

Khabazin Yu G, Pokhotelov O A and Porotov A V 1979 The role of proton injection processes in the nonlinear evolution of pearls *Planet. Space Sci.* **27** 165

Kikuchi H and Taylor H A Jr 1972 Irregular structure of thermal ion plasma near the plasmapause observed from OGO-3 and Pc1 measurements *J. Geophys. Res.* **77** 131

Kimura I 1974 Interrelation between VLF and ULF Emissions *Space Sci. Rev.* **16** 389

King J H 1989 *Interplanetary Medium Data Book NSSDS, WDC–A–R&S* (Greenbelt, MD: Goddard Space Flight Center)

Kiselev B V, Kozlovskiy A E and Pilipenko V A 1991 Non-linear distortion of the ULF waveform *Planet. Space Sci.* **39** 1119

Kitamura T 1965 Geomagnetic pulsations and the exosphere, part III, ion density distribution in the exosphere *Rep. Ionos. Space Res. Japan* **19** 21

Kitamura T and Jacobs J A 1968 Ray paths of Pc1 waves in the magnetosphere *Planet. Space Sci.* **16** 863

Kitamura T, Saka O and Watanabe T 1981 The extent of Pc1 source region in high latitudes *Can. J. Phys.* **59** 1097

Kivelson M G and Southwood D J 1985 Resonant ULF waves: a new interpretation *Geophys. Res. Lett.* **12** 49

Knaflich H B and Kenney J F 1967 Ipdp events and their generation in the magnetosphere *Earth Planet. Sci. Lett.* **2** 453

Knopoff L 1955 The interaction between elastic wave motions and a magnetic field in electrical conductors *J. Geophys. Res.* **60** 617

Knox F B 1962 A possible hydromagnetic duct in relation to high-altitude explosions *NZ J. Geology and Geophys.* **5** 1016

Kovach R L and Ben-Menahem A 1966 Nuclear explosion induced micropulsations *J. Geophys. Res.* **71** 1427

Kozhevnikov A A, Mikhailovsky A B and Pokhotelov O A 1976 The role of protons of the radiation belts in the generation of Pc3-5 *Planet. Space Sci.* **24** 465

Kozlowski M 1963 Guiding-center approximation in the diamagnetic ring current *J. Geophys. Res.* **68** 4421

Krasnushkin P E and Yablochkin N A 1963 *Theory of the Propagation of Ultralong Waves* (Moscow: AN SSSR)

Kravtzov Yu A and Orlov Yu I 1980 *Geometric Optics of Inhomogeneous Media* (Moscow: Nauka)

Krimigis S M and Dassoulas J 1984 Active experiments in the distant magnetosphere: the AMPTE program *Preprint* 84-01, APL Johns Hopkins University, USA, 1984

Krylov A L and Fedorov E N 1976 On eigen oscillations of the bounded volume of magnetized cold plasma *Dokl. AN SSSR* **231** 68

Krylov A L, Lifshits A E and Fedorov E N 1979 On the resonance features of a plasma in curvilinear magnetic field *Dokl. AN SSSR* **247** 1094

Krylov A L, Lifshits A E and Fedorov E N 1981 On resonance features of the magnetosphere *Izv. AN SSSR Ser. Fiz. Zhemli* **6** 49

Kuckes A F 1973 Relations between electrical conductivity of a mantle and fluctuating magnetic field *Geophys. J. R Astron. Soc.* **32** 119

Kurchashov Yu P, Petviashvili N V, Pokhotelov O A and Feygin F Z 1987 Cyclotron instability of magnetospheric plasma in longitudinally inhomogeneous plasma *Geomagn. Aeron.* **27** 448

Kuwashima M, Toya T, Kawamura M, Hirasawa T, Fukunishi H and Ayukawa M 1981 Comparative study of magnetic Pc1 pulsations between low latitudes and high latitudes: Statistical study *Mem. Nat. Inst. Polar Res.* special issue **18** 101

Lacourly S 1969 Evaluation de certains paramètres de la magnètosphere a partir des propriètès des pulsations hydromagnètiques irrègulières (SIP et IPDP) *Ann. Gèophys.* **25** 651

Lamb H 1932 *Hydrodynamics* 6th edn (Cambridge: Cambridge University Press)

Landau L D and Lifshitz E M 1963 *Quantum Mechanics* (Moscow: Nauka)

Landau L D and Lifshitz E M 1970 *Theory of Elasticity* 2nd edn (Oxford: Pergamon)

Landau L D and Lifshitz E M 1984 *Electrodynamics of Continuous Media* (Oxford: Pergamon)

Landau L D and Lifshitz E M 1988a *Hydrodynamics* (Moscow: Nauka)

Landau L D and Lifshitz E M 1988b *Theory of Field* (Moscow: Nauka)

Lanzerotti L J, Hasegawa A and Maclennan C G 1969 Drift mirror instability in the magnetosphere: particle and field oscillations and electron heating *J. Geophys. Res.* **74** 5565

Lanzerotti L J 1974 Magnetohydrodynamic waves in the magnetosphere and the photon rest mass *Geophys. Res. Lett.* **1** 229

Lanzerotti L J and Fukunishi H 1974 Modes of magnetohydrodynamic waves in the magnetosphere *Rev. Geophys. Space Phys.* **12** 724

Lanzerotti L J, Fukunishi H and Chen L 1974 ULF pulsation evidence of the plasmapause, 3, Interpretation of polarization and spectral amplitude studies of Pc3 and Pc4 pulsations near $L = 4$ *J. Geophys. Res.* **79** 4648

Lanzerotti L J 1976 Hydromagnetic waves *Proc. Int. Symp. Solar–terrestrial Physics* (Boulder, CO, 1976)

Lanzerotti L J 1979 Geomagnetic influences on man-made systems *J. Atmos. Terr. Phys.* **41** 787

Lanzerotti L J and Southwood D J 1979 Hydromagnetic waves *Solar System Plasma Physics* ed L J Lanzerotti *et al* (Amsterdam: North-Holland) ch 3, p 111

Lanzerotti L J, Medford L V, Maclennan C G and Hasegawa A 1981 Polarization characteristics of hydromagnetic waves at low latitudes *J. Geophys. Res.* A **86** 5500

Lanzerotti L J 1983 Geomagnetic induction effects in ground-based systems *Space Sci. Rev.* **34** 347

Lanzerotti L J and Medford L V 1984 Local night, impulsive (Pi2-type) hydromagnetic wave polarization at low latitudes *Planet. Space Sci.* **32** 135

Lanzerotti L J and Wolfe A 1988 Comment on the research note by Miyake *et al J. Geomagn. Geoelectr.* **40** 1407

Lanzerotti L J, Russell C T, Wolfe A, Maclennan C G, Medford L V and Lepping R P 1989 Propagation of magnetosheath hydromagnetic fluctuations into the magnetosphere *J. Geophys. Res.* **94** 6933

Lavrentiev M A and Shabat B V 1973 *Problems of Hydrodynamics and their Mathematical Models* (Moscow: Nauka)

Le G, Russell C T and Smith E J 1989 Discrete wave packets upstream from the Earth and comets *J. Geophys. Res.* A **94** 3755

Leonovich A S and Mazur V A 1991a On electromagnetic field, induced in the ionosphere and atmosphere and on the Earth's surface by low-frequency Alfvén oscillations of the magnetosphere: general theory *Planet. Space Sci.* **39** 529

Leonovich A S and Mazur V A 1991b An electromagnetic field, induced on the Earth's surface by standing Alfvén waves in the magnetosphere *Planet. Space Sci.* **39** 547

Leontovich M A 1948 On the approximate boundary conditions for the electromagnetic field on the surface of the conducting bodies *The Study of the Radiowave Propagation* ed B A Vvedensky (Moscow–Leningrad: AN SSSR) part 2, p 5

Levin J, Levshenko V T and Sadovsky A M 1988 On some features of earthquake registration by the inertialess seismometer *Dokl. AN SSSR* **300** 326

Liemohn H B 1967 Cyclotron-resonance amplification of VLF and ULF whistlers *J. Geophys. Res.* **72** 39

Liemohn H B, Kenney J F and Knaflich H B 1967 Proton densities in the magnetosphere from pearl dispersion measurements *Earth Planet. Sci. Lett.* **2** 360

Lifshitz E M and Pitaevsky L P 1979 *Physical Kinetics* (Moscow: Nauka)

Lifshitz A E 1980 On the oscillations of anisotropic resonators *Funct. Anal. Appl.* **14** 63

Lilley F E M and Sloane M N 1976 On estimating electrical conductivity using gradient data from magnetometer arrays *J. Geomagn. Geoelectr.* **28** 321

Liperovsky V A, Pokhotelov O A and Shalimov S L 1992 *Ionospheric Earthquake Precursors* (Moscow: Nauka)

Lokken J E, Shaud J A and Wright C S 1963 Some characteristics of electromagnetic background signals in the vicinity of one cycle per second *J. Geophys. Res.* **68** 789

Long L T and Rivers W K 1975 Field measurements of the electroseismic effect *Geophysics* **40** 233

Ludlow G R, Cornilleau-Wehrlin N and Hughes W J 1989 Simultaneous observation of a Pc1 pulsation by the Air Force geophysical laboratory magnetometer network and GEOS-1 *J. Geophys. Res.* **94** 6633

Lundin R 1988 Acceleration/heating of plasma on auroral field lines: preliminary results of the Viking satellite *Ann. Geophys.* **6** 143

Lundin R and Hultqvist B 1989 Ionospheric plasma escape by high-amplitude electric fields: magnetic moment pumping *J. Geophys. Res.* **94** 6665

Lundquist S 1952 Studies in magneto-hydrodynamics *Ack. Fys.* **5** 297

MacDonald G 1961 Spectrum of hydromagnetic waves in the exosphere *J. Geophys. Res.* **66** 3639

Maclure K S, Hafer R A and Weaver J T 1964 Magnetic variations produced by ocean swell *Nature* **204** 1290

Maltseva N F, Guglielmi A V and Vinogradova V N 1970 Effect of 'western frequency drift' in the intervals of IPDP pulsations *Geomagn. Aeron.* **10** 939

Maltseva N F, Troitskaya V A, Schepetnov R, Pokhotelov O A, Gokhberg M B, Pilipenko V A, McPherron R and Barfield J 1982 Pc4-Pc1 magnetic pulsations at synchronous orbit and their relation to pulsations on the ground *J. Geophys. Res.* **87** 439

Manchester R N 1966 Propagation of Pc1 micropulsations from high to low latitudes *J. Geophys. Res.* **71** 3749

Marchenko V A, Nezlina Yu M and Pokhotelov O A 1988 Drift field-swelling instability in anisotropic plasmas *Plasma Phys. Controlled Fusion* **30** 957

Martner S T and Sparks N R 1959 The electroseismic effect *Geophysics* **24** 297

Matveeva E T and Troitskaya V A 1965 General features of the oscillation regime of pearl type pulsations *Geomagn. Aeron.* **5** 1078

Matveeva E T, Gnevyshev M N and Troitskaya V A 1968 On the connection of 11-year cyclic variations of Pc1 with solar activity *Geomagn. Aeron.* **8** 973

Matveeva E T, Troitskaya V A and Feygin F Z 1976 Intervals of pulsations with rising periods (IPRP) in polar caps *Planet. Space Sci.* **24** 673

Matveeva E T, Troitskaya V A and Feygin F Z 1978 Isolated bursts of type Pc1b geomagnetic pulsations at high latitudes *Geomagn. Aeron.* **18** 75

Mauk B M 1982 Helium resonance and dispersion effects on geostationary Alfvén/ion cyclotron waves *J. Geophys. Res.* **87** 9107

McKenzie J F 1970 Hydromagnetic oscillation of the geomagnetic tail and plasma sheet *J. Geophys. Res.* **75** 5331

McKenzie J F 1971 Hydromagnetic wave coupling between the solar wind and the plasma sheet *J. Geophys. Res.* **76** 2958

McKenzie J F 1991 Solar corona and wind *J. Geomagn. Geoelectr.* **43** 45

Meerson B I, Mikhailovskii A B and Pokhotelov O A 1978 Excitation of Alfvén waves by fast particles in a finite pressure plasma of adiabatic traps *J. Plasma Phys.* **2** 137

Menk F W 1985 Stimulation of Pc1 geomagnetic pulsations by HF radio transmissions *J. Atmos. Terr. Phys.* **47** 713

Mikhailovskii A B 1974 *Theory of Plasma Instabilities* (New York: Plenum)

Mikhailovskii A B and Pokhotelov O A 1975a Influence of whistlers and ion cyclotron waves on the growth of Alfvén waves in the magnetospheric plasma *Sov. J. Plasma Phys.* **1** 548

Mikhailovskii A B and Pokhotelov O A 1975b New mechanism for generation of geomagnetic pulsations by fast particles *Sov. J. Plasma Phys.* **1** 430

Mikhailovskii A B and Pokhotelov O A 1976 Electromagnetic trapped-electron instability in the magnetosphere *Sov. J. Plasma Phys.* **2** 515

Mikhailovskii A B, Pokhotelov O A, Ryzhov N M and Suprunenko V A 1976 Drift-mirror instability in finite-β plasma with hot electrons *Sov. J. Plasma Phys.* **2** 46

Mikhailovskii A B and Pokhotelov O A 1977 Excitation of Alfvén waves by fast ions in a finite-β plasma *Sov. Phys. Tech. Phys.* **22** 779

Miletits J Cz, Verö J and Stuart W F 1988 Dynamic spectra of pulsation events at L~19 and L~33 *J. Atmos. Terr. Phys.* **50** 649

Miletits J Cz, Verö J, Szendöi J, Ivanova P, Best A, and Kivinen M 1990 Pulsation periods at mid-latitudes—a seven-station study *Planet. Space Sci.* **38** 85

Mio K, Ogino T, Minami K and S Takeda 1976 Modified nonlinear Schrödinger equation for Alfvén waves propagating along the magnetic field in cold plasma *J. Phys. Soc. Japan* **41** 265, 667

Miyake W, Mukai T, Yumoto K, Saito T and Hirao K 1987 A correlation study between the solar wind speed observed by Suisei and the amplitude of Pc3 geomagnetic pulsations *J. Geomagn. Geoelectr.* **39** 159

Mjølhus E 1976 On the modulational instability of hydromagnetic waves parallel to the magnetic field *J. Plasma Phys.* **16** 321

Moffatt H K 1978 *Magnetic Field Generation in Electrically Conducting Fluids* (Cambridge: Cambridge University Press)

Mogi K 1985 *Earthquake Prediction* (Tokyo: Academic)

Mond M, Hameiri E and Hu N 1990 Coupling of magnetohydrodynamic waves in inhomogeneous magnetic field configurations *J. Geophys. Res.* A **95** 89

Moore G 1964 Magnetic disturbances preceding the 1964 Alaska earthquake *Nature* **203** No 4944, 508

Morris R J, Cole K D, Matveeva E T and Troitskaya V A 1982 Hydromagnetic 'whistlers' at the dayside cusps IPRP events *Planet. Space Sci.* **30** 113

Morris R J and Cole K D 1987 'Serpentine emission' at the high latitude Antarctic station, Davis *Planet. Space Sci.* **35** 313

Morrison K 1991 On the nature of Pc3 pulsations at $L = 4$ and their solar wind dependence *Planet. Space Sci.* **39** 1017

Mursula K, Kangas J and Pikkarainen T 1991 Pc1 micropulsations at a high-latitude station: a study over nearly four solar cycles *J. Geophys. Res.* A **96** 17651

Mursula K, Blomberg L G, Lindqvist P-A, Marklund G T, Bräysy T, Rasinkangas R and Tanskanen P 1994 Dispersive Pc1 bursts observed by Freja *Geophys. Res. Lett.* **21** 1851

Nakaryakov V M and Fainshtein S M 1991 Generation of the second harmonics of Alfvén wave in plane plasma waveguide *Izv. Vyssh. Uchebn. Zaved. Ser. Radiofizika* **34** 211

Nambu M 1974 Wave–particle interactions between the ring current particles and micropulsations associated with the plasmapause *Space Sci. Rev.* **16** 427

Newton R S, Southwood D J and Hughes W J 1978 Damping of geomagnetic pulsations by the ionosphere *Planet. Space Sci.* **26** 201

Nezlina Yu M, Pokhotelov O A and Khabazin Yu G 1984 The role of injection processes in the nonlinear evolution of monochromatic Alfvén waves in the Earth's magnetosphere *Geomagn. Aeron.* **24** 784

Nikolaenko A P and Rabinovich L M 1982 On the global electromagnetic resonances on the planets of solar system *Kosmich. Issled.* **20** 82

Nishida A 1964 Ionospheric screening effect and storm sudden commencement *J. Geophys. Res.* **69** 1861

Nishida A 1978 *Geomagnetic Diagnosis of the Magnetosphere* (New York: Springer)

Obayashi T 1958 Geomagnetic pulsations and the Earth's outer atmosphere *Ann. Geophys.* **14** 464

Obayashi T 1965 Hydromagnetic whistlers *J. Geophys. Res.* **70** 1069

Ochadlick A R Jr 1989 Measurement of the magnetic fluctuation associated with ocean swell compared with Weaver's theory *J. Geophys. Res.* **94** 16237

Ochadlick A R Jr 1990 Time series and correlation of pulsations observed simultaneously by two aircraft *Geophys. Res. Lett.* **17** 1889

Odera T J 1984 Control of the solar wind parameters on the ground Pc3,4 pulsations *Acta Geodaet. Geophys. Mont. Hung.* **19** 305

Offen R J 1972 Geoelectric field strengths as deduced from the mid-frequency slopes of Pc1 hydromagnetic whistlers *Planet. Space Sci.* **20** 135

Ogawa T, Tanaka Y and Yasuhara M 1969 Schumann resonances and worldwide thunderstorm activity *Planetary Electrodynamics* vol 2, ed S Coronity and W J Hughes (New York: Gordon and Breach)

Olson J V, Struckman P E and Price C P 1991 Correlation of cusp region Pc3 pulsations with solar wind parameters *ANARE Res. Notes 80, Antarctic Div. Australia* 93

Orr D and Matthew J A D 1971 The variation of geomagnetic micropulsation period with latitude and the plasmapause *Planet. Space Sci.* **19** 897

Orr D 1973 Magnetic pulsations within the magnetosphere *J. Atmos. Terr. Phys.* **35** 1

Ostrovskii A A and Potapov A I 1988 *Modulated Waves in Linear Media with Dispersion* (Gorky: Gorky State University)

Ovchinnikov A O 1991 Analytical study of the waveguide propagation in the ionospheric MHD-duct *Izv. Vyssh. Ucheb. Zaved. Ser. Radiofiz.* **34** 123

Pain H J 1976 *The Physics of Vibrations and Waves* (London: Wiley)

Pajunpää K 1988 Application of the horizontal spatial gradient technique to magnetometer array data in Finland *Rep. Geophys. University Oulu* No 15, 1

Parker E N 1963 *Interplanetary Dynamical Processes* (New York: Wiley–Interscience)

Parkhomenko E I 1967 *Electrical Properties of Rocks* (New York: Plenum)

Paschmann G, Sckopke N, Asbridge J R, Bame S J and Gosling J T 1980 Energization of solar wind ions by reflection from the Earth's bow shock *J. Geophys. Res.* A **85** 4689

Paschmann G, Sckopke N, Papamastorakis I, Asbridge J R, Bame S J and Gosling J T 1981 Characteristics of reflected and diffuse ions upstream from the Earth's bow shock *J. Geophys. Res.* A **86** 4355

Patel V L 1965 Structure of the equations of cosmic electrodynamics and the photon rest mass *Phys. Rev. Lett.* **14** 105

Patel V L 1968a Origin of long period micropulsations *Nature* **218** 857

Patel V L 1968b Magnetospheric tail as a hydromagnetic waveguide *Phys. Lett.* A **26** 596

Perraut S, Gendrin R, Roux A and de Villedary C 1984 Ion cyclotron waves: direct comparison between ground-based measurements and observations in the source region *J. Geophys. Res.* A **89** 195

Petersen R A and Poehls K A 1982 Model spectrum of magnetic induction caused by ambient internal waves *J. Geophys. Res.* C **87** 433

Petviashvili V I and Pokhotelov O A 1977 Alfvén and magnetosonic vortices in a plasma *Sov. Phys.–JETP* **46** 260

Petviashvili V I, Pokhotelov O A and Chudin N V 1982 Solitary toroidal vortices *Sov. Phys.–JETP* **55** 1056

Petviashvili V I and Pokhotelov O A 1985 Dipole Alfvén vortices *Sov. Phys.–JETP Lett.* **42** 54

Petviashvili V I and Pokhotelov O A 1986 Solitary vortices in plasmas *Sov. J. Plasma Phys.* **12** 651

Petviashvili V I, Pokhotelov O A and Stenflo L 1986 Toroidal Alfvén solitons in a space plasma *Sov. J. Plasma Phys.* **12** 545

Petviashvili V I and Pokhotelov O A 1992 *Solitary Waves in Plasmas and in the Atmosphere* (Philadelphia: Gordon and Breach)

Pikkarainen T 1987 Statistical results of Ipdp pulsations recorded in Finland during 1975–1979 *Geophysica* **23** 1

Pikkarainen T 1989 *Characteristics, Sources and Generation Mechanisms of type IPDP magnetic pulsations* (Oulu, Finland: University of Oulu Press)

Pitaevsky L A 1961 Electric forces in a transparent dispersive medium *Sov. Phys.–JETP* **12** 1009

Plyasova-Bakounina T A, Golikov Yu V and Troitskaya V A 1978 Pulsations in the solar wind and on the ground *Planet. Space Sci.* **26** 547

Plyasova-Bakounina T A and Münch J W 1991 Ground-based monitoring of IMF magnitude *Acta Geod. Geophys. Mont. Hung.* **26** 263

Pokhotelov O A and Pilipenko V A 1976 Contribution to the theory of the drift mirror instability of the magnetospheric plasma *Geomagn. Aeron.* **16** 296

Pokhotelov O A and Khabazin Yu G 1979 Nonlinear evolution of ion–cyclotron waves in the magnetosphere during fast-proton injection *Sov. J. Plasma Physics* **5** 183

Pokhotelov O A, Pilipenko V A and Amata E 1985 Drift anisotropy instability of a finite β magnetospheric plasma *Planet. Space Sci.* **33** 1229

Pokhotelov O A, Pilipenko V A, Nezlina Yu M, Woch J, Kremser G, Korth A and Amata E 1986a Excitation of high-β plasma instabilities at the geostationary orbit: theory and observations *Planet. Space Sci.* **34** 695

Pokhotelov O A, Nezlina Yu M and Pilipenko V A 1986b Anisotropic drift instability of the ring current *Trans. USSR Acad. Sci. (Earth Sci. Sect.)* **289** 18

Polk C 1969 Relation of ELF noise and Schumann resonances to thunderstorm activity *Planetary Electrodynamics* vol 2, ed S Coronity and J Hughes (New York: Gordon and Breach)

Polk C 1982 Schumann resonances *CRC Handbook of Atmospherics* vol 1, ed H Volland (Boca Raton, FL: Chemical Rubber Company)

Polyakov S V and Rapoport V O 1980 Parametric excitation of ionospheric Alfvén resonator *Geomagn. Aeron.* **20** 1114

Polyakov S V, Rapoport V O and Trakhtengertz V Yu 1983 Alfvén sweep-maser *Sov. J. Plasma Phys.* **9** 371

Potapov A S 1973 Dissipation of hydromagnetic waves in the plasma sheet of geomagnetic tail *Geomagn. Aeron.* **13** 377

Potapov A S and Polyushkina T N 1982 Geomagnetic pulsations and interplanetary medium *Prediction the Evolution of Natural Phenomena* ed I P Druzhinin and V P Kukushkin (Novosibirsk: Nauka) p 12

Potapov A S and Polyushkina T N 1992 A phenomenological study of the D_{st} storm variation *Planet. Space Sci.* **40** 731

Price A T 1949 The induction of electric current in non-uniform thin sheets and shells *Q. J. Mech. Appl. Math.* **2** 283

Prikner K 1968 Resonance of a plane HM-wave in a horizontally stratified lower magnetosphere in middle and lower geomagnetic latitudes and Pc pulsations with period of 5 to 40 secs observed during geomagnetic storms *Geofys. Sbornik* No 297, 189

Prince C E Jr and Bostick F X Jr 1964 Ionospheric transmission of transversely propagated plane waves of micropulsation frequencies and theoretical power spectrums *J. Geophys. Res.* **69** 3213

Rabinovich L M 1986 On the influence of day-night inhomogeneity on the ELF fields *Izv. Vyssh. Ucheb. Zaved. Ser. Radiofiz.* **29** 635

Radoski H R and Carovillano R L 1966 Axisymmetric plasmasphere resonances: toroidal mode *Phys. Fluids* **9** 285

Radoski H R 1967 Poloidal axisymmetric resonances: a separable case *J. Geomagn. Geoelectr.* **19** 1

Radoski H R and McClay J F 1967 Hydromagnetic toroidal resonance *J. Geophys. Res.* **72** 4899

Radoski H R 1971 A note on the problem of hydromagnetic resonances in the magnetosphere *Planet. Space Sci.* **19** 1012

Rao M S V G and Booker H G 1963 Guiding of electromagnetic waves along a magnetic field in a plasma *J. Geophys. Res.* **68** 387

Richter C F 1958 *Elementary Seismology* (San Francisco: Freeman)

Risbeth H and Garriott O K 1969 *Introduction to Ionospheric Physics* (New York: Academic)

Roederer J G 1970 *Dynamics of Geomagnetically Trapped Radiation* (New York: Springer)

Rokityansky I I 1982 *Geoelectric Investigation of the Earth's Crust and Mantle* (Berlin: Springer)

Rostoker G 1979 Geomagnetic micropulsations *Fund. Cosmic Phys.* **4** 211

Roux A, Gendrin R, Wehrlin N, Pellat R and Welti R 1973 Fine structure of Pc1 pulsations. 2. Theoretical interpretation *J. Geophys. Res.* **78** 3176

Russell C T, Holzer R E and Smith E J 1970 OGO-3 observations of ELF noise in the magnetosphere. 2. The nature of the equatorial noise *J. Geophys. Res.* **75** 755

Russell C T, Childers D D and Coleman P J Jr 1971 OGO-5 observations of upstream waves in the interplanetary medium: discrete waves packets *J. Geophys. Res.* **76** 845

Russell C T and Fleming B K 1976 Magnetic pulsations as a probe of the interplanetary magnetic field: a test of the Borok *B*-index *J. Geophys. Res.* **81** 5882

Russell C T and Hoppe M M 1981 The dependence of upstream wave periods on the interplanetary magnetic field strength *Geophys. Res. Lett.* **8** 615

Russell C T and Hoppe M M 1983 Upstream waves and particles *Space Sci. Rev.* **34** 155

Russell C T, Luhmann J G, Odera T J and Stuart W F 1983 The rate of occurrence of dayside Pc3,4 pulsations: the L-value dependence of the IMF cone angle effect *Geophys. Res. Lett.* **10** 663

Russell C T 1986 Solar wind control of magnetospheric configuration *Solar Wind–Magnetosphere Coupling* ed Y Kamide and J A Slavin (Tokyo: TERRAPUB) p 209

Rytov S M 1940 Evaluation of skin-effect by perturbation method *JETP* **10** 180

Sagdeev R Z and Shafranov V D 1960 On plasma instability with anisotropic velocity distribution in the magnetic field *JETP* **39** 181

Saito T 1964 A new index of geomagnetic pulsations and its relation to solar M-regions *Rep. Ionos. Space Res. Japan* **18** 260

Saito T 1969 Geomagnetic pulsations *Space Sci. Rev.* **10** 319

Sakurai T, Tonegawa Y and Kato Y 1981 Magnetic pulsations in the period range from 40 to 170 sec observed at synchronous orbit *Mem. Nat. Inst. Polar Res.* Spec. No 18 189

Samson J C, Jacobs J A and Rostoker G 1971 Latitude dependent characteristics of long period geomagnetic micropulsations *J. Geophys. Res.* **76** 3675

Samson J C and Rostoker G 1972 Latitude-dependent characteristics of high-latitude Pc4 and Pc5 micropulsations *J. Geophys. Res.* **77** 6133

Samson J C and Harrold B G 1985 Characteristic time constant and velocities of high-latitude Pi2's *J. Geophys. Res.* **90** 12173

Samson J C, Wallis D D, Hughes T J, Creutzberg F, Ruohoniemi J M and Greenwald R A 1992 Substorm intensification and field line resonances in the nightside magnetosphere *J. Geophys. Res.* **97** 8495

Savin M G, Izraelskii Yu G and Aplakov P A 1984 Directional analysis of the wave packets of magnetotelluric field *Geomagn. Aeron.* **24** 472

Schmucker U 1970 An introduction to induction anomalies *J. Geomagn. Geoelectr.* **22** 9

Schrödinger E 1943 The earth's and the sun's permanent magnetic field in the unitary field theory *Proc. R. Ir. Acad.* A **49** 135

Schrödinger E 1950 *Space–Time Structure* (Cambridge: Cambridge University Press)

Schulze-Berge S, Cowley S and Chen L 1992 Theory of field line resonances of standing shear Alfvén waves in three-dimensional inhomogeneous plasma *J. Geophys. Res.* A **97** 3219

Schumann W O 1952a On the radiation free self-oscillations of a conducting sphere, which is surrounded by an air layer and ionospheric shell *Z. Naturforsch.* a **7** 149

Schumann W O 1952b On the damping of electromagnetic self-oscillations of the system earth–air–ionosphere *Z. Naturforsch.* a **7** 250

Sentman D D 1987 Magnetic elliptical polarization of Schumann resonances *Radio Sci.* **22** 595

Shafranov V D 1963 Electromagnetic waves in a plasma *Reviews of Plasma Physics* vol 3, ed M A Leontovich (New York: Consultants Bureau) p 3

Sheinman S M 1947 On stabilization of electric fields in the Earth *Appl. Geophys.* 3

Siebert M 1964 Geomagnetic pulsations with latitude dependent periods and their relation to the structure of the magnetosphere *Planet. Space Sci.* **12** 137

Singer H J, Russell C T and Kivelson M G 1977 Evidence for the control of Pc3,4 magnetic pulsations by the solar wind velocity *Geophys. Res. Lett.* **4** 377

Singer H J, Southwood D J, Walker R J and Kivelson M G 1981 Alfvén waves resonances in a realistic magnetospheric magnetic field geometry *J. Geophys. Res.* A **86** 4589

Siscoe G L 1969 Resonant compressional waves in the geomagnetic tail *J. Geophys. Res.* **74** 6482

Slawinski R, Venkatesan D, Wolfe A, Lanzerotti L J and Maclennan C G 1988 Transmission of solar wind hydromagnetic energy into the terrestrial magnetosphere *Geophys. Res. Lett.* **15** 1275

Smagin V P and Savchenko V N 1986 Perturbation of the geomagnetic field by knoidal waves *Geomagn. Aeron.* **26** 347

Smith R L, Helliwell R A and Jabroff A 1960 A theory of trapping whistlers in field-aligned columns of enhanced ionization *J. Geophys. Res.* **65** 815

Smith R L 1961 Propagation characteristics of whistlers trapped in field-aligned columns of enhanced ionization *J. Geophys. Res.* **66** 3699

Smith R L, Brice N M, Katsufracis J, Gurnett D A, Shawhan S D, Belrose J S and Barrington R E 1964 An ion gyrofrequency phenomenon observed in satellites *Nature* **204** 274

Smith P H and Hoffman R A 1974 Storm time ring current particles *J. Geophys. Res.* **79** 966

Snyder C W, Neugebauer M and Rao U R 1963 The solar wind velocity and its correlation with cosmic ray variations and with solar and geomagnetic activity *J. Geophys. Res.* **68** 6361

Sonnerup B U O 1969 Acceleration of particles reflected at a shock front *J. Geophys. Res.* **74** 1301

Sorenson W R 1968 Investigation of possible ionospheric heating by hydromagnetic waves *J. Geophys. Res.* **73** 287

Søraas F, Lundblad J Å, Maltseva N F, Troitskaya V A and Selivanov V 1980 A comparison between simultaneous IPDP ground based observations and observations of energetic protons obtained by satellites *Planet. Space Sci.* **28** 387

Southwood D J 1974a Some features of field line resonances in the magnetosphere *Planet. Space Sci.* **22** 483

Southwood D J 1974b Recent studies in micropulsation theory *Space Sci. Rev.* **16** 413

Southwood D J and Hughes W J 1983 Theory of hydromagnetic waves in the magnetosphere *Space Sci. Rev.* **35** 301

Southwood D J and Kivelson M G 1986 The effect of parallel inhomogeneity in the magnetospheric hydromagnetic wave coupling *J. Geophys. Res.* **91** 6871

Southwood D J and Kivelson M G 1990 The magnetohydrodynamic response of the magnetospheric cavity to changes in solar wind pressure *J. Geophys. Res.* A **95** 2301

Southwood D J and Kivelson M G 1993 Mirror instability: 1 physical mechanism of linear instability *J. Geophys. Res.* **98** 9181

Stix T H 1962 *The Theory of Plasma Waves* (New York: McGraw-Hill)

Storey L R O 1953 An investigation of whistling atmospherics *Phil. Trans. R. Soc.* A **246** 113

Streltsov A V, Chmyrev V M, Pokhotelov O A, Marchenko V A and Stenflo L 1990 The formation and nonlinear evolution of convective cells in auroral plasma *Phys. Scr.* **40** 535

Stubbe P and Kopka H 1981 Generation of Pc5 pulsations by polar electrojet modulation: first experimental evidence *J. Geophys. Res.* A **86** 1606

Sucksdorff E 1936 Occurrences of rapid micropulsations at Sodankylä during 1932 to 1935 *J. Geophys. Res.* **41** 337

Sugiura M 1961 Some evidence of hydromagnetic waves in the Earth's magnetic field *Phys. Rev. Lett.* **6** 255

Sugiura M and Wilson C R 1964 Oscillation of the geomagnetic field lines and associated magnetic perturbations at conjugate points *J. Geophys. Res.* **69** 1211

Sugiura M 1965 Propagation of hydromagnetic waves in the magnetosphere *Radio Sci.* D **69** 1133

Sugiura M 1972 The ring current *Goddard Space Flight Center, Report* X-645-72-176

Synge J L 1960 *Relativity: The General Theory* (Amsterdam: North-Holland)

Takahashi K, McPherron R L, Greenstadt E W and Neeley C A 1981 Factors controlling the occurrence of Pc3 magnetic pulsations at synchronous orbit *J. Geophys. Res.* **86** 5472

Takahashi K, Lopez R E, McEntire R W, Zanetti L J, Kistler L M and Ipavich F M 1987 An eastward propagating compressional Pc5 waves observed by AMPTE/CCE in the postmidnight sector *J. Geophys. Res.* A **92** 13472

Takahashi K, McEntire R W, Lui A T and Potemra T A 1990a Ion flux oscillations associated with a radially polarized transverse Pc5 magnetic pulsation *J. Geophys. Res.* A **95** 3717

Takahashi K, Anderson B J and Strangeway R J 1990b AMPTCCE E observations of Pc3-4 pulsations at $L = 2$–6 *J. Geophys. Res.* A **95** 17179

Takahashi K and Anderson B J 1992 Distribution of ULF energy ($f < 80$ mHz) in the inner magnetosphere: a statistical analysis of AMPTCCE E magnetic field data *J. Geophys. Res.* A **97** 10751

Tepley L R 1965 Regular oscillations near 1 c/sec observed at middle and low latitudes *Radio Sci.* D **69** 1089

Tepley L and Landshoff R K 1966 Waveguide theory for ionospheric propagation of hydromagnetic emissions *J. Geophys. Res.* **71** 1499

Thorne R M 1974 The consequences of micropulsations on geomagnetically trapped particles *Space Sci. Rev.* **16** 443

Tian M, Yeoman T K, Lester M and Jones T B 1991 Statistics of Pc5 pulsation events observed by SABRE *Planet. Space Sci.* **39** 1239

Troitskaya V A and Melnikova M V 1959 On the characteristic intervals of oscillations, diminishing over the period (10–1 s) *Dokl. AN SSSR* **128** 917

Troitskaya V A 1961 Pulsation of the Earth's electromagnetic field with periods of 1 to 15 s and their connection with phenomena in the high atmosphere *J. Geophys. Res.* **66** 5

Troitskaya V A 1964 Rapid variations in the electromagnetic field of the Earth *Research in Geophysics* ed H Odishaw (Cambridge, MA: MIT) p 485

Troitskaya V A and Guglielmi A V 1967 Geomagnetic micropulsations and diagnostics of the magnetosphere *Space Sci. Rev.* **7** 689

Troitskaya V A and Maltseva N F 1968 On the possibility of influence of the change in the properties of the lower exosphere resonance cavity on the formation of the Ipdp *Ann. Geophys.* **24** 617

Troitskaya V A and Guglielmi A V 1969 Geomagnetic pulsations and diagnostics of the magnetosphere *Sov. Phys. Usp.* **12** 195

Troitskaya V A, Matveeva E T and Guglielmi A V 1969 Amplification and damping of pearls in isolated series *Izv. AN SSSR, Fiz. Zhemli* No 9, 113

Troitskaya V A and Guglielmi A V 1970 Hydromagnetic diagnostics of plasma in the magnetosphere *Ann. Geophys.* **26** 893

Troitskaya V A, Plyasova-Bakounina T A and Guglielmi A V 1971 Correlation of Pc2-4 pulsations with IMF *Dokl. AN SSSR* **197** 1312

Tverskoy B A 1967 A stability of radiation belts *Geomagn. Aeron.* **7** 226

Verö J and Hollo L 1978 Connections between interplanetary magnetic field and geomagnetic pulsations *J. Atmos. Terr. Phys.* **40** 857

Verö J, Hollo L, Potapov A S and Polyushkina T N 1985 Connection between solar wind parameters and day-side geomagnetic pulsations *J. Atmos. Terr. Phys.* **47** 557

Vinogradov P A 1956 Measurements of the vertical component of electro-telluric field in the Baikal lake *Izv. AN SSSR Ser. Geophys.* No 1, 83

Vinogradov P A and Parkhomov V A 1975 MHD waves in solar wind as the source of geomagnetic pulsations *Geomagn. Aeron.* **15** 135

Vladimirsky B M, Narmansky Ya V and Temurjants N A 1994 *Cosmic Rhythms: In the Earth, in the Magnetosphere–ionosphere, in the Environment, in the Biosphere–noosphere* (Simferopol)

Vlasov A A 1938 On the vibration properties of an electron gas *Sov. Phys.–JETP* **8** 291

Von Neumann J 1961 *Collected Works* vol 1 (New York: Pergamon)

Vörös Z 1994 The magnetosphere as a nonlinear system *Studia Geoph. Geod.* **38** 168

Vörös Z, Verö J and Kristek J 1994 Nonlinear time series analysis of geomagnetic pulsations *Nonlinear Processes in Geophys.* **1** 145

Wait J R 1962 *Electromagnetic Waves in Stratified Media* (New York: MacMillan)

Wait J R 1982 *Geo-electromagnetism* (New York: Academic)

Walker A D M, Greenwald R A, Stuart W F and Green C A 1979 STARE auroral radar observations of Pc5 geomagnetic pulsations *J. Geophys. Res.* **84** 3373

Walker A D M and Greenwald R A 1981 Statistics of occurrence of hydromagnetic oscillations in the Pc5 range observed by the STARE auroral radar *Planet. Space Sci.* **29** 293

Warburton F and Caminiti R 1964 The induced magnetic field of sea waves *J. Geophys. Res.* **69** 4311

Washimi H and Karpman V I 1976 The ponderomotive force of a high-frequency electromagnetic field in a dispersive medium *Sov. Phys.–JETP* **44** 528

Watanabe T 1959 Hydromagnetic oscillation of the outer ionosphere and geomagnetic pulsation *J. Geomagn. Geoelectr.* **10** 195

Watanabe T 1961 On the origin of geomagnetic pulsations *Sci. Rep. Tohoku Univ. Ser.* **5** 127

Watanabe T 1965 Determination of the electron distribution in the magnetosphere using hydromagnetic whistlers *J. Geophys. Res.* **70** 5839

Watanabe T 1966 Quasi-linear theory of transverse plasma instabilities with applications to hydromagnetic emissions from the magnetosphere *Can. J. Phys.* **44** 815

Watanabe S and Ondoh T 1975 Deuteron cyclotron whistler and transequatorial propagation of whistlers *J. Radio Res. Lab.* **22** 63

Weaver J T 1965 Magnetic variations associated with ocean waves and swell *J. Geophys. Res.* **70** 1921

Webb D and Orr D 1976 Geomagnetic pulsations (5–50 mHz) and interplanetary magnetic field *J. Geophys. Res.* **81** 5941

Webb D C, Lanzerotti L J and Park C G 1977 A comparison of ULF and VLF measurements of magnetospheric cold plasma densities *J. Geophys. Res.* **82** 5063

Weinberg S 1962 Eikonal method in magnetohydrodynamics *Phys. Rev.* **126** 1899

Wentworth R C 1964 Enhancement of hydromagnetic emissions after geomagnetic storms *J. Geophys. Res.* **69** 2291

Wentworth R C, Tepley L R, Amundsen K D and Heacock R R 1966 Intra- and interhemisphere difference in occurrence time of hydromagnetic emissions *J. Geophys. Res.* **71** 1492

Wentzel D G 1961 Hydromagnetic waves and the trapped radiation *J. Geophys. Res.* **66** 359

Westphal K O and Jacobs J A 1962 Oscillations of the earth's outer atmosphere and micropulsations *J. R. Astron. Soc.* **6** 360

Williams E and Park D 1971 Photon mass and galactic magnetic field *Phys. Rev. Lett.* **26** 1651

Whitham G B 1974 *Linear and Nonlinear Waves* (New York: Wiley–Interscience)

Woch J, Kremser G, Korth A, Pokhotelov O A, Pilipenko V A, Nezlina Yu M and Amata E 1988 Curvature-driven drift mirror instability in the magnetosphere *Planet. Space Sci.* **36** 383

Woch J, Kremser G and Korth A 1990 A comprehensive investigation of compressional ULF waves observed in the ring current *J. Geophys. Res.* A **95** 15113

Wolfe A, Lanzerotti L J and Maclennan C G 1980 Dependence of hydromagnetic energy spectra on solar wind velocity and interplanetary magnetic field direction *J. Geophys. Res.* **85** 114

Wolfe A 1980 Dependence of mid-latitude hydromagnetic energy spectra on solar wind speed and interplanetary magnetic field direction *J. Geophys. Res.* **85** 5977

Wolfe A, Lanzerotti L J, Maclennan C G, Slawinski R and Venkatesan D 1989 *Transmission of Solar Wind Hydromagnetic Energy into the High-latitude Magnetosphere, Electromagnetic Coupling in the Polar Clefts and Caps* (Deventer: Kluwer) p 203

Wolfe A, Uberoi C, Russell C T, Lanzerotti L J, Maclennan C G and Medford L V 1989 Penetration of hydromagnetic energy deep into the magnetosphere *Planet. Space Sci.* **37** 1317

Wolfe A, Venkatesan D, Slawinski R and Maclennan C G 1990 A conjugate area study of Pc3 pulsations near cusp latitudes *J. Geophys. Res.* A **95** 10695

Yanagihara K 1963 Geomagnetic micropulsations with periods from 003 to 10 seconds in the auroral zones with special reference to conjugate-point studies *J. Geophys. Res.* **68** 3383

Yumoto K and Saito T 1983 Relation of compressional HM waves at GOES-2 to low-latitude Pc3 magnetic pulsations *J. Geophys. Res.* A **88** 10041

Yumoto K, Saito T, Tsurutani B T, Smith E J and Akasofu S U 1984 Relationship between the IMF magnitude and Pc3 magnetic pulsations in the magnetosphere *J. Geophys. Res.* A **89** 9731

Yumoto K 1985 Low frequency upstream wave as a probable source of low latitude Pc3-4 magnetic pulsations *Planet. Space Sci.* **33** 239

Yumoto K, Saito T, Tsurutani B T, Smith E J and Akasofu S I 1985 Effects of the interplanetary magnetic field on the characteristics of Pc3-4 pulsations at globally coordinated stations *Mem. Nat. Inst. Polar Res.* No 36, 15

Yumoto K and Miyake W 1988 Reply to Lanzerotti LJ and Wolfe A *J. Geomagn. Geoelectr.* **40** 1411

Zakharov V E, Manakov S V, Novikov S P and Pitaevskii L P 1980 *The Theory of Solitons* (Moscow: Nauka)

Zotov O D and Kalisher A L 1979 Effects of artificial influence on the ionosphere *Effects of Powerful Radioemissions on the Ionosphere* (Apatity) p 150

Index